P9-CBL-742

Problem Books in Mathematics

Edited by P. R. Halmos

Problem Books in Mathematics

Series Editor: P.R. Halmos

Polynomials
by *Edward J. Barbeau*

Problems in Geometry
by *Marcel Berger, Pierre Pansu, Jean-Pic Berry, and Xavier Saint-Raymond*

Problem Book for First Year Calculus
by *George W. Bluman*

Exercises in Probability
by *T. Cacoullos*

An Introduction to Hilbet Space and Quantum Logic
by *David W. Cohen*

Unsolved Problems in Geometry
by *Hallard T. Croft, Kenneth J. Falconer, and Richard K. Guy*

Problems in Analysis
by *Bernard R. Gelbaum*

Problems in Real and Complex Analysis
by *Bernard R. Gelbaum*

Theorems and Counterexamples in Mathematics
by *Bernard R. Gelbaum and John M.H. Olmsted*

Exercises in Integration
by *Claude George*

Algebraic Logic
by *S.G. Gindikin*

Unsolved Problems in Number Theory (2nd ed)
by *Richard K. Guy*

An Outline of Set Theory
by *James M. Henle*

Demography Through Problems
by *Nathan Keyfitz and John A. Beekman*

(continued after index)

Unsolved Problems in Intuitive Mathematics
Volume I

Richard K. Guy

Unsolved Problems in Number Theory

Second Edition

With 18 figures

Springer-Verlag
New York Berlin Heidelberg London Paris
Tokyo Hong Kong Barcelona Budapest

Richard K. Guy
Department of Mathematics and Statistics
The University of Calgary
Calgary, Alberta
Canada, T2N 1N4

AMS Classification (1991): 11-01

Library of Congress Cataloging-in-Publication Data
Guy, Richard K.
Unsolved problems in number theory / Richard K. Guy.
p. cm. -- (Problem books in mathematics)
Includes bibliographical references and index.
ISBN 0-387-94289-0
1. Number theory. I. Title. II. Series.
QA241.G87 1994
512''.7---dc20 94-3818

Production managed by Karen Phillips; manufacturing coordinated by Vincent Scelta.
Photocomposed pages prepared from the author's TEX files.
Printed and bound by R.R. Donnelley & Sons, Harrisonburg, VA.
Printed in the United States of America

9 8 7 6 5 4 3 2 1

ISBN 0-387-94289-0 Springer-Verlag New York Berlin Heidelberg
ISBN 3-540-94289-0 Springer-Verlag Berlin Heidelberg New York

Preface to the First Edition

To many laymen, mathematicians appear to be problem solvers, people who do "hard sums". Even inside the profession we classify ouselves as either theorists or problem solvers. Mathematics is kept alive, much more than by the activities of either class, by the appearance of a succession of unsolved problems, both from within mathematics itself and from the increasing number of disciplines where it is applied. Mathematics often owes more to those who ask questions than to those who answer them. The solution of a problem may stifle interest in the area around it. But "Fermat's Last Theorem", because it is not yet a theorem, has generated a great deal of "good" mathematics, whether goodness is judged by beauty, by depth or by applicability.

To pose good unsolved problems is a difficult art. The balance between triviality and hopeless unsolvability is delicate. There are many simply stated problems which experts tell us are unlikely to be solved in the next generation. But we have seen the Four Color Conjecture settled, even if we don't live long enough to learn the status of the Riemann and Goldbach hypotheses, of twin primes or Mersenne primes, or of odd perfect numbers. On the other hand, "unsolved" problems may not be unsolved at all, or may be much more tractable than was at first thought.

Among the many contributions made by Hungarian mathematician Erdős Pál, not least is the steady flow of well-posed problems. As if these were not incentive enough, he offers rewards for the first solution of many of them, at the same time giving his estimate of their difficulty. He has made many payments, from \$1.00 to \$1000.00.

One purpose of this book is to provide beginning researchers, and others who are more mature, but isolated from adequate mathematical stimulus, with a supply of easily understood, if not easily solved, problems which they can consider in varying depth, and by making occasional partial progress, gradually acquire the interest, confidence and persistence that are essential to successful research.

But the book has a much wider purpose. It is important for students and teachers of mathematics at all levels to realize that although they are not yet capable of research and may have no hopes or ambitions in that direction, there are plenty of unsolved problems that are well within their comprehension, some of which will be solved in their lifetime. Many amateurs have been attracted to the subject and many successful researchers first gained their confidence by examining problems in euclidean geometry, in number theory, and more recently in combinatorics and graph theory, where it is possible to understand questions and even to formulate them and obtain original results without a deep prior theoretical knowledge.

The idea for the book goes back some twenty years, when I was impressed by the circulation of lists of problems by the late Leo Moser and co-author Hallard Croft, and by the articles of Erdős. Croft agreed to let me help him amplify his collection into a book, and Erdős has repeatedly encouraged and prodded us. After some time, the Number Theory chapter swelled into a volume of its own, part of a series which will contain a volume on Geometry, Convexity and Analysis, written by Hallard T. Croft, and one on Combinatorics, Graphs and Games by the present writer.

References, sometimes extensive bibliographies, are collected at the end of each problem or article surveying a group of problems, to save the reader from turning pages. In order not to lose the advantage of having all references collected in one alphabetical list, we give an Index of Authors, from which particular papers can easily be located provided the author is not too prolific. Entries in this index and in the General Index and Glossary of Symbols are to problem numbers instead of page numbers.

Many people have looked at parts of drafts, corresponded and made helpful comments. Some of these were personal friends who are no longer with us: Harold Davenport, Hans Heilbronn, Louis Mordell, Leo Moser, Theodor Motzkin, Alfred Rényi and Paul Turán. Others are H. L. Abbott, J. W. S. Cassels, J. H. Conway, P. Erdős, Martin Gardner, R. L. Graham, H. Halberstam, D. H. and Emma Lehmer, A. M. Odlyzko, Carl Pomerance, A. Schinzel, J. L. Selfridge, N. J. A. Sloane, E. G. Straus, H. P. F. Swinnerton-Dyer and Hugh Williams. A grant from the National Research Council of Canada has facilitated contact with these and many others. The award of a Killam Resident Fellowship at the University of Calgary was especially helpful during the writing of a final draft. The technical typing was done by Karen McDermid, by Betty Teare and by Louise Guy, who also helped with the proof-reading. The staff of Springer-Verlag in New York has been courteous, competent and helpful.

In spite of all this help, many errors remain, for which I assume reluctant responsibility. In any case, if the book is to serve its purpose it will start becoming out of date from the moment it appears; it has been becoming out of date ever since its writing began. I would be glad to hear from readers. There must be many solutions and references and problems which I don't know about. I hope that people will avail themselves of this clearing house. A few good researchers thrive by rediscovering results for themselves, but many of us are disappointed when we find that our discoveries have been anticipated.

Calgary 81-08-13 Richard K. Guy

Preface to the Second Edition

Erdős recalls that Landau, at the International Congress in Cambridge in 1912, gave a talk about primes and mentioned four problems (see **A1**, **A5**, **C1** below) which were unattackable in the present state of science, and says that they still are. On the other hand, since the first edition of this book, some remarkable progress has been made. Fermat's last theorem (modulo some holes that are expected to be filled in), the Mordell conjecture, the infinitude of Carmichael numbers, and a host of other problems have been settled.

The book is perpetually out of date; not always the 1700 years of one statement in **D1** in the first edition, but at least a few months between yesterday's entries and your reading of the first copies off the press. To ease comparison with the first edition, the numbering of the sections is still the same. Problems which have been largely or completely answered are **B47**, **D2**, **D6**, **D8**, **D16**, **D26**, **D27**, **D28**, **E15**, **F15**, **F17** & **F28**. Related open questions have been appended in some cases, but in others they have become exercises, rather than problems.

Two of the author's many idiosyncrasies are mentioned here: the use of the ampersand (&) to denote joint work and remove any possible ambiguity from phrases such as '... follows from the work of Gauß and Erdős & Guy'; and the use of the notation

¿ . . . . ?

borrowed from the Hungarians, for a conjectural or hypothetical statement. This could have alleviated some anguish had it been used by the well intentioned but not very well advised author of an introductory calculus text. A student was having difficulty in finding the derivative of a product. Frustrated myself, I asked to see the student's text. He had highlighted a displayed formula stating that the derivative of a product was the product of the derivatives, without noting that the context was 'Why is ... not the right answer?'

The threatened volume on *Unsolved Problems in Geometry* has appeared, and is already due for reprinting or for a second edition.

It will be clear from the text how many have accepted my invitation to use this as a clearing house and how indebted I am to correspondents. Extensive though it is, the following list is far from complete, but I should at least offer my thanks to Harvey Abbott, Arthur Baragar, Paul Bateman, T. G. Berry, Andrew Bremner, John Brillhart, R. H. Buchholz, Duncan Buell, Joe Buhler, Mitchell Dickerman, Hugh Edgar, Paul Erdős,

Steven Finch, Aviezri Fraenkel, David Gale, Sol Golomb, Ron Graham, Sid Graham, Andrew Granville, Heiko Harborth, Roger Heath-Brown, Martin Helm, Gerd Hofmeister, Wilfrid Keller, Arnfried Kemnitz, Jeffery Lagarias, Jean Lagrange, John Leech, Dick & Emma Lehmer, Hendrik Lenstra, Hugh Montgomery, Peter Montgomery, Shigeru Nakamura, Richard Nowakowski, Andrew Odlyzko, Richard Pinch, Carl Pomerance, Aaron Potler, Herman te Riele, Raphael Robinson, Øystein Rødseth, K. R. S. Sastry, Andrzej Schinzel, Reese Scott, John Selfridge, Ernst Selmer, Jeffery Shallit, Neil Sloane, Stephane Vandemergel, Benne de Weger, Hugh Williams, Jeff Young and Don Zagier. I particularly miss the impeccable proof-reading, the encyclopedic knowledge of the literature, and the clarity and ingenuity of the mathematics of John Leech.

Thanks also to Andy Guy for setting up the electronic framework which has made both the author's and the publisher's task that much easier. The Natural Sciences and Engineering Research Council of Canada continue to support this and many other of the author's projects.

Calgary 94-01-08 Richard K. Guy

Contents

Preface to the First Edition v

Preface to the Second Edition vii

Glossary of Symbols xiii

Introduction 1

A. Prime Numbers 3
A1. Prime values of quadratic functions. *4* **A2.** Primes connected with factorials. *7* **A3.** Mersenne primes. Repunits. Fermat numbers. Primes of shape $k \cdot 2^n + 2$. *8* **A4.** The prime number race. *13* **A5.** Arithmetic progressions of primes. *15* **A6.** Consecutive primes in A.P. *17* **A7.** Cunningham chains. *18* **A8.** Gaps between primes. Twin primes. *19* **A9.** Patterns of primes. *23* **A10.** Gilbreath's conjecture. *25* **A11.** Increasing and decreasing gaps. *26* **A12.** Pseudoprimes. Euler pseudoprimes. Strong pseudoprimes. *26* **A13.** Carmichael numbers. *30* **A14.** "Good" primes and the prime number graph. *32* **A15.** Congruent products of consecutive numbers. *33* **A16.** Gaussian primes. Eisenstein-Jacobi primes. *33* **A17.** Formulas for primes. *36* **A18.** The Erdős-Selfridge classification of primes. *41* **A19.** Values of n making $n - 2^k$ prime. Odd numbers not of the form $\pm p^a \pm 2^b$. *42*

B. Divisibility 44
B1. Perfect numbers. *44* **B2.** Almost perfect, quasi-perfect, pseudoperfect, harmonic, weird, multiperfect and hyperperfect numbers. *45* **B3.** Unitary perfect numbers. *53* **B4.** Amicable numbers. *55* **B5.** Quasi-amicable or betrothed numbers. *59* **B6.** Aliquot sequences. *60* **B7.** Aliquot cycles or sociable numbers. *62* **B8.** Unitary aliquot sequences. *63* **B9.** Superperfect numbers. *65* **B10.** Untouchable numbers. *66* **B11.** Solutions of $m\sigma(m) = n\sigma(n)$. *67* **B12.** Analogs with $d(n)$, $\sigma_k(n)$. *67* **B13.** Solutions of $\sigma(n) = \sigma(n+1)$. *68* **B14.** Some irrational series. *69* **B15.** Solutions

of $\sigma(q) + \sigma(r) = \sigma(q + r)$. *69* **B16.** Powerful numbers. *70* **B17.** Exponential-perfect numbers. *73* **B18.** Solutions of $d(n) = d(n + 1)$. *73* **B19.** $(m, n + 1)$ and $(m + 1, n)$ with same set of prime factors. *75* **B20.** Cullen numbers. *77* **B21.** $k \cdot 2^n + 1$ composite for all n. *77* **B22.** Factorial n as the product of n large factors. *79* **B23.** Equal products of factorials. *79* **B24.** The largest set with no member dividing two others. *80* **B25.** Equal sums of geometic progressions with prime ratios. *81* **B26.** Densest set with no l pairwise coprime. *81* **B27.** The number of prime factors of $n + k$ which don't divide $n + i$, $0 \leq i < k$. *82* **B28.** Consecutive numbers with distinct prime factors. *83* **B29.** Is x determined by the prime divisors of $x + 1$, $x + 2$, ..., $x + k$? *83* **B30.** A small set whose product is square. *84* **B31.** Binomial coefficients. *84* **B32.** Grimm's conjecture. *85* **B33.** Largest divisor of a binomial coefficient. *87* **B34.** If there's an i such that $n - i$ divides $\binom{n}{k}$. *89* **B35.** Products of consecutive numbers with the same prime factors. *89* **B36.** Euler's totient function. *90* **B37.** Does $\phi(n)$ properly divide $n - 1$? *92* **B38.** Solutions of $\phi(m) = \sigma(n)$. *93* **B39.** Carmichael's conjecture. *94* **B40.** Gaps between totatives. *95* **B41.** Iterations of ϕ and σ. *96* **B42.** Behavior of $\phi(\sigma(n))$ and $\sigma(\phi(n))$. *99* **B43.** Alternating sums of factorials. *99* **B44.** Sums of factorials. *100* **B45.** Euler numbers. *101* **B46.** The largest prime factor of n. *101* **B47.** When does $2^a - 2^b$ divide $n^a - n^b$? *102* **B48.** Products taken over primes. *102* **B49.** Smith numbers. *103*

C. Additive Number Theory 105

C1. Goldbach's conjecture. *105* **C2.** Sums of consecutive primes. *107* **C3.** Lucky numbers. *108* **C4.** Ulam numbers. *109* **C5.** Sums determining members of a set. *110* **C6.** Addition chains. Brauer chains. Hansen chains. *111* **C7.** The money-changing problem. *113* **C8.** Sets with distinct sums of subsets. *114* **C9.** Packing sums of pairs. *115* **C10.** Modular difference sets and error correcting codes. *118* **C11.** Three-subsets with distinct sums. *121* **C12.** The postage stamp problem. *123* **C13.** The corresponding modular covering problem. Harmonious labelling of graphs. *127* **C14.** Maximal sum-free sets. *128* **C15.** Maximal zero-sum-free sets. *129* **C16.** Nonaveraging sets. Nondividing sets. *131* **C17.** The minimum overlap problem. *132* **C18.** The n queens problem. *133* **C19.** Is a weakly independent sequence the finite union of strongly independent ones? *135* **C20.** Sums of squares. *136*

D. Diophantine Equations 139

D1. Sums of like powers. Euler's conjecture. *139* **D2.** The Fermat problem. *144* **D3.** Figurate numbers. *146* **D4.** Sums of l kth powers. *150* **D5.** Sum of four cubes. *151* **D6.** An elementary solution

of $x^2 = 2y^4 - 1$. *152* **D7.** Sum of consecutive powers made a power. *153*
D8. A pyramidal diophantine equation. *154* **D9.** Difference of two powers. *155* **D10.** Exponential diophantine equations. *157*
D11. Egyptian fractions. *158* **D12.** Markoff numbers. *166* **D13.** The equation $x^x y^y = z^z$. *168* **D14.** $a_i + b_j$ made squares. *169*
D15. Numbers whose sums in pairs make squares. *170* **D16.** Triples with the same sum and same product. *171* **D17.** Product of blocks of consecutive integers not a power. *172* **D18.** Is there a perfect cuboid? Four squares whose sums in pairs are square. Four squares whose differences are square. *173* **D19.** Rational distances from the corners of a square. *181* **D20.** Six general points at rational distances. *185*
D21. Triangles with integer sides, medians and area. *188*
D22. Simplexes with rational contents. *190* **D23.** Some quartic equations. *191* **D24.** Sum equals product. *193* **D25.** Equations involving factorial n. *193* **D26.** Fibonacci numbers of various shapes. *194* **D27.** Congruent numbers. *195* **D28.** A reciprocal diophantine equation. *197*

E. Sequences of Integers 199
E1. A thin sequence with all numbers equal to a member plus a prime. *199* **E2.** Density of a sequence with l.c.m. of each pair less than x. *200* **E3.** Density of integers with two comparable divisors. *201*
E4. Sequence with no member dividing the product of r others. *201*
E5. Sequence with members divisible by at least one of a given set. *202*
E6. Sequence with sums of pairs not members of a given sequence. *203*
E7. A series and a sequence involving primes. *203* **E8.** Sequence with no sum of a pair a square. *203* **E9.** Partitioning the integers into classes with numerous sums of pairs. *204* **E10.** Theorem of van der Waerden. Szemerédi's theorem. Partitioning the integers into classes; at least one contains an A.P. *204* **E11.** Schur's problem. Partitioning integers into sum-free classes. *209* **E12.** The modular version of Schur's problem. *211*
E13. Partitioning into strongly sum-free classes. *213* **E14.** Rado's generalizations of van der Waerden's and Schur's problems. *213* **E15.** A recursion of Göbel. *214* **E16.** Collatz's sequence. *215*
E17. Permutation sequences. *218* **E18.** Mahler's Z-numbers. *219*
E19. Are the integer parts of the powers of a fraction infinitely often prime? *220* **E20.** Davenport-Schinzel sequences. *220* **E21.** Thue sequences. *222* **E22.** Cycles and sequences containing all permutations as subsequences. *224* **E23.** Covering the integers with A.P.s. *224*
E24. Irrationality sequences. *225* **E25.** Silverman's sequence. *225*
E26. Epstein's Put-or-Take-a-Square game. *226* **E27.** Max and mex sequences. *227* **E28.** B_2-sequences. *228* **E29.** Sequence with sums and products all in one of two classes. *229* **E30.** MacMahon's prime numbers of measurement. *230* **E31.** Three sequences of Hofstadter. *231*
E32. B_2-sequences formed by the greedy algorithm. *232*

E33. Sequences containing no monotone A.P.s. *233* **E34.** Happy numbers. *234* **E35.** The Kimberling shuffle. *235* **E36.** Klarner-Rado sequences. *237* **E37.** Mousetrap. *237* **E38.** Odd sequences. *238*

F. None of the Above 240
F1. Gauß's lattice point problem. *240* **F2.** Lattice points with distinct distances. *241* **F3.** Lattice points, no four on a circle. *241* **F4.** The no-three-in-line problem. *242* **F5.** Quadratic residues. Schur's conjecture. *244* **F6.** Patterns of quadratic residues. *245* **F7.** A cubic analog of a Pell equation. *248* **F8.** Quadratic residues whose differences are quadratic residues. *248* **F9.** Primitive roots *248* **F10.** Residues of powers of two. *249* **F11.** Distribution of residues of factorials. *250* **F12.** How often are a number and its inverse of opposite parity? *251* **F13.** Covering systems of congruences. *251* **F14.** Exact covering systems. *253* **F15.** A problem of R. L. Graham. *256* **F16.** Products of small prime powers dividing n. *256* **F17.** Series associated with the ζ-function. *257* **F18.** Size of the set of sums and products of a set. *258* **F19.** Partitions into distinct primes with maximum product. *258* **F20.** Continued fractions. *259* **F21.** All partial quotients one or two. *259* **F22.** Algebraic numbers with unbounded partial quotients. *260* **F23.** Small differences between powers of 2 and 3. *261* **F24.** Squares with just two different decimal digits. *262* **F25.** The persistence of a number. *262* **F26.** Expressing numbers using just ones. *263* **F27.** Mahler's generalization of Farey series. *263* **F28.** A determinant of value one. *265* **F29.** Two congruences, one of which is always solvable. *266* **F30.** A polynomial whose sums of pairs of values are all distinct. *266* **F31.** An unusual digital problem. *266*

Index of Authors Cited 268

General Index 280

Glossary of Symbols

A.P.	arithmetic progression, $a, a+d, \ldots a+kd, \ldots$.	A5, A6, E10, E33
$a_1 \equiv a_2 \bmod b$	a_1 congruent to a_2, modulo b; $a_1 - a_2$ divisible by b.	A3, A4, A12, A15, B2, B4, B7, ...
$A(x)$	number of members of a sequence not exceeding x; e.g. number of amicable numbers not exceeding x	B4, E1, E2, E4
c	a positive constant (not always the same!)	A1, A3, A8, A12, B4, B11, ...
d_n	difference between consecutive primes; $p_{n+1} - p_n$	A8, A10, A11
$d(n)$	the number of (positive) divisors of n; $\sigma_0(n)$	B, B2, B8, B12, B18, ...
$d \mid n$	d divides n; n is a multiple of d; there is an integer q such that $dq = n$	B, B17, B32, B37, B44, C20, D2, E16
$d \nmid n$	d does not divide n	B, B2, B25, E14, E16, ...
e	base of natural logarithms; 2.718281828459045...	A8, B22, B39, D12, ...
E_n	Euler numbers; coefficients in series for $\sec x$	B45
exp{..}	exponential function	A12, A19, B4, B36, B39, ...
F_n	Fermat numbers; $2^{2^n} + 1$	A3, A12

Symbol	Meaning	Sections
$f(x) \sim g(x)$	$f(x)/g(x) \to 1$ as $x \to \infty$. $(f,\ g > 0)$	A1, A3, A8, B33, B41, C1, C17, D7, E2, E30, F26
$f(x) = o(g(x))$	$f(x)/g(x) \to 0$ as $x \to \infty$. $(g > 0)$	A1, A18, A19, B4, C6, C9, C11, C16, C20, D4, D11, E2, E14, F1
$f(x) = O(g(x))$	there is a c such that $\lvert f(x)\rvert < cg(x)$ for all sufficiently large x.	A19, B37, C8, C9, C10, C12, C16, D4, D12, E4, E8, E20, E30, F1, F2, F16
$f(x) \ll g(x)$	(as above)	A4, B4, B18, B32, B40, C9, C14, D11, E28, F4
$f(x) = \Omega(g(x))$	there is a $c > 0$ such that there are arbitrarily large x with $\lvert f(x)\rvert \geq cg(x)$ $(g(x) > 0)$.	D12, E25
$f(x) \asymp g(x)$	there are c_1, c_2 such that $c_1 g(x) \leq f(x) \leq c_2 g(x)$ $(g(x) > 0)$ for all sufficiently large x.	B18
$f(x) = \Theta(g(x))$	(as above)	E20
i	square root of -1; $i^2 = -1$	A16
$\ln x$	natural logarithm of x	A1, A2, A3, A5, A8, A12, ...
(m, n)	g.c.d. (greatest common divisor) of m and n; h.c.f. (highest common factor) of m and n	A, B3, B4, B5 B11, D2
$[m, n]$	l.c.m. (least common multiple) of m and n.	B35, E2, F14
	Also the block of consecutive integers, m, $m+1$, ..., n	B24, B26, B32, C12, C16
$m \perp n$	m, n coprime; $(m, n) = 1$; m prime to n.	A, A4, B3, B4, B5, B11, D2
M_n	Mersenne numbers; $2^n - 1$	A3, B11, B38

$n!$	factorial n; $1 \times 2 \times 3 \times \cdots \times n$	A2, B12, B14, B22 B23, B43, ...
$!n$	$0! + 1! + 2! + \ldots + (n-1)!$	B44
$\binom{n}{k}$	n choose k; the binomial coefficient $n!/k!(n-k)!$	B31, B33, C10, D3
$\left(\frac{p}{q}\right)$	Legendre (or Jacobi) symbol	see F5 (A1, A12, F7)
$p^a \Vert n$	p^a divides n, but p^{a+1} does not divide n	B, B8, B37, F16
p_n	the nth prime, $p_1 = 2$, $p_2 = 3$, $p_3 = 5$, ...	A2, A5, A14, A17 E30
$P(n)$	largest prime factor of n	B32, B46
$\mathbb{Q}$	the field of rational numbers	D2, F7
$r_k(n)$	least number of numbers not exceeding n, which must contain a k-term A.P.	see E10
$s(n)$	sum of aliquot parts (divisors of n other than n) of n; $\sigma(n) - n$	B, B1, B2, B8, B10, ...
$s^k(n)$	kth iterate of $s(n)$	B, B6, B7
$s^*(n)$	sum of unitary aliquot parts of n	B8
$S \bigcup T$	union of sets S and T	E7
$W(k,l)$	van der Waerden number	see E10
$\lfloor x \rfloor$	floor of x; greatest integer not greater than x.	A1, A5, C7, C12, C15, ...
$\lceil x \rceil$	ceiling of x; least integer not less than x.	B24
$\mathbb{Z}$	the integers $\ldots, -2, -1, 0, 1, 2, \ldots$	F14
$\mathbb{Z}_n$	the ring of integers, 0, 1, 2, ..., $n-1$ (modulo n)	E8
γ	Euler's constant; 0.577215664901532...	A8

ϵ	arbitrarily small positive constant.	A8, A18, A19, B4, B11, ...
ζ_p	p-th root of unity.	D2
$\zeta(s)$	Riemann zeta-function; $\sum_{n=1}^{\infty}(1/n^s)$	D2
π	ratio of circumference of circle to diameter; 3.141592653589793...	F1, F17
$\pi(x)$	number of primes not exceeding x	A17, E4
$\pi(x; a, b)$	number of primes not exceeding x and congruent to a modulo b	A4
$\prod$	product	A1, A2, A3, A8 A15, ...
$\sigma(n)$	sum of divisors of n; $\sigma_1(n)$	B, B2, B5, B8, B9, ...
$\sigma_k(n)$	sum of kth powers of divisors of n	B, B12, B13, B14
$\sigma^k(n)$	kth iterate of $\sigma(n)$	B9
$\sigma^*(n)$	sum of unitary divisors of n	B8
Σ	sum	A5, A8, A12, B2 B14, ...
$\phi(n)$	Euler's totient function; number of numbers not exceeding n and prime to n	B8, B11, B36, B38, B39, ...
$\phi^k(n)$	kth iterate of $\phi(n)$	B41
ω	complex cube root of 1 $\omega^3 = 1, \omega \neq 1$, $\omega^2 + \omega + 1 = 0$	A16
$\omega(n)$	number of distinct prime factors of n	B2, B8, B37
$\Omega(n)$	number of prime factors n, counting repetitions	B8
¿ $\cdots$?	conjectural or hypothetical statement	A1, A9, B37, C6 E10, E28, F2, F18

Introduction

Number theory has fascinated both the amateur and the professional for a longer time than any other branch of mathematics, so that much of it is now of considerable technical difficulty. However, there are more unsolved problems than ever before, and though many of these are unlikely to be solved in the next generation, this probably won't deter people from trying. They are so numerous that they have already filled more than one volume: the present book is just a personal sample.

Some good sources of problems in number theory were listed in the Introduction to the first edition, some of which are repeated here, along with more recent references.

Paul Erdős, Problems and results in combinatorial number theory III, *Springer Lecture Notes in Math.*, **626**(1977) 43–72;*MR* **57** #12442.

Paul Erdős, A survey of problems in combinatorial number theory, in *Combinatorial Mathematics, Optimal Designs and their Applications* (Proc. Symp. Colo. State Univ. 1978) *Ann. Discrete Math.*, **6**(1980) 89–115.

P. Erdős & R. L. Graham, *Old and New Problems and Results in Combinatorial Number Theory*, Monographies de l'Enseignement Math. No. 28, Geneva, 1980.

Pál Erdős & András Sárközy, Some solved and unsolved problems in combinatorial number theory, *Math. Slovaca*, **28**(1978) 407–421; *MR* **80i**:10001.

P. Erdős, Problems and results in number theory, in Halberstam & Hooley (eds) Recent Progress in Analytic Number Theory, Vol. 1, Academic Press, 1981, 1–13.

H. Fast & S. Świerczkowski, *The New Scottish Book*, Wrocław, 1946–1958.

Heini Halberstam, Some unsolved problems in higher arithmetic, in Ronald Duncan & Miranda Weston-Smith (eds.) *The Encyclopaedia of Ignorance*, Pergamon, Oxford & New York, 1977, 191–203.

Victor Klee & Stan Wagon, *Old and New Unsolved Problems in Plane Geometry and Number Theory*, Math. Assoc. of Amer. Dolciani Math. Expositions, **11**(1991).

Proceedings of Number Theory Conference, Univ. of Colorado, Boulder, 1963.

Report of Institute in the Theory of Numbers, Univ. of Colorado, Boulder, 1959.

Joe Roberts, *Lure of the Integers*, Math. Assoc. of America, Spectrum Series, 1992.

Daniel Shanks, *Solved and Unsolved Problems in Number Theory*, Chelsea, New York, 2nd ed. 1978; *MR* **80e**:10003.

W. Sierpiński, *A selection of Problems in the Theory of Numbers*, Pergamon, 1964.

Robert D. Silverman, A perspective on computational number theory, in Computers and Mathematics, *Notices Amer. Math. Soc.*, **38**(1991) 562–568.

S. Ulam, *A Collection of Mathematical Problems*, Interscience, New York, 1960.

Throughout this volume, "number" means natural number, i.e.,

$$0, 1, 2, \ldots$$

and c is an absolute positive constant, not necessarily taking the same value at each appearance. We use Donald Knuth's now familiar "floor" ($\lfloor\, \rfloor$) and "ceiling" ($\lceil\, \rceil$) symbols for "the greatest integer not greater than" and "the least integer not less than." A less familiar symbol may be "$m \perp n$" for "m is prime to n" or "$\gcd(m, n) = 1$."

The book is partitioned, somewhat arbitrarily at times, into six sections:

A. Prime numbers
B. Divisibility
C. Additive number theory
D. Diophantine equations
E. Sequences of integers
F. None of the above.

A. Prime Numbers

We can partition the positive integers into three classes:

the unit 1

the primes 2, 3, 5, 7, 11, 13, 17, 19, 23, 29, 31, 37, ...

the composite numbers 4, 6, 8, 9, 10, 12, 14, 15, 16, ...

A number greater than 1 is **prime** if its only positive divisors are 1 and itself; otherwise it's **composite**. Primes have interested mathematicians at least since Euclid, who showed that there are infinitely many.

Denote the n-th prime by p_n, e.g. $p_1 = 2$, $p_2 = 3$, $p_{99} = 523$; and the number of primes not greater than x by $\pi(x)$, e.g., $\pi(2) = 1$, $\pi(3\frac{1}{2}) = 2$, $\pi(1000) = 168$, $\pi(4 \cdot 10^{16}) = 1075292778753150$. The greatest common divisor (gcd) of m and n is denoted by (m, n), e.g., $(36, 66) = 6$, $(14, 15) = 1$, $(1001, 1078) = 77$. If $(m, n) = 1$, we say that m and n are **coprime** and write $m \perp n$; for example $182 \perp 165$.

Dirichlet's theorem tells us that there are infinitely many primes in any **arithmetic progression**,

$$a, \quad a+b, \quad a+2b, \quad a+3b, \quad \ldots$$

provided $a \perp b$. An article, giving a survey of problems about primes and a number of further references, is

A. Schinzel & W. Sierpiński, Sur certains hypothèses concernant les nombres premiers, *Acta Arith.*, **4**(1958) 185–208; erratum **5**(1959) 259; *MR* **21** #4936; and see **7**(1961) 1–8.

Table 7 (**D 27**) can be used as a table of primes < 1000; an entry 1, 3, 5 or 7 indicates a prime in that residue class (see **A4**) modulo 8.

The general problem of determining whether a large number is prime or composite, and in the latter case of determining its factors, has fascinated number theorists down the ages. With the advent of high speed computers, considerable advances have been made, and a special stimulus has recently been provided by the application to cryptanalysis. Some other references appear after Problem **A3** and in the first edition of this book.

William Adams & Daniel Shanks, Strong primality tests that are not sufficient, *Math. Comput.*, **39**(1982) 255–300.

Richard K. Guy, How to factor a number, *Congressus Numerantium XVI*, Proc. 5th Manitoba Conf. Numer. Math., Winnipeg, 1975, 49–89; *MR* **53** #7924.

Wilfrid Keller, Woher kommen die größten derzeit bekannten Primzahlen? *Mitt. Math. Ges. Hamburg*, **12**(1991) 211–229; *MR* **92j**:11006.

Arjen K. Lenstra & Mark S. Manasse, Factoring by electronic mail, in Advances in Cryptology—EUROCRYPT'89, *Springer Lect. Notes in Comput. Sci.*, **434**(1990) 355–371; *MR* **91i**:11182.

Hendrik W. Lenstra, Factoring integers with elliptic curves, *Ann. of Math.*(2), **126**(1987) 649–673; *MR* **89g**:11125.

Hendrik W. Lenstra & Carl Pomerance, A rigorous time bound for factoring integers, *J. Amer. Math. Soc.*, **5**(1992) 483–916; *MR* **92m**:11145.

G. L. Miller, Riemann's hypothesis and tests for primality, *J. Comput. System Sci.*, **13**(1976) 300–317; *MR* **58** #470ab.

Peter Lawrence Montgomery, An FFT extension of the elliptic curve method of factorization, PhD dissertation, UCLA, 1992.

J. M. Pollard, Theorems on factoring and primality testing, *Proc. Cambridge Philos. Soc.*, **76**(1974) 521–528; *MR* **50** #6992.

J. M. Pollard, A Monte Carlo method for factorization, *BIT*, **15**(1975) 331–334; *MR* **52** #13611.

Carl Pomerance, Recent developments in primality testing, *Math. Intelligencer*, **3**(1980/81) 97–105.

Carl Pomerance, Notes on Primality Testing and Factoring, *MAA Notes* **4**(1984) Math. Assoc. of America, Washington DC.

Carl Pomerance (editor), Cryptology and Computational Number Theory, *Proc. Symp. Appl. Math.*, **42** Amer. Math. Soc., Providence, 1990; *MR* **91k**: 11113.

Paulo Ribenboim, *The Book of Prime Number Records*, Springer-Verlag, New York, 1988.

Paulo Ribenboim, *The Little Book of Big Primes*, Springer-Verlag, New York, 1991.

Hans Riesel, Wie schnell kann man Zahlen in Faktoren zerlegen? *Mitt. Math. Ges. Hamburg*, **12**(1991) 253–260.

R. Rivest, A. Shamir & L. Adleman, A method for obtaining digital signatures and public key cryptosystems, *Communications A.C.M.*, Feb. 1978.

R. Solovay & V. Strassen, A fast Monte-Carlo test for primality, *SIAM J. Comput.*, **6**(1977) 84–85; erratum **7**(1978) 118; *MR* **57** #5885.

Jonathan Sorenson, Counting the integers cyclotomic methods can factor, *Comput. Sci. Tech. Report*, **919**, Univ. of Wisconsin, Madison, March 1990.

H. C. Williams & J. S. Judd, Some algorithms for prime testing using generalized Lehmer functions, *Math. Comput.*, **30**(1976) 867–886.

A1 Prime values of quadratic functions.

Are there infinitely many primes of the form $a^2 + 1$? Probably so, and in fact Hardy and Littlewood (their conjecture E) guessed that the number,

$P(n)$, of such primes less than n, was asymptotic to $c\sqrt{n}/\ln n$,

$$¿ \quad P(n) \sim c\sqrt{n}/\ln n \quad ?$$

i.e., that the ratio of $P(n)$ to $\sqrt{n}/\ln n$ tends to c as n tends to infinity. The constant c is

$$c = \prod \left\{ 1 - \frac{\left(\frac{-1}{p}\right)}{p-1} \right\} = \prod \left\{ 1 - \frac{(-1)^{(p-1)/2}}{p-1} \right\} \approx 1.3727$$

where $\left(\frac{-1}{p}\right)$ is the Legendre symbol (see **F5**) and the product is taken over all odd primes. They make similar conjectures, differing only in the value of c, for the number of primes represented by more general quadratic expressions. But we don't know of any integer polynomial, of degree greater than one, for which it has been proved that it takes an infinity of prime values. Is there even one prime a^2+b for each $b>0$? Sierpiński has shown that for every k there is a b such that there are more than k primes of the form a^2+b.

Iwaniec has shown that there are infinitely many n for which n^2+1 is the product of at most two primes, and his results extend to other irreducible quadratics.

If $P(n)$ is the largest prime factor of n, Maurice Mignotte has shown that $P(a^2+1) \geq 17$ if $a \geq 240$. Note that $239^2+1 = 2 \cdot 13^4$ (yet another property of 239). It has been known for 50 years that $P(a^2+1) \to \infty$ with a.

Ulam and others noticed that the pattern formed by the prime numbers when the sequence of numbers is written in a "square spiral" seems to favor diagonals which correspond to certain "prime-rich" quadratic polynomials. For example the main diagonal of Figure 1 corresponds to Euler's famous formula n^2+n+41.

There are some results for expressions (*not* polynomials!) of degree greater than 1, starting with that of Pyateckii-Šapiro, who proved that the number of primes of the form $\lfloor n^c \rfloor$ in the range $1<n<x$ is $(1+o(1))x/(1+c)\ln x$ if $1 \leq c \leq \frac{12}{11}$. This range has been successively extended to $\frac{10}{9}$, $\frac{69}{62}$, $\frac{755}{662}$, $\frac{39}{34}$ and $\frac{15}{13}$ by Kolesnik, Graham and Leitmann independently, Heath-Brown, Kolesnik again, and by Liu & Rivat.

Gilbert W. Fung & Hugh Cowie Williams, Quadratic polynomials which have a high density of prime values, *Math. Comput.*, **55**(1990) 345–353; *MR* **90j**:11090.

Martin Gardner, The remarkable lore of prime numbers, *Scientific Amer.*, **210** #3 (Mar. 1964) 120–128.

G. H. Hardy & J. E. Littlewood, Some problems of 'partitio numerorum' III: on the expression of a number as a sum of primes, *Acta Math.*, **44**(1922) 1–70.

421	420	**419**	418	417	416	415	414	413	412	411	410	**409**	408	407	406	405	404	403	402
422	**347**	346	345	344	343	342	341	340	339	338	**337**	336	335	334	333	332	**331**	330	**401**
423	348	**281**	280	279	278	**277**	276	275	274	273	272	**271**	270	**269**	268	267	266	329	400
424	**349**	282	**223**	222	221	220	219	218	217	216	215	214	213	212	**211**	210	265	328	399
425	350	**283**	224	**173**	172	171	170	169	168	**167**	166	165	164	**163**	162	209	264	327	398
426	351	284	225	174	**131**	130	129	128	**127**	126	125	124	123	122	161	208	**263**	326	**397**
427	352	285	226	175	132	**97**	96	95	94	93	92	91	90	121	160	207	262	325	396
428	**353**	286	**227**	176	133	98	**71**	70	69	68	**67**	66	**89**	120	159	206	261	324	395
429	354	287	228	177	134	99	72	**53**	52	51	50	65	88	119	158	205	260	323	394
430	355	288	**229**	178	135	100	**73**	54	**43**	42	49	64	87	118	**157**	204	259	322	393
431	356	289	230	**179**	136	**101**	74	55	44	**41**	48	63	86	117	156	203	258	321	392
432	357	290	231	180	**137**	102	75	56	45	46	**47**	62	85	116	155	202	**257**	320	391
433	358	291	232	**181**	138	**103**	76	57	58	**59**	60	**61**	84	115	154	201	256	319	390
434	**359**	292	**233**	182	**139**	104	77	78	**79**	80	81	82	**83**	114	153	200	255	318	**389**
435	360	**293**	234	183	140	105	106	**107**	108	**109**	110	111	112	**113**	152	**199**	254	**317**	388
436	361	294	235	184	141	142	143	144	145	146	147	148	**149**	150	**151**	198	253	616	387
437	362	295	236	185	186	187	188	189	190	**191**	192	**193**	194	195	196	**197**	252	315	386
438	363	296	237	238	**239**	240	**241**	242	243	244	245	246	247	248	249	250	**251**	314	385
439	364	297	298	299	300	301	302	303	304	305	306	**307**	308	309	310	**311**	312	**313**	384
440	365	366	**367**	368	369	370	371	372	**373**	374	375	376	377	378	**379**	380	381	382	**383**

Figure 1. Primes (in **bold**) Form Diagonal Patterns.

D. R. Heath-Brown, The Pyateckii-Šapiro prime number theorem, *J. Number Theory*, **16**(1983) 242–266.

D. R. Heath-Brown, Zero-free regions for Dirichlet L-functions, and the least prime in an arithmetic progression, Proc. London Math. Soc.(3) **64**(1992) 265–338.

Henryk Iwaniec, Almost-primes represented by quadratic polynomials, *Invent. Math.*, **47**(1978) 171–188; *MR* **58** #5553.

G. A. Kolesnik, The distribution of primes in sequences of the form $[n^c]$, *Mat. Zametki*(2), **2**(1972) 117–128.

G. A. Kolesnik, Primes of the form $[n^c]$, *Pacific J. Math.*(2), **118**(1985) 437–447.

D. Leitmann, Abschätzung trigonometrischer Summen, *J. reine angew. Math.*, **317**(1980) 209–219.

D. Leitmann, Durchschnitte von Pjateckij-Shapiro-Folgen, *Monatsh. Math.*, **94**(1982) 33–44.

H. Q. Liu & J. Rivat, On the Pyateckii-Šapiro prime number theorem, *Bull. London Math. Soc.*, **24**(1992) 143–147.

Maurice Mignotte, $P(x^2+1) \geq 17$ si $x \geq 240$, *C. R. Acad. Sci. Paris Sér. I Math.*, **301**(1985) 661–664; *MR* **87a**:11026.

Carl Pomerance, A note on the least prime in an arithmetic progression, *J. Number Theory*, **12**(1980) 218–223.

I. I. Pyateckii-Šapiro, On the distribution of primes in sequences of the form $[f(n)]$ (Russian), *Mat. Sbornik N.S.*, **33**(1953) 559–566; *MR* **15**, 507.

Daniel Shanks, On the conjecture of Hardy and Littlewood concerning the number of primes of the form $n^2 + a$, *Math. Comput.*, **14**(1960) 321–332.

W. Sierpiński, Les binômes $x^2 + n$ et les nombres premiers, *Bull. Soc. Roy. Sci. Liège*, **33**(1964) 259–260.

E. R. Sirota, Distribution of primes of the form $p = [n^c] = [t^d]$ in arithmetic progressions (Russian), *Zap. Nauchn. Semin. Leningrad Otdel. Mat. Inst. Steklova*, **121**(1983) 94–102; *Zbl.* **524**.10038.

A2 Primes connected with factorials.

Are there infinitely many primes of the form $n! \pm 1$ or of the form

$$X_k = 1 + \prod_{i=1}^{k} p_i$$

or of the form $X_k - 2$? Buhler, Crandall & Penk have shown that $n! + 1$ is prime for $n = 1$, 2, 3, 11, 27, 37, 41, 73, 77, 116, 154, 320, 340, 399 and 427; that $n! - 1$ is prime for $n = 3$, 4, 6, 7, 12, 14, 30, 32, 33, 38, 94, 166, 324, 379 and 469 and for no other $n < 546$; that X_k is prime for $p_k = 2$, 3, 5, 7, 11, 31, 379, 1019, 1021 and 2657 and for no other $p_k < 3088$; and that $X_k - 2$ is prime for $p_k = 3$, 5, 11, 13, 41, 89, 317, 337, 991, 1873 and 2053, and is a probable prime (i.e. final tests had not been carried out) for $p_k = 2377$. Harvey Dubner has discovered the primes $872! + 1$ and $1477! + 1$ and shown that X_k is prime for $p_k = 3229$, 4547 and 4787. It is also prime for $p_k = 11549$ and 13649.

Let q_k be the least prime greater than X_k. Then R. F. Fortune conjectures that $q_k - X_k + 1$ is prime for all k. It is clear that it is not divisible by the first k primes, and Selfridge observes that the truth of the conjecture would follow from Schinzel's formulation of Cramer's conjecture, that for $x > 8$ there is always a prime between x and $x + (\ln x)^2$. Stan Wagon has calculated the first 100 fortunate primes:

3 5 7 13 23 17 19 23 37 61 67 61 71 47 107 59 61 109 89 103
79 151 197 101 103 233 223 127 223 191 163 229 643 239 157 167 439 239 199 191
199 383 233 751 313 773 607 313 383 293 443 331 283 277 271 401 307 331 379 491
331 311 397 331 353 419 421 883 547 1381 457 457 373 421 409 1061 523 499 619 727
457 509 439 911 461 823 613 617 1021 523 941 653 601 877 607 631 733 757 877 641

under the assumption that the very large probable primes involved are genuine primes. The answers to the questions are probably "yes", but it does not seem conceivable that such conjectures will come within reach either of computers or of analytical tools in the foreseeable future. Schinzel's conjecture has been attributed to Cramér, but Cramér conjectured (see reference at **A8**)

$$¿ \quad \limsup_{n\to\infty} \frac{p_{n+1} - p_n}{(\ln p_n)^2} = 1 \quad ?$$

Schinzel notes that this doesn't imply the existence of a prime between x and $x + (\ln x)^2$, even for sufficiently large x.

More hopeful, but still difficult, is the following conjecture of Erdős and Stewart: are $1! + 1 = 2$, $2! + 1 = 3$, $3! + 1 = 7$, $4! + 1 = 5^2$, $5! + 1 = 11^2$ the only cases where $n! + 1 = p_k^a p_{k+1}^b$ and $p_{k-1} \le n < p_k$? [Note that $(a, b) = (1, 0)$, $(1, 0)$, $(0, 1)$, $(2, 0)$ and $(0, 2)$ in these five cases.]

Erdős also asks if there are infinitely many primes p for which $p - k!$ is composite for each k such that $1 \le k! < p$; for example, $p = 101$ and

$p = 211$. He suggests that it may be easier to show that there are infinitely many integers n $(l! < n \leq (l+1)!)$ all of whose prime factors are greater than l, and for which all the numbers $n - k!$ $(1 \leq k \leq l)$ are composite.

I. O. Angell & H. J. Godwin, Some factorizations of $10^n \pm 1$, *Math. Comput.*, **28**(1974) 307–308.

Alan Borning, Some results for $k! \pm 1$ and $2 \cdot 3 \cdot 5 \cdots p \pm 1$, *Math. Comput.*, **26**(1972) 567–570.

J. P. Buhler, R. E. Crandall & M. A. Penk, Primes of the form $n! \pm 1$ and $2 \cdot 3 \cdot 5 \cdots p \pm 1$, *Math. Comput.*, **38**(1982) 639–643; corrigendum, Wilfrid Keller, **40**(1983) 727; *MR* **83c**:10006, **85b**:11119.

Harvey Dubner, Factorial and primorial primes, *J. Recreational Math.*, **19** (1987) 197–203.

Martin Gardner, Mathematical Games, *Sci. Amer.*, **243**#6(Dec. 1980) 18–28.

Solomon W. Golomb, The evidence for Fortune's conjecture, *Math. Mag.*, **54**(1981) 209–210.

S. Kravitz & D. E. Penney, An extension of Trigg's table, *Math. Mag.*, **48**(1975) 92–96.

Mark Templer, On the primality of $k! + 1$ and $2 * 3 * 5 * \cdots * p + 1$, *Math. Comput.*, **34**(1980) 303–304.

A3 Mersenne primes. Repunits. Fermat numbers. Primes of shape $k \cdot 2^n + 2$.

Primes of special form have been of perennial interest, especially the **Mersenne primes** $2^p - 1$. Here p is necessarily prime, but that is *not* a sufficient condition! $2^{11} - 1 = 2047 = 23 \cdot 89$. They are connected with perfect numbers (see **B1**).

The powerful Lucas-Lehmer test, in conjunction with successive generations of computers, and more sophisticated techniques in using them, continues to add to the list of primes p for which $2^p - 1$ is also prime:

2, 3, 5, 7, 13, 17, 19, 31, 61, 89, 107, 127, 521, 607, 1279, 2203,
2281, 3217, 4253, 4423, 9689, 9941, 11213, 19937, 21701,
23209, 44497, 86243, 110503, 132049, 216091, 756839, 859433, ...

The number of Mersenne primes is undoubtedly infinite, but proof is again hopelessly beyond reach. Suppose $M(x)$ is the number of primes $p \leq x$ for which $2^p - 1$ is prime. Find a convincing heuristic argument for the size of $M(x)$. Gillies gave one suggesting that $M(x) \sim c \ln x$. H. W. Lenstra, Pomerance and Wagstaff all believe this and in fact suggest that

$$¿ \qquad M(x) \sim e^\gamma \log x \qquad ?$$

where the log is to base 2.

The largest known prime is usually a Mersenne prime, but for a while the record was $391581 \cdot 2^{216193} - 1$, discovered by J. Brown, L.C. Noll, B. Parady, G. Smith, J. Smith & S. Zarantonello. In late March, 1992 this was beaten by the penultimate item in the above list, discovered by Slowinski & Gage.

D. H. Lehmer puts $S_1 = 4$, $S_{k+1} = S_k^2 - 2$, supposes that $2^p - 1$ is a Mersenne prime, notes that $S_{p-2} \equiv 2^{(p+1)/2}$ or $-2^{(p+1)/2} \bmod 2^p - 1$ and asks: which?

Selfridge conjectures that if n is a prime of the form $2^k \pm 1$ or $2^{2k} \pm 3$, then $2^n - 1$ and $(2^n + 1)/3$ are either both prime or both composite. Moreover if both are prime, then n is of one of those forms. Is this an example of the Strong Law of Small Numbers? Dickson, on p. 28 of Vol. 1 of his *History*, says:

> In a letter to Tannery (*l'Intermédiaire des math.*, **2**(1895) 317) Lucas stated that Mersenne (1644, 1647) implied that a necessary and sufficient condition that $2^p - 1$ be a prime is that p be a prime of one of the forms $2^{2n} + 1$, $2^{2n} \pm 3$, $2^{2n+1} - 1$. Tannery expressed his belief that the theorem was empirical and due to Frenicle, rather than to Fermat.

If p is a prime, is $2^p - 1$ always **squarefree** (does it never contain a repeated factor)? This seems to be another unanswerable question. It is safe to conjecture that the answer is "No!" This *could* be settled by computer if you were lucky. As D. H. Lehmer has said about various factoring methods, "Happiness is just around the corner". Selfridge puts the computational difficulties in perspective by proposing the problem: find fifty more numbers like 1093 and 3511. [Fermat's theorem tells us that if p is prime, then p divides $2^p - 2$; the primes 1093 and 3511 are the only ones less than $6 \cdot 10^9$ for which p^2 divides $2^p - 2$.] It is not known if there are infinitely many primes p for which p^2 divides $2^p - 2$. It is not even known if there are infinitely many p for which p^2 *does not* divide $2^p - 2$ — although this can be deduced from the very powerful "ABC conjecture" (see **B17**).

The so-called **repunits**, $(10^p - 1)/9$, are prime for $p = 2, 19, 23, 317$ and 1031. Repunits other than 1 are known never to be squares and Rotkiewicz has shown that they are not cubes. When are they squarefree? The primes 3, 487 and 56598313 are the only ones less than 2^{32} for which p^2 divides $10^p - 10$. Peter Montgomery lists cases where p^2 divides $a^{p-1} - 1$ for $a < 100$ and $p < 2^{32}$.

Selfridge asks if the sequence (in decimal notation)

$$1,\ 12,\ 123,\ 1234,\ 12345,\ 123456,\ 1234567,\ 12345678,\ 123456789,$$
$$12345678910,\ 1234567891011,\ 123456789101112,\ \ldots$$

contains infinitely many primes. The question can also be asked for other scales of notation; for example

$$12345610111213_7 = 131870666077_{10} \quad \text{is prime.}$$

Wagstaff observes that the only primes < 180 for which $(p^p - 1)/(p-1)$ is prime are $p = 2$, 3, 7, 19 and 31; for $(p^p + 1)/(p+1)$ they are $p = 3$, 5, 17 and 157.

The **Fermat numbers**, $F_n = 2^{2^n} + 1$ are also of continuing interest; they are prime for $0 \le n \le 4$ and composite for $5 \le n \le 21$ and for many larger values of n. Hardy & Wright give a heuristic argument which suggests that only a finite number of them are prime. Selfridge would like to see this strengthened to support the conjecture that all the rest are composite.

It has been conjectured that the Fermat numbers are squarefree. It was verified by Gostin & McLaughlin that 82 of the 85 then known factors of the 71 known composite Fermat numbers (factors of F_m for $m = 3310$, 4724 and 6537 were *not* so tested) were not repeated. Wilfrid Keller and Hiromi Suyama have found several new factors of Fermat numbers. The table of 88 prime factors $k \cdot 2^n + 1$ of $2^{2^m} + 1$ given on p. lx of the Introduction of "the Cunningham project" (see the Brillhart *et al.* ref. below) expanded to 114 entries in the second edition, and has at least 150 at the time of writing. Lenstra, Lenstra, Manasse & Pollard have factored the ninth, and R. P. Brent the eleventh, Fermat numbers completely. People interested in factoring large numbers should make contact with Samuel S. Wagstaff.

Because of their special interest as potential factors of Fermat numbers, and because proofs of their primality are comparatively easy, numbers of the form $k \cdot 2^n + 1$ have received special attention, at least for small values of k. For example, large primes were found by Harvey Dubner and Wilfrid Keller, the record on 84-09-05 for a non-Mersenne prime being $(k, n) = (5, 23473)$ by Keller. Another of his discoveries, $(k, n) = (289, 18502)$ is amusing in that it may be written as (18496,18496), a Cullen prime (**B20**) and as $(17 \cdot 2^{9251})^2 + 1$, a prime of shape $a^2 + 1$ (**A1**).

As we mentioned, the record has since been beaten with a prime of shape $k \cdot 2^n - 1$ with $(k, n) = (391581, 216193)$. See also **B21**.

Hugh Williams has found, for $r = 3$, 5, 7 and 11, all values of $n \le 500$ for which $(r-1)r^n - 1$ is prime:

$r = 3$ $\quad n = 1, 2, 3, 7, 8, 12, 20, 23, 27, 35, 56, 62, 68, 131, 222, 384, 387$
$r = 5$ $\quad n = 1, 3, 9, 13, 15, 25, 39, 69, 165, 171, 209, 339$
$r = 7$ $\quad n = 1, 2, 7, 18, 55, 69, 87, 119, 141, 189, 249, 354$
$r = 11$ $\quad n = 1, 3, 37, 119, 255, 355, 371, 497$

We are very unlikely to know for sure that the **Fibonacci sequence**

1,1,**2**,**3**,**5**,8,**13**,21,34,55,**89**,144,**233**,377,610,987,**1597**,..., where $u_1 = u_2 = 1$ and $u_{n+1} = u_n + u_{n-1}$, contains infinitely many primes. (Hugh Williams has found the large prime Fibonacci number u_{2971}.) Similarly for the related **Lucas sequence**

$$1,\mathbf{3},4,\mathbf{7},\mathbf{11},18,\mathbf{29},\mathbf{47},76,123,\mathbf{199},322,\mathbf{521},843,1364,\ldots,$$

and most other Lucas-Lehmer sequences (with $u_1 \perp u_2$) defined by second-order recurrence relations. However, Graham has shown that the sequence with

$$u_0 = 1786\ 772701\ 928802\ 632268\ 715130\ 455793$$

$$u_1 = 1059\ 683225\ 053915\ 111058\ 165141\ 686995$$

contains no primes at all! Knuth notes that Graham's numbers should have been given as

$$u_0 = 331\,635635\,998274\,737472\,200656\,430763$$

$$u_1 = 1510\,028911\,088401\,971189\,590305\,498785$$

and gives the smaller example

$$u_1 = 49463\,435743\,205655,\ u_2 = 62638\,280004\,239857$$

Raphael Robinson considers the Lucas sequence (sometimes called the **Pell sequence**) $u_0 = 0$, $u_1 = 1$, $u_{n+1} = 2u_n + u_{n-1}$ and defines the **primitive part**, L_n, by

$$u_n = \prod_{d|n} L_d$$

He notes that $L_7 = 13^2$ and $L_{30} = 31^2$ and asks if there is any larger n for which L_n is a square.

R. C. Archibald, *Scripta Math.*, **3**(1935) 117.

A. O. L. Atkin & N. W. Rickert, Some factors of Fermat numbers, *Abstracts Amer. Math. Soc.*, **1**(1980) 211.

P. T. Bateman, J. L. Selfridge & S. S. Wagstaff, The new Mersenne conjecture, *Amer. Math. Monthly*, **96**(1989) 125–128; *MR* **90c**:11009.

Wieb Bosma, Explicit primality criteria for $h \cdot 2^k \pm 1$, *Math. Comput.*, **61**(1993) 97–109.

R. P. Brent, Factorization of the eleventh Fermat number, *Abstracts Amer. Math. Soc.*, **10**(1989) 89T-11-73.

R. P. Brent & J. M. Pollard, Factorization of the eighth Fermat number, *Math. Comput.*, **36**(1981) 627–630; *MR* **83h**:10014.

John Brillhart & G. D. Johnson, On the factors of certain Mersenne numbers, I, II *Math. Comput.*, **14**(1960) 365–369, **18**(1964) 87–92; *MR* **23**#A832, **28**#2992.

John Brillhart, D. H. Lehmer, J. L. Selfridge, Bryant Tuckerman & Samuel S. Wagstaff, Factorizations of $b^n \pm 1$, $b = 2, 3, 5, 6, 7, 10, 11, 12$ up to high powers, *Contemp. Math.*, **22**. Amer. Math. Soc., Providence RI, 1983, 1988; *MR* **84k**:10005, **90d**:11009.

John Brillhart, Peter L. Montgomery & Robert D. Silverman, Tables of Fibonacci and Lucas factorizations, *Math. Comput.*, **50**(1988) 251–260 & S1–S15.

John Brillhart, J. Tonascia & P. Weinberger, On the Fermat quotient, in *Computers in Number Theory*, Academic Press, 1971, 213–222.

W. N. Colquitt & L. Welsh, A new Mersenne prime, *Math. Comput.*, **56**(1991) 867–870; *MR* **91h**:11006.

Harvey Dubner, Generalized Fermat numbers, *J. Recreational Math.*, **18** (1985–86) 279–280.

Harvey Dubner, Generalized repunit primes, *Math. Comput.*, **61** (1993) 927–930.

John R. Ehrman, The number of prime divisors of certain Mersenne numbers, *Math. Comput.*, **21**(1967) 700–704; *MR* **36**#6368.

Donald B. Gillies, Three new Mersenne primes and a statistical theory, *Math. Comput.*, **18**(1964) 93–97; *MR* **28**#2990.

Gary B. Gostin & Philip B. McLaughlin, Six new factors of Fermat numbers, *Math. Comput.*, **38**(1982) 645–649; *MR* **83c**:10003.

R. L. Graham, A Fibonacci-like sequence of composite numbers, *Math. Mag.*, **37**(1964) 322–324; *Zbl* **125**, 21.

Richard K. Guy, The strong law of small numbers, *Amer. Math. Monthly*, **95**(1988) 697–712; *MR* **90c**:11002 (see also *Math. Mag.*, **63**(1990) 3–20; *MR* **91a**:11001.

Wilfrid Keller, Factors of Fermat numbers and large primes of the form $k.2^n + 1$, *Math. Comput.*, **41**(1983) 661–673; *MR* **85b**:11117.

Wilfrid Keller, New factors of Fermat numbers, *Abstracts Amer. Math. Soc.*, **5**(1984) 391

Wilfrid Keller, The 17th prime of the form $5.2^n + 1$, *Abstracts Amer. Math. Soc.*, **6**(1985) 121.

Donald E. Knuth, A Fibonacci-like sequence of composite numbers, *Math. Mag.*, **63**(1990) 21–25; *MR* **91e**:11020.

M. Kraitchik, *Sphinx*, 1931, 31.

D. H. Lehmer, *Sphinx*, 1931, 32, 164.

D. H. Lehmer, On Fermat's quotient, base two, *Math. Comput.*, **36**(1981) 289–290; *MR* **82e**:10004.

A. K. Lenstra, H. W. Lenstra, M. S. Manasse & J. M. Pollard, The factorization of the ninth Fermat number, *Math. Comput.*, **61**(1993) 319–349; *MR* **93k**:11116.

Peter L. Montgomery, New solutions of $a^{p-1} \equiv 1 \bmod p^2$, *Math. Comput.*, **61**(1993) 361–363.

Thorkil Naur, New integer factorizations, *Math. Comput.*, **41**(1983) 687–695; *MR* **85c**:11123.

Rudolf Ondrejka, Titanic primes with consecutive like digits, *J. Recreational Math.*, **17**(1984-85) 268–274.

Herman te Riele, Walter Lioen & Dik Winter, Factorization beyond the googol with MPQS on a single computer, *CWI Quarterly*, **4**(1991) 69–72.

A. Rotkiewicz, Note on the diophantine equation $1 + x + x^2 + \ldots + x^n = y^m$, *Elem. Math.*, **42**(1987) 76.

Daniel Shanks & Sidney Kravitz, On the distribution of Mersenne divisors, *Math. Comput.*, **21**(1967) 97–100; *MR* **36**:#3717.

Hiromi Suyama, Searching for prime factors of Fermat numbers with a microcomputer (Japanese) *bit*, **13**(1981) 240-245; *MR* **82c**:10012.

Hiromi Suyama, Some new factors for numbers of the form $2^n \pm 1$, 82T-10-230 *Abstracts Amer. Math. Soc.*, **3**(1982) 257; IV, **5**(1984) 471.

Hiromi Suyama, The cofactor of F_{15} is composite, 84T-10-299 *Abstracts Amer. Math. Soc.*, **5**(1984) 271.

Samuel S. Wagstaff, Divisors of Mersenne numbers, *Math. Comput.*, **40**(1983) 385–397; *MR* **84j**:10052.

Samuel S. Wagstaff, The period of the Bell exponential integers modulo a prime, in *Math. Comput. 1943–1993* (Vancouver, 1993), Proc. Sympos. Appl. Math., Amer. Math. Soc., 1994.

H. C. Williams, The primality of certain integers of the form $2Ar^n - 1$, *Acta Arith.*, **39**(1981) 7–17; *MR* **84h**:10012.

H. C. Williams, How was F_6 factored? *Math. Comput.*, **61**(1993) 463–474.

Samuel Yates, Titanic primes, *J. Recreational Math.*, **16**(1983-84) 265–267.

Samuel Yates, Sinkers of the Titanics, *J. Recreational Math.*, **17**(1984-85) 268–274.

Samuel Yates, Tracking Titanics, in *The Lighter Side of Mathematics*, Proc. Strens Mem. Conf., Calgary 1986, Math. Assoc. of America, Washington DC, *Spectrum* series, 1993, 349–356.

Jeff Young & Duncan Buell, The twentieth Fermat number is composite, *Math. Comput.*, **50**(1988) 261–263.

A4 The prime number race.

A number a is said to be **congruent** to c, **modulo** a positive number b, written $a \equiv c \bmod b$, if b is a divisor of $a - c$. S. Chowla conjectured that if $a \perp b$, then there are infinitely many pairs of consecutive primes such that $p_n \equiv p_{n+1} \equiv a \bmod b$. The case $b = 4$, $a = 1$ follows from a theorem of Littlewood. Bounds between which such consecutive primes occur have been given in this case, and for $b = 4$, $a = 3$ by Knapowski & Turán. Turán observed that it would be of interest (in connexion with the Riemann hypothesis, for example) to discover long sequences of consecutive primes $\equiv 1 \bmod 4$. Den Haan found the nine primes

$$11593, 11597, 11617, 11621, 11633, 11657, 11677, 11681, 11689.$$

Four sequences of 10 such primes end at 373777, 495461, 509521 and 612217 and a sequence of 11 ends at 766373. Stephane Vandemergel has discovered no fewer than 16 consecutive primes of shape $4k + 1$; they are 207622000 + 273, 297, 301, 313, 321, 381, 409, 417, 421, 489, 501, 517, 537, 549, 553, 561.

Thirteen consecutive primes congruent to 3 mod 4 are 241000 + 603, 639, 643, 651, 663, 667, 679, 687, 691, 711, 727, 739 and 771.

If $p(b, a)$ is the least prime in the arithmetic progression $a + nb$, with $a \perp b$, then Linnik showed that there is a constant L, now called **Linnik's constant**, such that $p(b, a) \ll b^L$. Pan Cheng-Tung, Chen Jing-Run, Matti Jutila, Chen Jing-Run, Matti Jutila, S. Graham, Chen Jing-Run, Chen Jing-Run & Liu Jian-Min, and Wang Wei3 have successively improved the best known value of L to 5448, 777, 550, 168, 80, 36, 17, 13.5, and 8, and Heath-Brown has recently established the remarkableresult $L \leq 5.5$.

Elliott & Halberstam have shown that

$$p(b,a) < \phi(b)(\ln b)^{1+\delta}$$

almost always.

In the other direction it is known (see the papers of Prachar, Schinzel and Pomerance) that, given a, there are infinitely many values of b for which

$$p(b,a) > \frac{cb \ln b \ln \ln b \ln \ln \ln \ln b}{(\ln \ln \ln b)^2}$$

where c is an absolute constant.

Turán was particularly interested in the **prime number race**. Let $\pi(n; a, b)$ be the number of primes $p \le n$, $p \equiv a \bmod b$. Is it true that for every a and b with $a \perp b$, there are infinitely many values of n for which

$$\pi(n; a, b) > \pi(n : a_1, b)$$

for every $a_1 \not\equiv a \bmod b$? Knapowski & Turán settled special cases, but the general problem is wide open.

Chebyshev noted that $\pi(n; 1, 3) < \pi(n; 2, 3)$ and $\pi(n; 1, 4) \le \pi(n; 3, 4)$ for small values of n. Leech, and independently Shanks & Wrench, discovered that the latter inequality is reversed for $n = 26861$ and Bays & Hudson that the former is reversed for two sets, each of more than 150 million integers, between $n = 608981813029$ and $n = 610968213796$.

Carter Bays & Richard H. Hudson, The appearance of tens of billions of integers x with $\pi_{24,13}(x) < \pi_{24,1}(x)$ in the vicinity of 10^{12}, *J. reine angew. Math.*, **299/300**(1978) 234–237; *MR* **57** #12418.

Carter Bays & Richard H. Hudson, Details of the first region of integers x with $\pi_{3,2}(x) < \pi_{3,1}(x)$, *Math. Comput.*, **32**(1978) 571–576.

Carter Bays & Richard H. Hudson, Numerical and graphical description of all axis crossing regions for the moduli 4 and 8 which occur before 10^{12}, *Internat. J. Math. Sci.*, **2**(1979) 111–119; *MR* **80h**:10003.

Chen Jing-Run, On the least prime in an arithmetical progression, *Sci. Sinica*, **14**(1965) 1868–1871; *MR* **32** #5611.

Chen Jing-Run, On the least prime in an arithmetical progression and two theorems concerning the zeros of Dirichlet's L-functions, *Sci. Sinica*, **20**(1977) 529–562; *MR* **57** #16227.

Chen Jing-Run, On the least prime in an arithmetical progression and theorems concerning the zeros of Dirichlet's L-functions II, *Sci. Sinica*, **22**(1979) 859–889; *MR* **80k**:10042.

Chen Jing-Run & Liu Jian-Min, On the least prime in an arithmetic progression III, IV, *Sci. China Ser. A*, **32**(1989) 654–673, 792–807; *MR* **91h**:11090ab.

S. Graham, On Linnik's constant, *Acta Arith.*, **39**(1981) 163–179; *MR* **83d**: 10050.

Andrew Granville & Carl Pomerance, On the least prime in certain arithmetic progressions, *J. London Math. Soc.*(2), **41**(1990) 193–200; *MR* **91i**:11119.

D. R. Heath-Brown, Siegel zeros and the least prime in an arithmetic progression, *Quart. J. Math. Oxford Ser.*(2), **41**(1990) 405–418; *MR* **91m**:11073.

D. R. Heath-Brown, Zero-free regions for Dirichlet L-functions and the least prime in an arithmetic progression, *Proc. London Math. Soc.*(3), **64**(1992) 265–338; *MR* **93a**:11075.

Richard H. Hudson, A common combinatorial principle underlies Riemann's formula, the Chebyshev phenomenon, and other subtle effects in comparative prime number theory I, *J. reine angew. Math.*, **313**(1980) 133–150.

Richard H. Hudson, Averaging effects on irregularities in the distribution of primes in arithmetic progressions, *Math. Comput.*, **44**(1985) 561–571; *MR* **86h**:11064.

Matti Jutila, A new estimate for Linnik's constant, *Ann. Acad. Sci. Fenn. Ser. A I No.* 471 (1970), *8pp.*; *MR* **42** #5939.

Matti Jutila, On Linnik's constant, *Math. Scand.*, **41**(1977) 45–62; *MR* **57** #16230.

S. Knapowski & P. Turán, Über einige Fragen der vergleichenden Primzahltheorie, *Number Theory and Analysis*, Plenum Press, New York, 1969, 157–171.

S. Knapowski & P. Turán, On prime numbers $\equiv 1$ resp. 3 mod 4, *Number Theory and Algebra*, Academic Press, New York, 1977, 157–165; *MR* **57** #5926.

John Leech, Note on the distribution of prime numbers, *J. London Math. Soc.*, **32**(1957) 56–58.

U. V. Linnik, On the least prime in an arithmetic progression I. The basic theorem. II. The Deuring-Heilbronn phenomenon. *Rec. Math.* [*Mat. Sbornik*] *N.S.*, **15**(**57**)(1944) 139–178, 347–368; *MR* **6**, 260bc.

Pan Cheng-Tung, On the least prime in an arithmetic progression, *Sci. Record* (*N.S.*) **1**(1957) 311–313; *MR* **21** #4140.

Carl Pomerance, A note on the least prime in an arithmetic progression, *J. Number Theory*, **12**(1980) 218–223; *MR* **81m**:10081.

K. Prachar, Über die kleinste Primzahl einer arithmetischen Reihe, *J. reine angew. Math.*, **206**(1961) 3–4; *MR* **23** #A2399; and see Andrzej Schinzel, Remark on the paper of K. Prachar, **210**(1962) 121–122; *MR* **27** #118.

Daniel Shanks, Quadratic residues and the distribution of primes, *Math. Tables Aids Comput.*, **13**(1959) 272–284.

Wang Wei$_3$, On the least prime in an arithmetic progression, *Acta Math. Sinica*(**N.S.**), **7**(1991) 279–289; *MR* **93c**:11073.

A5 Arithmetic progressions of primes.

How long can an arithmetic progression be which consists only of primes? Table 1 shows progressions of n primes, a, $a+d$, $\ldots$, $a+(n-1)d$, discovered by James Fry, V.A. Golubev, Andrew Moran, Paul Pritchard, S.C. Root, W.N. Seredinskii, S. Weintraub and Jeff Young (see the first edition for earlier, smaller discoveries). Of course, the common difference must have every prime $p \le n$ as a divisor (unless $n = a$). It is conjectured that n can be as large as you like. This would follow if it were possible to improve Szemerédi's theorem (see **E10**).

More generally, Erdős conjectures that if $\{a_i\}$ is any infinite sequence of integers for which $\sum 1/a_i$ is divergent, then the sequence contains arbitrarily long arithmetic progressions. He offers $3000.00 for a proof or disproof of this conjecture.

Table 1. Long Arithmetic Progressions of Primes.

n	d	a	$a+(n-1)d$	source
12	30030	23143	353473	G, 1958
13	510510	766439	6892559	S, 1965
14	2462460	46883579	78895559	
16	9699690	53297929	198793279	
16	223092870	2236133941	5582526991	R, 1969
17	87297210	3430751869	4827507229	W, 1977
18	717777060	4808316343	17010526363	P
19	4180566390	8297644387	83547839407	P
19	13608665070	244290205469	489246176729	F, Mar 1987
20	2007835830	803467381001	841616261771	F, Mar 1987
20	7643355720	1140997291211	1286221049891	F, Mar 1987
20	18846497670	214861583621	572945039351	Y&F, 87-09-01
20	1140004565700	1845449006227	23505535754527	M&P, Nov 1990
20	19855265430	24845147147111	25222397190281	M&P, Nov 1990
21	1419763024680	142072321123	28537332814723	M&P, 90-11-30

Sierpiński defines $g(x)$ to be the maximum number of terms in a progression of primes not greater than x. The least x, $l(x)$, for which $g(x)$ takes the values

$$g(x) = 0 \quad 1 \quad 2 \quad 3 \quad 4 \quad 5 \quad 6 \quad 7 \quad 8 \quad 9 \quad 10 \quad \ldots$$

is

$$l(x) = 1 \quad 2 \quad 3 \quad 7 \quad 23 \quad 29 \quad 157 \quad 1307 \quad 1669 \quad 1879 \quad 2089 \quad \ldots$$

Günter Löh has searched for arithmetic progressions of primes with first term q and length q. Examples are $(q, d) = (7, 150), (11, 1536160080)$ and $(13, 9918821194590)$.

Pomerance produces the "prime number graph" by plotting the points (n, p_n) and shows that for every k we can find k primes whose points are collinear.

Grosswald has shown that there are long arithmetic progressions consisting only of **almost primes**, in the following sense. There are infinitely many arithmetic progressions of k terms, each term being the product of at most r primes, where

$$r \leq \lfloor k \ln k + 0.892k + 1 \rfloor.$$

He has also shown that the Hardy-Littlewood estimate is of the right order of magnitude for 3-term arithmetic progressions of primes.

P. D. T. A. Elliott & H. Halberstam, The least prime in an arithmetic progression, in *Studies in Pure Mathematics* (*Presented to Richard Rado*), Academic Press, London, 1971, 59–61; *MR* **42** #7609.

P. Erdős & P. Turán, On certain sequences of integers, *J. London Math. Soc.*, **11**(1936) 261–264.

J. Gerver, The sum of the reciprocals of a set of integers with no arithmetic progression of k terms, *Proc. Amer. Math. Soc.*, **62**(1977) 211–214.

Joseph L. Gerver & L. Thomas Ramsey, Sets of integers with no long arithmetic progressions generated by the greedy algorithm, *Math. Comput.*, **33**(1979) 1353–1359.

V. A. Golubev, Faktorisation der Zahlen der Form $x^3 \pm 4x^2 + 3x \pm 1$, *Anz. Oesterreich. Akad. Wiss. Math.-Naturwiss. Kl.*, **1969** 184–191 (see also 191–194; 297–301; **1970**, 106–112; **1972**, 19–20, 178–179).

Emil Grosswald, Long arithmetic progressions that consist only of primes and almost primes, *Notices Amer. Math. Soc.*, **26**(1979) A451.

Emil Grosswald, Arithmetic progressions of arbitrary length and consisting only of primes and almost primes, *J. reine angew. Math.*, **317**(1980) 200–208.

Emil Grosswald, Arithmetic progressions that consist only of primes, *J. Number Theory*, **14**(1982) 9–31.

Emil Grosswald & Peter Hagis, Arithmetic progressions consisting only of primes, *Math. Comput.*, **33**(1979) 1343–1352; *MR* **80k**:10054.

H. Halberstam, D. R. Heath-Brown & H.-E. Richert, On almost-primes in short intervals, in *Recent Progress in Analytic Number Theory*, Vol. 1, Academic Press, 1981, 69–101; *MR* **83a**:10075.

D. R. Heath-Brown, Three primes and an almost-prime in arithmetic progression, *J. London Math. Soc.*, (2) **23**(1981) 396–414.

D. R. Heath-Brown, Almost-primes in arithmetic progressions and short intervals, *Math. Proc. Cambridge Philos. Soc.*, **83**(1978) 357–375; *MR* **58** #10789.

Edgar Karst, 12–16 primes in arithmetical progression, *J. Recreational Math.*, **2**(1969) 214–215.

Edgar Karst, Lists of ten or more primes in arithmetical progression, *Scripta Math.*, **28**(1970) 313–317.

Edgar Karst & S. C. Root, Teilfolgen von Primzahlen in arithmetischer Progression, *Anz. Oesterreich. Akad. Wiss. Math.-Naturwiss. Kl.*, **1972**, 19–20 (see also 178–179).

Andrew Moran & Paul Pritchard, The design of a background job on a local area network, *Proc. 14th Austral. Comput. Sci. Conf.*,

Carl Pomerance, The prime number graph, *Math. Comput.*, **33**(1979) 399–408; *MR* **80d**:10013.

Paul Pritchard, Eighteen primes in arithmetic progression, *Math. Comput.*, **41** (1983) 697.

Paul Pritchard, Long arithmetic progressions of primes: some old, some new, *Math. Comput.*, **45**(1985) 263–267.

W. Sierpiński, Remarque sur les progressions arithmétiques, *Colloq. Math.*, **3** (1955) 44–49.

Sol Weintraub, Primes in arithmetic progression, *BIT* **17**(1977) 239–243.

K. Zarankiewicz, Problem 117, *Colloq. Math.*, **3**(1955) 46, 73.

A6 Consecutive primes in A.P.

It has even been conjectured that there are arbitrarily long arithmetic progressions of *consecutive* primes, such as

$$251, 257, 263, 269 \quad \text{and} \quad 1741, 1747, 1753, 1759.$$

Jones, Lal & Blundon discovered the sequence $10^{10} + 24493 + 30k$ $(0 \le k \le 4)$ of five consecutive primes, and Lander & Parkin, soon after, found six such primes, $121174811 + 30k$ $(0 \le k \le 5)$. They also established that $9843019 + 30k$ $(0 \le k \le 4)$ is the least progression of five terms, that there are 25 others less than $3 \cdot 10^8$, but no others of length six.

It is not known if there are infinitely many sets of three *consecutive* primes in arithmetic progression, but S. Chowla has demonstrated this without the restriction to consecutive primes.

Harry Nelson has collected the $100.00 prize that Martin Gardner offered to the first discoverer of a 3×3 magic square whose nine entries are consecutive primes. These are *not* in arithmetic progression, of course. The central prime is 1480028171 and the others are this ± 12, ± 18, ± 30 and ± 42. He found more than 20 other such squares.

S. Chowla, There exists an infinity of 3-combinations of primes in A.P., *Proc. Lahore Philos. Soc.*, **6** no. 2(1944) 15–16; *MR* **7**, 243.

P. Erdős & A. Rényi, Some problems and results on consecutive primes, *Simon Stevin*, **27**(1950) 115–125; *MR* **11**, 644.

M. F. Jones, M. Lal & W. J. Blundon, Statistics on certain large primes, *Math. Comput.*, **21**(1967) 103–107; *MR* **36** #3707.

L. J. Lander & T. R. Parkin, Consecutive primes in arithmetic progression, *Math. Comput.*, **21**(1967) 489.

H. L. Nelson, A consecutive prime 3×3 magic square, *J. Recreational Math.*

A7 Cunningham chains.

A common method for proving that p is a prime involves the factorization of $p-1$. If $p-1 = 2q$, where q is another prime, the size of the problem has only been reduced by a factor of 2, so it's interesting to observe **Cunningham chains** of primes with each member one more than twice the previous one. D. H. Lehmer found just three such chains of 7 primes with least member $< 10^7$:

1122659, 2245319, 4490639, 8981279, 17962559, 35925119, 71850239

2164229, 4328459, 8656919, 17313839, 34627679, 69255359, 138510719

2329469, 4658939, 9317879, 18635759, 37271519, 74543039, 149086079

and two others with least members 10257809 and 10309889. The factorization of $p+1$ can also be used to prove that p is prime. Lehmer found seven chains of length 7 based on $p+1=2q$. The first three had least members 16651, 67651 and 165901, but the second of these must be discarded, since the fifth member is $1082401 = 601 \cdot 1801$ (curiously, this is a divisor of $2^{25}-1$).

Lalout & Meeus found chains of length 8 of each kind, starting with 19099919 and 15514861, and these are the smallest of this length. Günter Löh has found many new chains: the least of length 9 start with 85864769 and 857095381; of length 10 with 26089808579 and 205528443121; of length 11 with 665043081119 and 1389122693971; of length 12 with 554688278429 and 216857744866621; and a chain of length 13 of the second kind starts with 758083947856951. A count of all chains of the first kind starting below 10^{11} and of length 6, 7, 8, 9, 10 gave the respective frequencies 19991, 2359, 257, 21, 2.

Claude Lalout & Jean Meeus, Nearly-doubled primes, *J. Recreational Math.*, **13** (1980/81) 30–35.

D. H. Lehmer, Tests for primality by the converse of Fermat's theorem, *Bull. Amer. Math. Soc.*, **33**(1927) 327–340.

D. H. Lehmer, On certain chains of primes, *Proc. London Math. Soc.*, **14A** (Littlewood 80 volume, 1965) 183–186.

Günter Löh, Long chains of nearly doubled primes, *Math. Comput.*, **53**(1989) 751–759; *MR* **90e**:11015.

A8 Gaps between primes. Twin primes.

There are many problems concerning the gaps between consecutive primes. Write $d_n = p_{n+1} - p_n$ so that $d_1 = 1$ and all other d_n are even. How large and how small can d_n be? Rankin has shown that

$$d_n > \frac{c \ln n \ln\ln n \ln\ln\ln\ln n}{(\ln\ln\ln n)^2}$$

for infinitely many n and Erdős offers $5,000 for a proof or disproof that the constant c can be taken arbitrarily large. Rankin's best value is $c = e^\gamma$ where γ is Euler's constant: Maier & Pomerance have improved this by a factor $k \approx 1.31256$, the root of the equation $4/k - e^{-4/k} = 3$, and Pintz has made a further improvement.

A very famous conjecture is the Twin Prime Conjecture, that $d_n = 2$ infinitely often. If $n > 6$, are there always twin primes between n and $2n$? Conjecture B of Hardy and Littlewood (cf. **A1**) is that $P_k(n)$, the number of pairs of primes less than n and differing by an even number k, is given asymptotically by

$$P_k(n) \sim \frac{2cn}{(\ln n)^2} \prod \left(\frac{p-1}{p-2}\right)$$

where the product is taken over all odd prime divisors of k (and so is empty and taken to be 1 when k is a power of 2) and $c = \prod(1 - 1/(p-1)^2)$ taken over all odd primes, so that $2c \approx 1.32032$. If $\pi_{1,2}(n)$ is the number of primes p such that $p+2$ has at most two prime factors, then Fouvry & Grupp have shown that

$$\pi_{1,2}(n) \geq 0.71 \times \frac{2cn}{(\ln n)^2}$$

and 0.71 has been improved to 1.015 by Liu and then to 1.05 by Wu.

The large twin primes $9 \cdot 2^{211} \pm 1$ were discovered by the Lehmers and independently by Riesel. Crandall & Penk found twin primes with 64, 136, 154, 203 and 303 digits, Williams found $156 \cdot 5^{202} \pm 1$, Baillie $297 \cdot 2^{546} \pm 1$, Atkin & Rickert

$$694503810 \cdot 2^{2304} \pm 1 \quad \text{and} \quad 1159142985 \cdot 2^{2304} \pm 1$$

and in 1989 Brown, Noll, Parady, Smith, Smith & Zarantonello found

$$663777 \cdot 2^{7650} \pm 1, \quad 571305 \cdot 2^{7701} \pm 1, \quad 1706595 \cdot 2^{11235} \pm 1.$$

On 93:08:16 Harvey Dubner announced a new record with

$$2^{4025} \cdot 3 \cdot 5^{4020} \cdot 7 \cdot 11 \cdot 13 \cdot 79 \cdot 223 \pm 1,$$

numbers with 4030 decimal digits.

Richard Brent counted 224376048 primes p less than 10^{11} for which $p+2$ is also prime; about 9% more than predicted by Conjecture B.

Bombieri & Davenport have shown that

$$\liminf \frac{d_n}{\ln n} < \frac{2+\sqrt{3}}{8} \approx 0.46650$$

(no doubt the real answer is zero; of course the truth of the Twin Prime Conjecture would imply this); G.Z. Pilt′yaĭ has improved the constant on the right to $(2\sqrt{2}-1)/4 \approx 0.45711$; Uchiyama to $(9-\sqrt{3})/16 \approx 0.454256$; Huxley to $(4\sin\theta + 3\theta)/(16\sin\theta) \approx 0.44254$, where $\theta + \sin\theta = \pi/4$, and later to 0.4394; and Helmut Maier to 0.248.

Huxley has also shown that

$$d_n < p_n^{7/12+\epsilon},$$

Heath-Brown & Iwaniec have improved the exponent to 11/20; Mozzochi to 0.548; and Lou & Yao to 6/11. Cramér proved, using the Riemann hypothesis, that

$$\sum_{n<x} d_n^2 < cx(\ln x)^4.$$

Erdős conjectures that the right-hand side should be $cx(\ln x)^2$, but thinks that there is no hope of a proof. The Riemann hypothesis implies that $d_n < p_n^{1/2+\epsilon}$.

Dorin Andrica conjectures that, for all natural n,

$$¿ \quad \sqrt{p_{n+1}} - \sqrt{p_n} < 1 \quad ?$$

Dan Grecu has verified this for $p_n < 10^6$. In *Amer. Math. Monthly,* **83**(1976) 61, it is given as a difficult unsolved problem that

$$¿ \quad \lim_{n\to\infty} (\sqrt{p_{n+1}} - \sqrt{p_n}) = 0 \quad ?$$

If true, this implies Andrica's conjecture for large enough n, which is comparable with that of Cramér, mentioned in **A2**, and with the following one of Shanks, who has given a heuristic argument which supports the conjecture that if $p(g)$ is the first prime that follows a gap of g between consecutive primes, then $\ln p(g) \sim \sqrt{g}$. Record gaps between consecutive primes have been observed by Lehmer, Lander & Parkin, Brent, Weintraub, Young & Potler and others. Table 2 illustrates Shanks's conjecture.

Table 2. Earliest large gaps between consecutive primes.

g	$p(g)$	$(\ln p)^2$	$g/(\ln p)^2$
456	25056082543	573.33	0.7953
464	42652618807	599.09	0.7745
468	127976335139	654.09	0.7155
474	182226896713	672.29	0.7051
486	241160624629	686.90	0.7075
490	297501076289	697.95	0.7021
500	303371455741	698.98	0.7153
514	304599509051	699.19	0.7351
516	416608696337	715.85	0.7208
532	461690510543	721.36	0.7375
534	614487454057	736.80	0.7247
540	738832928467	746.84	0.7230
582	1346294311331	779.99	0.7462
588	1408695494197	782.53	0.7514
602	1968188557063	801.35	0.7512
652	2614941711251	817.52	0.7975
674	7177162612387	876.27	0.7692
716	13829048560417	915.53	0.7821
766	19581334193189	936.70	0.8178
778	42842283926129	985.24	0.7897
804	90874329412297	1033.01	0.7783

The last entry is an unpublished result of Aaron Potler & Jeff Young, communicated Aug. 1993.

Chen Jing-Run showed that, for x large enough, there is always a number with at most two prime factors in the interval $[x - x^\alpha, x]$ for any value of $\alpha \geq 0.477$. Halberstam, Heath-Brown & Richert (see reference at **A5**) showed that in such an interval with $\alpha = 0.455$ there are at least $x^\alpha / 121 \ln x$ numbers with at most two prime factors, and Iwaniec & Laborde further reduced the exponent to $\alpha = 0.45$.

Victor Meally used the phrase **prime deserts**. He notes that below 373 the commonest gap is 2; below 467 there are 24 gaps of each of 2, 4 and 6; below 563 the commonest gap is 6, as it is between 10^{14} and $10^{14} + 10^8$ and probably also from 2 to 10^{14}. He asks: when does 30 take over as the commonest gap? Conway & Odlyzko call d a **champion** for x if it occurs most frequently as the difference between consective primes $\leq x$. There may be more than one champion for the same x: $C(135) = 4$, $C(100) = \{2, 4\}$. They suggest that the only champions are 4 and the prime factorials 2, 6, 30, 210, 2310, $\ldots$. Do champions $\to \infty$? Does each prime p divide all champions for $x \geq x_0(p)$?

A. O. L. Atkin & N. W. Rickert, On a larger pair of twin primes, Abstract 79T-A132, *Notices Amer. Math. Soc.*, **26**(1979) A-373.

E. Bombieri & H. Davenport, Small differences between prime numbers, *Proc. Roy. Soc. Ser. A*, **293**(1966)1–18; *MR* **33** #7314.

Richard P. Brent, The first occurrence of large gaps between successive primes, *Math. Comput.*, **27** (1973) 959–963; *MR* **48** #8360; (and see *Math. Comput.*, **35** (1980) 1435–1436.

Richard P. Brent, Irregularities in the distribution of primes and twin primes, *Math. Comput.*, **29**(1975) 43–56.

J. H. Cadwell, Large intervals between consecutive primes, *Math. Comput.*, **25** (1971) 909–913.

Chen Jing-Run, On the distribution of almost primes in an interval II, *Sci. Sinica*, **22**(1979) 253–275; *Zbl.*, 408.10030.

Chen Jing-Run & Wang Tian-Ze, , *Acta Math. Sinica*, **32**(1989) 712–718; *MR* **91e**:11108.

Chen Jing-Run & Wang Tian-Ze, On distribution of primes in an arithmetical progression, *Sci. China Ser. A*, **33**(1990) 397–408; *MR* **91k**:11078.

H. Cramér, On the order of magnitude of the difference between consecutive prime numbers, *Acta Arith.*, **2**(1937) 23–46.

É. Fouvry & F. Grupp, On the switching principle in sieve theory, *J. reine angew. Math.*, **370**(1986) 101–126; *MR* **87j**:11092.

J. B. Friedlander & J. C. Lagarias, On the distribution in short intervals of integers having no large prime factor, *J. Number Theory*, **25**(1987) 249–273.

D. A. Goldston, On Bombieri and Davenport's theorem concerning small gaps between primes, *Mathematika*, **39**(1992) 10–17; *MR* **93h**:11102.

Glyn Harman, Short intervals containing numbers without large prime factors, *Math. Proc. Cambridge Philos. Soc.*, **109**(1991) 1–5.

Jan Kristian Haugland, Large prime-free intervals by elementary methods, *Normat*, **39**(1991) 76–77.

Martin Huxley, An application of the Fouvry-Iwaniec theorem, *Acta Arith.*, **43**(1984) 441–443.

H. Iwaniec & M. Laborde, P_2 in short intervals, *Ann. Inst. Fourier(Grenoble)*, **31**(1981) 37–56; *MR* **83e**:10061.

Jia Chao-Hua, Three primes theorem in a short interval VI (Chinese), *Acta Math. Sinica*, **34**(1991) 832–850; *MR* **93h**:11104.

Liu Hong-Quan, On the prime twins problem, *Sci. China Ser. A*, **33**(1990) 281–298; *MR* **91i**:11125.

Lou Shi-Tuo & Qi Yao, Upper bounds for primes in intervals (Chinese), *Chinese Ann. Math. Ser. A*, **10**(1989) 255–262; *MR* **91d**:11112.

Helmut Maier, Small differences between prime numbers, *Michigan Math. J.*, **35**(1988) 323–344.

Helmut Maier, Primes in short intervals, *Michigan Math. J.*, **32**(1985) 221–225.

Helmut Maier & Carl Pomerance, Unusually large gaps between consecutive primes, *Théorie des nombres*, (Quebec, PQ, 1987), de Gruyter, 1989, 625–632; *MR* **91a**:11045: and see *Trans. Amer. Math. Soc.*, **322**(1990) 201–237; *MR* **91b**:11093.

C. J. Mozzochi, On the difference between consecutive primes, *J. Number Theory* **24** (1986) 181–187.

Bodo K. Parady, Joel F. Smith & Sergio E. Zarantonello, Largest known twin primes, *Math. Comput.*, **55**(1990) 381–382; *MR* **90j**:11013.

Г. З. Пильтяй, О величине разности между соседними простыми числами, Исследования по теории чисел, вып. 4, Издательство Саратовского университета 1972, 73–79.

Daniel Shanks, On maximal gaps between successive primes, *Math. Comput.*, **18**(1964) 646–651; *MR* **29** #4745.

S. Uchiyama, On the difference between consecutive prime numbers, *Acta Arith.*, **27** (1975) 153–157.

Jie Wu, Sur la suite des nombres premiers jumeaux, *Acta Arith.*, **55**(1990) 365–394; *MR* **91j**:11074.

Jeff Young & Aaron Potler, First occurrence prime gaps, *Math. Comput.*, **52**(1989) 221–224.

Alessandro Zaccagnini, A note on large gaps between consecutive primes in arithmetic progressions, *J. Number Theory*, **42**(1992) 100–102.

A9 Patterns of primes.

A conjecture more general than Chowla's (see **A4**) is that there are infinitely many sets of consecutive primes of any given pattern, provided that there are no congruence relations which rule them out. It seems likely, for example, that there are infinitely many triples of primes $\{6k-1, 6k+1, 6k+5\}$ and $\{6k+1, 6k+5, 6k+7\}$. This would be even harder to settle than the Twin Prime Conjecture, but its plausibility is of interest, since Hensley & Richards have shown that it is incompatible with the well-known conjecture (also due to Hardy & Littlewood)

$$¿¿¿ \qquad \pi(x+y) \le \pi(x) + \pi(y) \qquad ???$$

for all integers x, $y \ge 2$. We've put more queries than usual round this, since it is very likely to be false. Indeed, there's some hope of finding values of x and y which contradict it. However there's an alternative conjecture,

$$¿ \qquad \pi(x+y) \le \pi(x) + 2\pi(y/2) \qquad ?$$

that the Hensley-Richards method doesn't comment on.

Montgomery & Vaughan showed that

$$\pi(x+y) - \pi(x) \le 2y/\ln y$$

and Iwaniec observed that for each θ, $0 < \theta < 1$, there is an $\eta(\theta) > \theta$ such that

$$\pi(x + x^\theta) - \pi(x) < (2+\epsilon)x^\theta/(\eta(\theta)\ln x)$$

for sufficiently large x and he found that $\eta(\theta) = \frac{5}{3}\theta - \frac{2}{9}$ for $\theta > \frac{1}{3}$, and that $\eta(\theta) = (1+\theta)/2$ for $\theta > \frac{1}{2}$. Lou & Yao improve this in part by showing that $\eta(\theta) = (100\theta - 45)/11$ for $\frac{6}{11} < \theta \le \frac{11}{20}$.

C. W. Trigg reported that in 1978 M. A. Penk found four primes p, $p+2$, $p+6$ and $p+8$ where

$$p = 802359150003121605557551380867519560344356971.$$

H. F. Smith noted that the pattern 11, 13, 17, 19, 23, 29, 31, 37 is repeated at least three times, starting with the primes 15760091, 25658841 and 93625991. In none of these cases is the number corresponding to 41 a prime, although $n-11$, $n-13$, $\dots n-41$ are all primes for $n = 88830$ and 855750.

Leech gave as an unsolved problem to find 33 consecutive numbers greater than 11 which include 10 primes. In 1961 Herschel Smith found 20 such sets and also 5 examples of 37 consecutive numbers containing 11 primes. Smith writes that Selfridge noted some errors in his 1957 paper. Sten Säfholm found primes

$$\{n+11, \dots, n+43\} \quad \text{for} \quad n = 33081664140$$

and rediscovered Smith's first three examples, that each of

$$\{n-11, \dots, n-43\} \quad \text{is prime for} \quad n = 9853497780,$$

for $n = 21956291910$ and for $n = 22741837860$. Leech wondered why the latter sets seem to occur more readily than the former. My guess is that this is just a more complicated version of the prime number race (see **A4**) and that with much more high-powered telescopes we'd see the balance being redressed (infinitely often). Dimitrios Betsis & Sten Säfholm have found many more patterns, culminating in $\{n+11,\ldots,n+61\}$ for $n=$ 21817283854511250 and $\{n-11,\ldots,n-61\}$ for $n=$ 79287805466244270.

Erdős asks, for each k, what is the smallest l for which p_k, p_{k+1}, ..., p_{k+l-1} is the only set of l consecutive primes with this pattern. E.g., the pattern 3, 5, 7 cannot occur again. The pattern 5, 7, 11, 13, 17 repeats at 101, 103, 107, 109, 113 and no doubt occurs infinitely often, but considerations mod 5 show that the pattern 5, 7, 11, 13, 17, 19 does not occur again. $(p_k,l)=(2,2)$, (3,3), (5,6), ...

Antal Balog, The prime k-tuplets conjecture on the average, *Analytic Number Theory (Allerton Park, IL,* 1989), 47–75, *Progr. Math.*, **85**, Birkhäuser, Boston, 1990.

Paul Erdős & Ian Richards, Density functions for prime and relatively prime numbers, *Monatsh. Math.*, **83**(1977) 99–112; *Zbl.* **355**.10034.

John B. Friedlander & Andrew Granville, Limitations to the equi-distribution of primes I, IV, III, *Ann. of Math.*(2), **129**(1989) 363–382; *MR* **90e**:11125; *Proc. Roy. Soc. London Ser. A*, **435**(1991) 197–204; *MR* **93g**:11098; *Compositio Math.*, **81**(1992) 19–32.

John B. Friedlander, Andrew Granville, Adolf Hildebrandt & Helmut Maier, Oscillation theorems for primes in arithmetic progressions and for sifting functions, *J. Amer. Math. Soc.*, **4**(1991) 25–86; *MR* **92a**:11103.

Douglas Hensley & Ian Richards, On the incompatibility of two conjectures concerning primes, *Proc. Symp. Pure Math.*, (Analytic Number Theory, St. Louis, 1972) **24** 123–127.

H. Iwaniec, On the Brun-Titchmarsh theorem, *J. Math. Soc. Japan*, **34**(1982) 95–123; *MR* **83a**:10082.

John Leech, Groups of primes having maximum density, *Math. Tables Aids Comput.*, **12**(1958) 144–145; *MR* **20** #5163.

H. L. Montgomery & R. C. Vaughan, The large sieve, *Mathematika*, **20**(1973) 119–134; *MR* **51** #10260.

Ian Richards, On the incompatibility of two conjectures concerning primes; a discussion of the use of computers in attacking a theoretical problem, *Bull. Amer. Math. Soc.*, **80**(1974) 419–438.

Herschel F. Smith, On a generalization of the prime pair problem, *Math. Tables Aids Comput.*, **11**(1957) 249–254; *MR* **20** #833.

Charles W. Trigg, A large prime quadruplet, *J. Recreational Math.*, **14** (1981/82) 167.

Sheng-Gang Xie, The prime 4-tuplet problem (Chinese. English summary), *Sichuan Daxue Xuebao*, **26**(1989) 168–171; *MR* **91f**:11066.

A10 Gilbreath's conjecture.

Define d_n^k by $d_n^1=d_n$ and $d_n^{k+1}=|d_{n+1}^k-d_n^k|$, that is, the successive absolute differences of the sequence of primes (Figure 2). N. L. Gilbreath conjectured (and Hugh Williams notes that Proth long before claimed to have proved) that $d_1^k = 1$ for all k. This was verified for $k < 63419$ by Killgrove & Ralston. Odlyzko has checked it for primes up to $\pi(10^{13}) \approx 3 \cdot 10^{11}$; he only needed to examine the first 635 differences.

```
1  2  3  4   5  6  7  8  9 10 11 12 13 14 15 16 17 18 19 20 21 22 23 24
2  3  5  7  11 13 17 19 23 29 31 37 41 43 47 53 59 61 67 71 73 79 83 89
 1  2  2  4  2  4  2  4  6  2  6  4  2  4  6  6  2  6  4  2  6  4  6
   1  0  2  2  2  2  2  2  4  4  2  2  2  2  0  4  4  2  2  4  2  2
     1  2  0  0  0  0  0  2  0  2  0  0  0  2  4  0  2  0  2  2  0
       1  2  0  0  0  0  2  2  2  2  0  0  2  2  4  2  2  2  0  2
         1  2  0  0  0  2  0  0  0  2  0  2  0  2  2  0  0  2  2
           1  2  0  0  2  2  0  0  2  2  2  2  2  0  2  0  2  0
             1  2  0  2  0  2  0  2  0  0  0  0  2  2  2  2  2
               1  2  2  2  2  2  2  2  0  0  0  2  0  0  0  0
                 1  0  0  0  0  0  0  2  0  0  2  2  0  0  0
```

Figure 2. Successive Absolute Differences of the Sequence of Primes.

Hallard Croft and others have suggested that it has nothing to do with primes as such, but will be true for any sequence consisting of 2 and odd numbers, which doesn't increase too fast, or have too large gaps. Odlyzko discusses this.

R. B. Killgrove & K. E. Ralston, On a conjecture concerning the primes, *Math. Tables Aids Comput.*, **13**(1959) 121–122; *MR* **21** #4943.

Andrew M. Odlyzko, Iterated absolute values of differences of consecutive primes, *Math. Comput.*, **61**(1993) 373–380; *MR* **93k**:11119.

F. Proth, Sur la série des nombres premiers, *Nouv. Corresp. Math.*, **4**(1878) 236–240.

A11 Increasing and decreasing gaps.

Since the proportion of primes gradually decreases, albeit somewhat erratically, $d_m < d_{m+1}$ infinitely often, and Erdős & Turán have shown that the same is true for $d_n > d_{n+1}$. They have also shown that the values of n for which $d_n > d_{n+1}$ have positive lower density, but it is not known if there are infinitely many decreasing or increasing sets of *three* consecutive values of d_n. If there were not, then there is an n_0 so that for every i and $n > n_0$ we have $d_{n+2i} > d_{n+2i+1}$ and $d_{n+2i+1} < d_{n+2i+2}$. Erdős offers \$100.00 for a proof that such an n_0 does not exist. He and Turán could not even prove that for $k > k_0$, $(-1)^r(d_{k+r+1} - d_{k+r})$ can't always have the same sign.

P. Erdős, On the difference of consecutive primes, *Bull. Amer. Math. Soc.*, **54**(1948) 885–889; *MR* **10**, 235.

P. Erdős & P. Turán, On some new questions on the distribution of prime numbers, *Bull. Amer. Math. Soc.*, **54**(1948) 371–378; *MR* **9**, 498.

A12 Pseudoprimes. Euler pseudoprimes. Strong pseudoprimes.

Pomerance, Selfridge & Wagstaff call an odd composite n for which $a^{n-1} \equiv 1 \bmod n$ a **pseudoprime to base** a (psp(a)). This usage is introduced to avoid the clumsy "composite pseudoprime" which appears throughout the literature. Odd composite n which are psp(a) for every a prime to n are **Carmichael numbers** (see **A13**). An odd composite n is an **Euler pseudoprime to base** a (epsp(a)) if $a \perp n$ and $a^{(n-1)/2} \equiv (\frac{a}{n}) \bmod n$, where $(\frac{a}{n})$ is the Jacobi symbol (see **F5**). Finally, an odd composite n with $n-1 = d \cdot 2^s$, d odd, is a **strong pseuodoprime to base** a (spsp(a)) if $a^d \equiv 1 \bmod n$ (otherwise $a^{d \cdot 2^r} \equiv -1 \bmod n$ for some r, $0 \le r < s$). These definitions are illustrated by a Venn diagram (Figure 3) which displays the smallest member of each set.

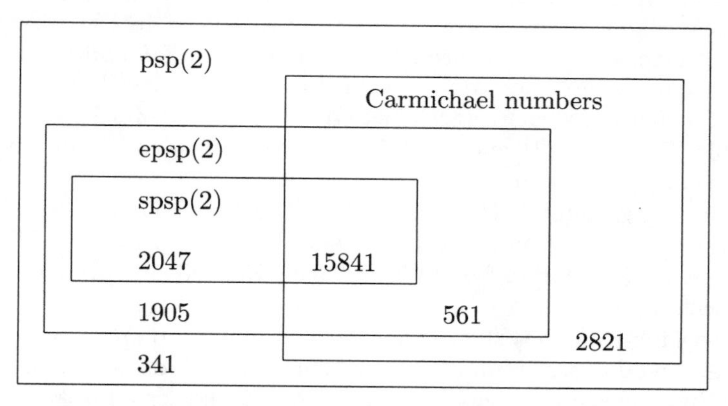

Figure 3. Relationships of Sets of psps with Least Element in Each Set.

The following values of $P_2(x)$, $E_2(x)$, $S_2(x)$ and $C(x)$ — the numbers of psp(2), epsp(2), spsp(2) and Carmichael numbers less than x, respectively — were given by Pomerance, Selfridge & Wagstaff:

x	10^3	10^4	10^5	10^6	10^7	10^8	10^9	10^{10}	$2.5 \cdot 10^{10}$
$P_2(x)$	3	22	78	245	750	2057	5597	14884	21853
$E_2(x)$	1	12	36	114	375	1071	2939	7706	11347
$S_2(x)$	0	5	16	46	162	488	1282	3291	4842
$C(x)$	1	7	16	43	105	255	646	1547	2163

Lehmer and Erdős showed that, for sufficiently large x,

$$c_1 \ln x < P_2(x) < x \exp\{-c_2(\ln x \ln \ln x)^{1/2}\}$$

and Pomerance improved these bounds to

$$\exp\{(\ln x)^{5/14}\} < P_2(x) < x\exp\{(-\ln x \ln\ln\ln x)/2\ln\ln x\}$$

and has a heuristic argument that the true estimate is the upper bound with the 2 omitted. The exponent 5/14 has been improved to 85/207 by Pomerance, using a result of Friedlander.

There are also examples of even numbers such that $2^n \equiv 2 \bmod n$. Lehmer found $161038 = 2 \cdot 73 \cdot 1103$ and Beeger showed that there are infinitely many. If F_n is the Fermat number $2^{2^n}+1$, Cipolla showed that $F_{n_1}F_{n_2}\cdots F_{n_k}$ is psp(2) if $k > 1$ and $n_1 < n_2 < \ldots < n_k < 2^{n_1}$.

If $P_n^{(a)}$ is the n-th psp(a), Szymiczek has shown that $\sum 1/P_n^{(2)}$ is convergent, while Mąkowski has shown that $\sum 1/\ln P_n^{(a)}$ is divergent. Rotkiewicz has a booklet on pseudoprimes which contains 58 problems and 20 conjectures.

For example, Problem #22 asks if there is a pseudoprime of form 2^N-2. Wayne McDaniel answers this affirmatively with $N = 465794$. Rotkiewicz has shown that the congruence $2^{n-2} \equiv 1 \bmod n$ has infinitely many composite solutions n. Shen Mok-Kong found five such less than a million, each of which ended in 7. McDaniel and Zhang Ming-Zhi have given the examples $73 \cdot 48544121$ and $524287 \cdot 13264529$ which each show that 3 is also a possible final digit.

Selfridge, Wagstaff & Pomerance offer \$500.00 + \$100.00 + \$20.00 for a composite $n \equiv 3$ or $7 \bmod 10$ which divides both $2^n - 2$ and the Fibonacci number u_{n+1} (see **A3**) or \$20.00 + \$100.00 + \$500.00 for a proof that there is no such n.

Shen Mok-Kong has shown that there are infinitely many k such that $2^{n-k} \equiv 1 \bmod n$ has infinitely many composite solutions n, and Kiss & Phong have shown that this is so for all $k \geq 2$ and for all $a \geq 2$ in place of 2.

Steven Arno, A note on Perrin pseudoprimes, *Math. Comput.*, **56**(1991) 371–376; *MR* **91k**:11011.

N. G. W. H. Beeger, On even numbers m dividing $2^m - 2$, *Amer. Math. Monthly*, **58**(1951) 553–555; *MR* **13**, 320.

R. D. Carmichael, On composite numbers P which satisfy the Fermat congruence $a^{P-1} \equiv 1 \bmod P$, *Amer. Math. Monthly*, **19**(1912) 22–27.

M. Cipolla, Sui numeri composti P che verificiano la congruenza di Fermat, $a^{P-1} \equiv 1 \pmod P$, *Annali di Matematica*, **9**(1904) 139–160.

P. Erdős, On the converse of Fermat's theorem, *Amer. Math. Monthly*, **56** (1949) 623–624; *MR* **11**, 331.

P. Erdős, On almost primes, *Amer. Math. Monthly*, **57**(1950) 404–407; *MR* **12**, 80.

P. Erdős, P. Kiss & A. Sárközy, A lower bound for the counting function of Lucas pseudoprimes, *Math. Comput.*, **51**(1988) 259–279; *MR* **89e**:11011.

P. Erdős & C. Pomerance, The number of false witnesses for a composite number, *Math. Comput.*, **46**(1986) 259–279; *MR* **87i**:11183.

J. B. Friedlander, Shifted primes without large prime factors, in *Number Theory and Applications*, (Proc. NATO Adv. Study Inst., Banff, 1988), Kluwer, Dordrecht, 1989, 393–401;

Daniel M. Gordon & Carl Pomerance, The distribution of Lucas and elliptic pseudoprimes, *Math. Comput.*, **57**(1991) 825–838; *MR* **92h**:11081; corrigendum **60**(1993) 877; *MR* **93h**:11108.

S. Gurak, Pseudoprimes for higher-order linear recurrence sequences, *Math. Comput.*, **55**(1990) 783–813.

Gerhard Jaeschke, On strong pseudoprimes to several bases, *Math. Comput.*, **61**(1993) 915–926.

I. Joó, On generalized Lucas pseudoprimes, *Acta Math. Hungar.*, **55**(1990) 315–322.

I. Joó & Phong Bui-Minh, On super Lehmer pseudoprimes, *Studia Sci. Math. Hungar.*, **25**(1990) 121–124; *MR* **92d**:11109.

Kim Su-Hee & Carl Pomerance, The probability that a random probable prime is composite, *Math. Comput.*, **53**(1989) 721–741; *MR* **90e**:11190.

Péter Kiss & Phong Bui-Minh, On a problem of A. Rotkiewicz, *Math. Comput.*, **48**(1987) 751–755; *MR* **88d**:11004.

D. H. Lehmer, On the converse of Fermat's theorem, *Amer. Math. Monthly*, **43**(1936) 347–354; II **56**(1949) 300–309; *MR* **10**, 681.

Andrzej Mąkowski, On a problem of Rotkiewicz on pseudoprime numbers, *Elem. Math.*, **29**(1974) 13.

A. Mąkowski & A. Rotkiewicz, On pseudoprime numbers of special form, *Colloq. Math.*, **20**(1969) 269–271; *MR* **39** #5458.

Wayne L. McDaniel, Some pseudoprimes and related numbers having special forms, *Math. Comput.*, **53**(1989) 407–409; *MR* **89m**:11006.

Carl Pomerance, A new lower bound for the pseudoprime counting function, *Illinois J. Math.*, **26**(1982) 4–9; *MR* **83h**:10012.

Carl Pomerance, On the distribution of pseudoprimes, *Math. Comput.*, **37** (1981) 587–593; *MR* **83k**:10009.

Carl Pomerance, Two methods in elementary analytic number theory, in *Number Theory and Applications*, (Proc. NATO Adv. Study Inst., Banff, 1988), Kluwer, Dordrecht, 1989, 135–161;

Carl Pomerance, John L. Selfridge & Samuel S. Wagstaff, The pseudoprimes to $25 \cdot 10^9$, *Math. Comput.*, **35**(1980) 1003–1026; *MR* **82g**:10030.

A. Rotkiewicz, *Pseudoprime Numbers and their Generalizations*, Student Association of the Faculty of Sciences, Univ. of Novi Sad, 1972; *MR* **48** #8373; *Zbl.* 324.10007.

A. Rotkiewicz, Sur les diviseurs composés des nombres $a^n - b^n$, *Bull. Soc. Roy. Sci. Liège*, **32**(1963) 191–195; *MR* **26** #3645.

A. Rotkiewicz, Sur les nombres pseudopremiers de la forme $ax + b$, *Comptes Rendus Acad. Sci. Paris*, **257**(1963) 2601–2604; *MR* **29** #61.

A. Rotkiewicz, Sur les formules donnant des nombres pseudopremiers, *Colloq. Math.*, **12**(1964) 69–72; *MR* **29** #3416.

A. Rotkiewicz, Sur les nombres pseudopremiers de la forme $nk+1$, *Elem. Math.*, **21**(1966) 32–33; *MR* **33** #112.

A. Rotkiewicz, On Euler-Lehmer pseudoprimes and strong Lehmer pseudoprimes with parameters L, Q in arithmetic progressions, *Math. Comput.*, **39** (1982) 239–247; *MR* **83k**:10004.

A. Rotkiewicz, On the congruence $2^{n-2} \equiv 1 (\text{mod } n)$, *Math. Comput.*, **43** (1984) 271–272; *MR* **85e**:11005.

Shen Mok-Kong, On the congruence $2^{n-k} \equiv 1 (\text{mod } n)$, *Math. Comput.*, **46**(1986) 715–716; *MR* **87e**:11005.

K. Szymiczek, On prime numbers p, q and r such that pq, pr and qr are pseudoprimes, *Colloq. Math.*, **13**(1964–65) 259–263; *MR* **31** #4757.

K. Szymiczek, On pseudoprime numbers which are products of distinct primes, *Amer. Math. Monthly*, **74**(1967) 35–37; *MR* **34** #5746.

S. S. Wagstaff, Pseudoprimes and a generalization of Artin's conjecture, *Acta Arith.*, **41**(1982) 151–161; *MR* **83m**:10004.

Masataka Yorinaga, Search for absolute pseudoprime numbers (Japanese), *Sûgaku*, **31** (1979) 374–376; *MR* **82c**:10008.

A13 Carmichael numbers.

The Carmichael numbers (psp(a) for all a prime to n, n composite) must be the product of at least three odd prime factors. As long ago as 1899 Korselt had given a necessary and sufficient condition for n to be a Carmichael number; namely that n be squarefree and such that $(p-1)|(n-1)$ for each p that divides n. The smallest example is $561 = 3 \cdot 11 \cdot 17$. More generally, if $p = 6k+1$, $q = 12k+1$ and $r = 18k+1$ are each prime, then pqr is a Carmichael number. It seems certain that there are infinitely many such triples of primes, but beyond our means to prove it. Alford, Granville & Pomerance have shown (by a different method!) that there are infinitely many Carmichael numbers, in fact, for sufficiently large x, more than x^β of them less than x, where

$$\beta = \frac{5}{12}\left(1 - \frac{1}{2\sqrt{e}}\right) > 0.290306 > \frac{2}{7}$$

Erdős had conjectured that $(\ln C(x))/\ln x$ tends to 1 as x tends to infinity and he improved a result of Knödel to show that

$$C(x) < x \exp\{-c \ln x \ln\ln\ln x / \ln\ln x\}.$$

Then Pomerance, Selfridge & Wagstaff (see **A12**) proved this with $c = 1-\epsilon$ and give a heuristic argument supporting the conjecture that the reverse inequality holds with $c = 2+\epsilon$.

They found 2163 Carmichael numbers $< 25 \cdot 10^9$ and Jaeschke finds 6075 more between that bound and 10^{12}; seven of these have eight prime factors. Richard Pinch has counted

	8241	19279	44706	105212	246683	such numbers
$<$	10^{12}	10^{13}	10^{14}	10^{15}	10^{16}	

A fair-sized specimen is

$$2013745337604001 = 17 \cdot 37 \cdot 41 \cdot 131 \cdot 251 \cdot 571 \cdot 4159.$$

J. R. Hill found the large Carmichael number pqr where $p = 5 \cdot 10^{19} + 371$, $q = 2p - 1$ and $r = 1 + (p-1)(q+2)/433$. Wagstaff produced a 321-digit example, and Woods & Huenemann one of 432 digits. Dubner has continued to beat this and his own records, a 3710-digit specimen being $N = PQR$ where $P = 6M + 1$, $Q = 12M + 1$ and $R = 1 + (PQ - 1)/X$ are primes given by $M = (TC - 1)^A/4$, T the product of the odd primes up to 47, $C = 141847$, $A = 41$ and $X = 123165$. But Günter Löh & Wolfgang Niebuhr have developed new algorithms which completely eclipse these by producing a Carmichael number with no fewer than 1101518 prime facors, a number of 16142049 decimal digits!

Alford, Granville & Pomerance proved that there are infinitely many Carmichael numbers n with the stronger requirement that $p|n$ implies $(p^2-1)|(n-1)$, but didn't know of any examples. Sid Graham found 18 such numbers, the smallest being

$$5893983289990395334700037072001 = 29 \cdot 31 \cdot 37 \cdot 43 \cdot 53 \cdot 67 \cdot 79 \cdot 89 \cdot 97 \cdot 151 \cdot 181 \cdot 191 \cdot 419 \cdot 881 \cdot 883$$

Richard Pinch had already found the smallest of all:

$$443372888629441 = 17 \cdot 31 \cdot 41 \cdot 43 \cdot 89 \cdot 97 \cdot 167 \cdot 331$$

Graham found 17 other numbers that satisfy the slightly weaker condition $\frac{p^2-1}{2}\,|(n-1)$.

W. Red Alford, Andrew Granville & Carl Pomerance, *Ann. of Math.* (to appear; 1992 preprint).

Robert Baillie & Samuel S. Wagstaff, Lucas pseudoprimes, *Math. Comput.*, **35** (1980) 1391–1417.

N. G. W. H. Beeger, On composite numbers n for which $a_{n-1} \equiv 1 \bmod n$ for every a prime to n, *Scripta Math.*, **16**(1950) 133–135.

R. D. Carmichael, Note on a new number theory function, *Bull. Amer. Math. Soc.*, **16**(1909–10) 232–238.

Harvey Dubner, A new method for producing large Carmichael numbers, *Math. Comput.*, **53**(1989) 411–414; *MR* **89m**:11013.

H. J. A. Duparc, On Carmichael numbers, *Simon Stevin*, **29**(1952) 21–24; *MR* **14**, 21f.

P. Erdős, On pseudoprimes and Carmichael numbers, *Publ. Math. Debrecen*, **4**(1956) 201–206; *MR* **18**, 18.

Andrew Granville, Prime testing and Carmichael numbers, *Notices Amer. Math. Soc.*, **39**(1992) 696–700.

D. Guillaume, Table de nombres de Carmichael inférieurs à 10^{12}, preprint, May 1991.

Jay Roderick Hill, Large Carmichael numbers with three prime factors, Abstract 79T-A136, *Notices Amer. Math. Soc.*, **26**(1979) A-374.

Gerhard Jaeschke, The Carmichael numbers to 10^{12}, *Math. Comput.*, **55** (1990) 383–389; *MR* **90m**:11018.

I. Joó & Phong Bui-Minh, On super Lehmer pseudoprimes, *Studia Sci. Math. Hungar.*, **25**(1990) 121–124.

W. Keller, The Carmichael numbers to 10^{13}, *Abstracts Amer. Math. Soc.*, **9**(1988) 328–329.

W. Knödel, Eine obere Schranke für die Anzahl der Carmichaelschen Zahlen kleiner als x, *Arch. Math.*, **4**(1953) 282–284; *MR* **15**, 289 (and see *Math. Nachr.*, **9**(1953) 343–350).

A. Korselt, Problème chinois, *L'intermédiaire des math.*, **6**(1899) 142–143.

D. H. Lehmer, Strong Carmichael numbers, *J. Austral. Math. Soc. Ser. A*, **21** (1976) 508–510.

G. Löh, Carmichael numbers with a large number of prime factors, *Abstracts Amer. Math. Soc.*, **9**(1988) 329; II (with W. Niebuhr) **10**(1989) 305.

Günter Löh & Wolfgang Niebuhr, New algorithms for constructing large Carmichael numbers, (92-11-01 preprint).

R. G. E. Pinch, The Carmichael numbers up to 10^{15}, *Math. Comput.*, **61** (1993) 381–391; *MR* **93m**:11137.

A. J. van der Poorten & A. Rotkiewicz, On strong pseudoprimes in arithmetic progressions, *J. Austral. Math. Soc. Ser. A*, **29**(1980) 316–321.

S. S. Wagstaff, Large Carmichael numbers, *Math. J. Okayama Univ.*, **22** (1980) 33–41; *MR* **82c**:10007.

H. C. Williams, On numbers analogous to Carmichael numbers, *Canad. Math. Bull.*, **20**(1977) 133–143.

Dale Woods & Joel Huenemann, Larger Carmichael numbers, *Comput. Math. Appl.*, **8**(1982) 215–216; *MR* **83f**:10017.

Masataka Yorinaga, Numerical computation of Carmichael numbers, I, II, *Math. J. Okayama Univ.*, **20**(1978) 151–163, **21**(1979) 183–205; *MR* **80d**:10026, **80j**:10002.

Masataka Yorinaga, Carmichael numbers with many prime factors, *Math. J. Okayama Univ.*, **22**(1980) 169–184; *MR* **81m**:10018.

Zhang Ming-Zhi, A method for finding large Carmichael numbers, *Sichuan Daxue Xuebao*, **29**(1992) 472–479; *MR* **93m**:11009.

A14 "Good" primes and the prime number graph.

Erdős and Straus called the prime p_n **good** if $p_n^2 > p_{n-1}p_{n+1}$ for all i, $1 \le i \le n-1$; for example, 5, 11, 17 and 29. Pomerance used the "prime number graph" (see **A5**) to show that there are infinitely many good primes. He asks the following questions. Is it true that the set of n for which p_n is good has density 0? Are there infinitely many n with $p_n p_{n+1} > p_{n-i}p_{n+1+i}$ for all i, $1 \le i \le n-1$? Are there infinitely many n with $p_n + p_{n+1} > p_{n-i} + p_{n+1+i}$ for all i, $1 \le i \le n-1$? Does the set of n for which $2p_n < p_{n-i} + p_{n+i}$ for all i, $1 \le i \le n-1$ have density 0? (Pomerance proved that there were infinitely many such n.) Is $\limsup\{\min_{0<i<n}(p_{n-i} + p_{n+i}) - 2p_n\} = \infty$?

A15 Congruent products of consecutive numbers.

Erdős, in a letter dated 79-10-31, observes that $3 \cdot 4 \equiv 5 \cdot 6 \cdot 7 \equiv 1 \bmod 11$ and asks for the least prime p such that there are integers a, k_1, k_2, k_3 and

$$\prod_{i=1}^{k_1}(a+i) \equiv \prod_{i=1}^{k_2}(a+k_1+i) \equiv \prod_{i=1}^{k_3}(a+k_1+k_2+i) \equiv 1 \bmod p.$$

He suggests that such primes p exist for any number of such congruent products.

Mąkowski sends examples corresponding to rows $n = 5$ and 6 in the table below [compare **F11**] and says that tables of indices can be used to find others. W. Narkiewicz also sends these examples, together with those in rows $n = 7$, 8 and 9 below. Landon Noll & Chuck Simmons generalize the problem slightly by asking for solutions of

$$q_1! \equiv q_2! \equiv \ldots \equiv q_n! \bmod p$$

and they give the least prime p for which there is a solution with n terms.

n	p	q_1	q_2	q_3	q_4	q_5	q_6	q_7	q_8	q_9	q_{10}	q_{11}
1	2	0										
2	2	0	1									
3	5	0	1	3								
4	17	0	1	5	11							
5	17	0	1	5	11	15						
6	23	0	1	4	8	11	21					
7	71	8	10	20	52	62	64	71				
8	599	29	51	123	184	251	290	501	540			
9	599	29	51	123	184	251	290	501	540	556		
10	3011	0	1	611	723	749	805	2205	2261	2287	2399	
11	3011	0	1	611	723	749	805	2205	2261	2287	2399	3009

Andrzej Mąkowski, On a number-theoretic problem of Erdős, *Elem. Math.*, **38** (1983) 101–102.

A16 Gaussian primes. Eisenstein-Jacobi primes.

Prime numbers can be defined in fields other than the rational field. In the complex number field they are called **Gaussian primes**. Many problems on ordinary primes can be reformulated for Gaussian primes.

Gaussian integers $a+bi$, where a, b are integers and $i^2 = -1$, behave like ordinary integers in the sense that there is **unique factorization** (apart from order, **units** $(\pm 1, \pm i)$ and **associates**; the associates of 7, for example, are 7, -7, $7i$ and $-7i$). Primes of shape $4k-1$ (3, 7, 11, 19, 23,

...) are still primes in the ring of Gaussian integers, but the other ordinary primes can be factored into Gaussian primes:

$$2 = (1+i)(1-i), \qquad 5 = (2+i)(2-i) = -(2i-1)(21+1), \text{ etc.}$$

$$13 = (2+3i)(2-3i), \qquad 17 = (4+i)(4-i), \qquad 29 = (5+2i)(5-2i), \ \ldots .$$

The Gaussian primes $\pm 1 \pm i$, $\pm 1 \pm 2i$, $\pm 2 \pm i$, ± 3, $\pm 3i$, $\pm 2 \pm 3i$, $\pm 3 \pm 2i$, $\pm 4 \pm i$, $\pm 1 \pm 4i$, $\pm 5 \pm 2i$, $\pm 2 \pm 5i$, ... make a pleasing pattern (Figure 4) when drawn on an Argand diagram, which has been used for tiling floors and weaving tablecloths.

Motzkin and Gordon asked if one can "walk" from the origin to infinity using the Gaussian primes as "stepping stones" and taking steps of bounded length. Presumably not. Jordan & Rabung have shown that steps of length at least 4 are necessary.

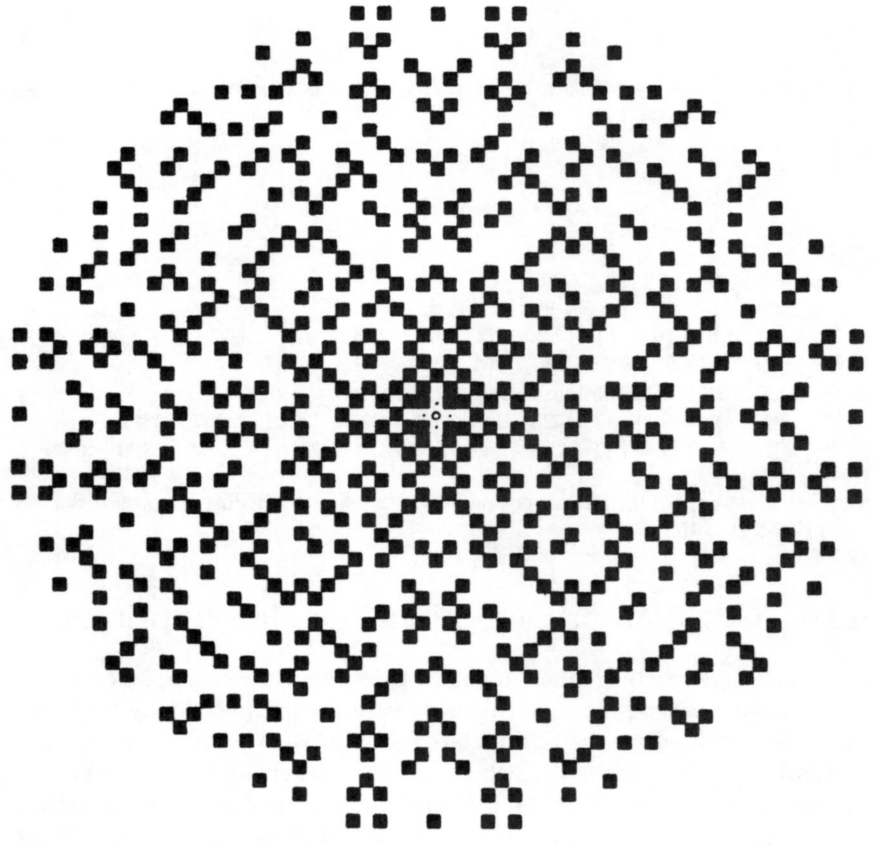

Figure 4. The Gaussian Primes with Norm Less Than 1000.

The **Eisenstein-Jacobi integers** $a + b\omega$, where a, b are integers and ω is a complex cube root of unity, $\omega^2 + \omega + 1 = 0$, also enjoy unique factorization. The primes again form a pattern (Figure 5), this time with hexagonal symmetry, because there are six units, ±1, $\pm\omega$, $\pm\omega^2$. The prime 2 and those of shape $6k-1$ (5, 11, 17, 23, 29, 41, ...,) are still Eisenstein-Jacobi primes, but 3 and those of shape $6k+1$ can be factored:

$$3 = (1-\omega)(1-\omega^2), \qquad 7 = (2-\omega)(2-\omega^2), \qquad 13 = (3-\omega)(3-\omega^2),$$

$$19 = (3-2\omega)(3-2\omega^2), \quad 31 = (5-\omega)(5-\omega^2), \quad 37 = (4-3\omega)(4-3\omega^2), \; \ldots .$$

John Leech asks for long arithmetic progressions of Gaussian primes and also of Eisenstein-Jacobi primes. He finds nine in Figure 4 and twelve in Figure 5. He later found the arithmetic progression

$$-8-13i, \; -3-8i, \; 2-3i, \; 7+2i, \; \ldots, \; 37+32i$$

of ten Gaussian primes, the last three of which are outside Figure 4.

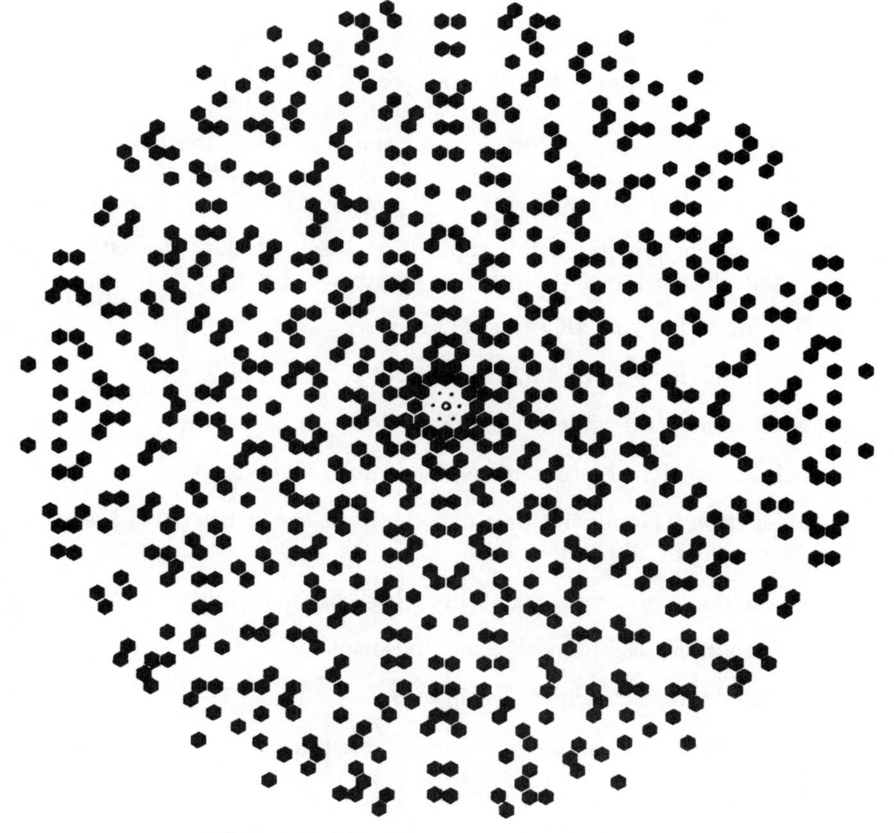

Figure 5. The Eisenstein-Jacobi Primes.

J. H. Jordan & J. R. Rabung, A conjecture of Paul Erdős concerning Gaussian primes, *Math. Comput.*, **24**(1970) 221–223.

A17 Formulas for primes.

Perhaps the philosopher's stone of number theory is a formula for p_n, or for $\pi(x)$, or for a necessary and sufficient condition for primality. Wilson's theorem seems to be unique (is Vantieghem's result, that $p > 2$ is prime if and only if $\prod_{d=1}^{p-1}(2^d - 1) \equiv p \bmod 2^p - 1$ equivalent to it?); but even that is useless for computation. C. P. Willans and C. P. Wormell used it to give formulas which use only elementary functions, but which are too clumsy to print here. The Mann-Shanks algorithm is another curiosity, of even less practical value. Matiyasevich and other logicians have used Wilson's theorem and their solution of Hilbert's tenth problem to produce polynomials the positive part of whose range is exactly the set of primes.

Three theorems of Boris Stechkin may be worth recording. They are based on the function

$$S(n) = \#\left\{m \;:\; 2 \le m \le n,\ (m-1) \left| \left\lfloor \frac{n(m-1)}{m} \right\rfloor \right.\right\}$$

(1) $n-1$ is prime just if $S(n) = d(n)$, the number of divisors of n,
(2) $n \pm 1$ are twin primes just if $S(n) + S(n+1) = 2d(n)$.
(3) $p < q$ odd primes implies $S(q) - S(q-1) + S(q-2) - \ldots - S(p+1) = 0$.

The numerous papers on this topic vary widely in their sophistication and in their aim. It seems desirable to distinguish between

1. A formula for $\pi(x)$ as a function of x.
2. A formula for p_n as a function of n.
3. A necessary and sufficient condition for n to be prime.
4. A function that is prime for each member of its domain.
5. A function (the positive part of) whose range consists only of primes, or consists of all of the primes.
6. A function whose range contains a high density of primes.
7. A formula for the largest prime divisor of n.
8. A formula for the prime factors of n.
9. A formula for the smallest prime greater than n.
10. A formula for p_{n+1} in terms of $p_1, p_2, \ldots, p_n$.
11. An algorithm for generating the primes. And so on ....

Examples of each can be found in the references. We have already mentioned (in **A1**) Euler's famous formula n^2+n+41. In some sense this is best possible, but quadratic expressions with positive discriminant can yield even longer sequences of prime values (though some of them may be negative). Gilbert Fung gives $47n^2-1701n+10181$, $0 \le n \le 42$, $\Delta = 979373$ and Russell Ruby $36n^2-810n+2753$, $0 \le n \le 44$, $\Delta = 2^2 3^2 7213$.

The first 1000 values of Euler's formula include 581 primes. Edgar Karst beats this with 598 values of $2n^2-199$ and in a 91-01-01 letter, Stephen Williams announces 602 prime values of $2n^2-1000n-2609$. The corresponding numbers among the first 10000 values are 4148, 4373 and 4151. However, what is significant is not the actual density over the first so many values, which clearly has to tend to zero in all cases, but the **asymptotic** density, which, if we believe Hardy & Littlewood (see **A1**), is always $c\sqrt{n}/\ln n$, and the best that can be done is to make the value of c as large as possible. Shanks has calculated $c = 3.3197732$ for Euler's formula and $c = 3.6319998$ for a polynomial $x^2+x+27941$ found by Beeger. Fung & Williams (see reference at **A1** and the references they give) have achieved $c = 5.0870883$ with the formula $x^2+x+132874279528931$. If Δ is the discriminant of the quadratic, then the Legendre symbol (see **F5**) $\left(\frac{\Delta}{p}\right)$ takes the value 1 for very few of the small primes, p.

Sierpiński observes that it follows from Fermat's theorem that if n is prime, then n divides

$$1^{n-1}+2^{n-1}+\ldots+(n-1)^{n-1}+1.$$

Is the converse true? Giuga so conjectured and verified it for $n \le 10^{1000}$ and Bedocchi verified it to $n \le 10^{1700}$. Giuga observed that a counterexample would be a Carmichael number (**A12**, **A13**), that $p|n$ would imply that $(p-1)|(n-1)$ and that

$$\sum_{p|n} \frac{1}{p} - \frac{1}{n}$$

must be an integer, so that n contains at least eight distinct prime factors. An equivalent conjecture is that

$$nB_{n-1} \equiv -1 \bmod n$$

where the **Bernoulli numbers** B_k are the coefficients in the expansion of $x/(e^x-1) = \sum_{k\ge0} B_k x^k/k!$ (compare **D2**).

Ulrich Abel & Hartmut Siebert, Sequences with large numbers of prime values, *Amer. Math. Monthly*, **100**(1993) 167–169.

William W. Adams, Eric Liverance & Daniel Shanks, Infinitely many necessary and sufficient conditions for primality, *Bull. Inst. Combin. Appl.*, **3**(1991) 69–76; *MR* **90e**:11011.

A. R. Ansari, On prime representing function, *Ganita*, **2**(1951) 81–82; *MR* **15**, 11.

Thøger Bang, A function representing all prime numbers, *Norsk Mat. Tidsskr.*, **34**(1952) 117–118; *MR* **14**, 621.

V. I. Baranov & B. S. Stechkin, *Extremal Combinatorial Problems and their Applications*, Kluwer, 1993, Problem 2.22.

Paul T. Bateman & Roger A. Horn, A heuristic asymptotic formula concerning the distribution of prime numbers, *Math. Comput.*, **16**(1962) 363–367.

Christoph Baxa, Über Gandhis Primzahlformel, *Elem. Math.*, **47**(1992) 82–84; *MR* **93h**: 11007.

E. Bedocchi, Nota ad una congettura sui numeri primi, *Riv. Mat. Univ. Parma*(4), **11**(1985) 229–236.

J. Braun, Das Fortschreitungsgesetz der Primzahlen durch eine transcendente Gleichung exakt dargestelt, *Wiss. Beilage Jahresber. Fr. W. Gymn. Trier*, 1899, 96 pp.

R. Creighton Buck, Prime representing functions, *Amer. Math. Monthly*, **53**(1946) 265.

John H. Conway, Problem 2.4, *Math. Intelligencer*, **3**(1980) 45.

L. E. Dickson, *History of the Theory of Numbers*, Carnegie Institute, Washington, 1919, 1920, 1923; reprinted Stechert, New York, 1934; Chelsea, New York, 1952, 1966, Vol. I, Chap. XVIII.

Underwood Dudley, History of a formula for primes, *Amer. Math. Monthly*, **76**(1969) 23–28; *MR* **38** #4270.

Underwood Dudley, Formulas for primes, *Math. Mag.*, **56**(1983) 17–22.

D. D. Elliott, A prime generating function, *Two-Year Coll. Math. J.*, **14**(1983) 57.

David Ellis, Some consequences of Wilson's theorem, *Univ. Nac. Tucumán Rev. Ser. A*, **12**(1959) 27–29; *MR* **21** #7179.

Reijo Ernvall, A formula for the least prime greater than a given integer, *Elem. Math.*, **30**(1975) 13–14; *MR* **54** #12616.

Robin Forman, Sequences with many primes, *Amer. Math. Monthly*, **99**(1992) 548–557; *MR* **93e**:11104.

Gilbert W. Fung & Hugh Cowie Williams, Quadratic polynomials which have a high density of prime values, *Math. Comput.*, (199)

J. M. Gandhi, Formulae for the n-th prime, *Proc. Washington State Univ. Conf. Number Theory*, Pullman, 1971, 96–106; *MR* **48** #218.

Betty Garrison, Polynomials with large numbers of prime values, *Amer. Math. Monthly*, **97**(1990) 316–317; *MR* **91i**:11124.

Giuseppe Giuga, Sopra alcune proprietà caratteristiche dei numeri primi, *Period. Math.* (4), **23**(1943) 12–27; *MR* **8**, 11.

Giuseppe Giuga, Su una presumibile proprietà caratteristica dei numeri primi, *Ist. Lombardo Sci. Lett. Rend. Cl. Sci. Mat. Nat.*(3), **14(83)**(1950) 511–528; *MR* **13**, 725.

P. Goetgheluck, On cubic polynomials giving many primes, *Elem. Math.*, **44**(1989) 70–73; *MR* **90j**:11014.

Solomon W. Golomb, A direct interpretation of Gandhi's formula, *Amer. Math. Monthly*, **81**(1974) 752–754; *MR* **50** #7003.

Solomon W. Golomb, Formulas for the next prime, *Pacific J. Math.*, **63**(1976) 401–404; *MR* **53** #13094.

R. L. Goodstein, Formulae for primes, *Math. Gaz.*, **51**(1967) 35–36.

H. W. Gould, A new primality criterion of Mann and Shanks, *Fibonacci Quart.*, **10**(1972) 355–364, 372; *MR* **47** #119.

Richard K. Guy, Conway's prime producing machine, *Math. Mag.*, **56**(1983) 26–33; *MR* **84j**:10008.

G. H. Hardy, A formula for the prime factors of any number, *Messenger of Math.*, **35**(1906) 145–146.

V. C. Harris, A test for primality, *Nordisk Mat. Tidskr.*, **17**(1969) 82; *MR* **40** #4197.

E. Härtter, Über die Verallgemeinerung eines Satzes von Sierpiński, *Elem. Math.*, **16** (1961) 123–127; *MR* **24** #A1869.

Olga Higgins, Another long string of primes, *J. Recreational Math.*, **14** (1981/82) 185.

C. Isenkrahe, Ueber eine Lösung der Aufgabe, jede Primzahl als Function der vorhergehenden Primzahlen durch einen geschlossenen Ausdruck darzustellen, *Math. Ann.*, **53**(1900) 42–44.

James P. Jones, Formula for the n-th prime number, *Canad. Math. Bull.*, **18** (1975) 433–434; *MR* **57** #9641.

James P. Jones, Daihachiro Sato, Hideo Wada & Douglas Wiens, Diophantine representation of the set of prime numbers, *Amer. Math. Monthly*, **83**(1976) 449–464; *MR* **54** #2615.

James P. Jones & Yuri V. Matiyasevich, Proof of recursive unsolvability of Hilbert's tenth problem, *Amer. Math. Monthly*, **98**(1991) 689–709; *MR* **92i**:03050.

Steven Kahan, On the smallest prime greater than a given positive integer, *Math. Mag.*, **47**(1974) 91–93; *MR* **48** #10964.

E. Karst, The congruence $2^{p-1} \equiv 1 \bmod p^2$ and quadratic forms with a high density of primes, *Elem. Math.*, **22**(1967) 85–88.

John Knopfmacher, Recursive formulae for prime numbers, *Arch. Math. (Basel)*, **33** (1979/80) 144–149; *MR* **81j**:10008.

Masaki Kobayashi, Prime producing quadratic polynomials and class-number one problem for real quadratic fields, *Proc. Japan Acad. Ser. A Math. Sci.*, **66**(1990) 119–121; *MR* **91i**:11140.

L. Kuipers, Prime-representing functions, *Nederl. Akad. Wetensch. Proc.*, **53** (1950) 309–310 = *Indagationes Math.*, **12**(1950) 57–58; *MR* **11**, 644.

J. C. Lagarias, V. S. Miller & A. M. Odlyzko, Computing $\pi(x)$: the Meissel-Lehmer method, *Math. Comput.*, **44**(1985) 537–560.

Klaus Langmann, Eine Formel für die Anzahl der Primzahlen, *Arch. Math. (Basel)*, **25**(1974) 40; *MR* **49** #4951.

D. H. Lehmer, On the function x^2+x+A, *Sphinx*, **6**(1936) 212–214; **7**(1937) 40.

S. Louboutin, R. A.Mollin & H. C. Williams, Class numbers of real quadratic fields, continued fractions, reduced ideals, prime-producing polynomials and quadratic residue covers, *Canad. J. Math.*, **44**(1992) 1–19.

H. B. Mann & Daniel Shanks, A necessary and sufficient condition for primality and its source, *J. Combin. Theory Ser. A*, **13**(1972) 131–134; *MR* **46** #5225.

J.-P. Massias & G. Robin, Effective bounds for some functions involving prime numbers, Preprint, Laboratoire de Théorie des Nombres et Algorithmique, 123 rue A. Thomas, 87060 Limoges Cedex, France.

Yuri V. Matiyasevich, Primes are enumerated by a polynomial in 10 variables, *Zap. Naučn. Sem. Leningrad. Otdel. Mat. Inst. Steklov*,**68**(1977) 62–82, 144–145; *MR* **58** #21534; English translation: *J. Soviet Math.*, **15**(1981) 33–44.

W. H. Mills, A prime-representing function, *Bull. Amer. Math. Soc.*, **53**(1947) 604; *MR* **8**, 567.

Richard A. Mollin, Prime valued polynomials and class numbers of quadratic fields, *Internat. J. Math. Math. Sci.*, **13**(1990) 1–11; *MR* **91c**:11060.

Richard A. Mollin, Ambiguous classes in quadratic fields, *Math. Comput.*, **61** (1993) 355–360.

Richard A. Mollin & Hugh Cowie Williams, Quadratic nonresidues and prime-producing polynomials, *Canad. Math. Bull.*, **32**(1989) 474–478; *MR* **91a**:11009. [see also *Number Theory*, de Gruyter, 1989, 654–663 and *Nagoya Math. J.*, **112**(1988) 143–151.]

Leo Moser, A prime-representing function, *Math. Mag.*, **23**(1950) 163–164.

K. S. Namboodiripad, A note on formulae for the n-th prime, *Monatsh. Math.*, **75**(1971) 256–262; *MR* **46** #126.

T. B. M. Neill & M. Singer, The formula for the Nth prime, *Math. Gaz.*, **49**(1965) 303.

Ivan Niven, Functions which represent prime numbers, *Proc. Amer. Math. Soc.*, **2**(1951) 753–755; *MR* **13**, 321a.

O. Ore, On the selection of subsequences, *Proc. Amer. Math. Soc.*, **3**(1952) 706–712; *MR* **14**, 256.

Joaquin Ortega Costa, The explicit formula for the prime number function $\pi(x)$, *Revista Mat. Hisp.-Amer.*(4), **10**(1950) 72–76; *MR* **12**, 392b.

Makis Papadimitriou, A recursion formula for the sequence of odd primes, *Amer. Math. Monthly*, **82**(1975) 289; *MR* **52** #246.

Carlos Raitzin, The exact count of the prime numbers that do not exceed a given upper bound (Spanish), *Rev. Ingr.*, **1**(1979) 37–43; *MR* **82e**:10074.

Stephen Regimbal, An explicit formula for the k-th prime number, *Math. Mag.*, **48**(1975) 230–232; *MR* **51** #12676.

J. B. Rosser & L. Schoenfeld, Approximate formulas for some functions of prime numbers, *Illinois J. Math.*, **6**(1962) 64–94.

Michael Rubinstein, A formula and a proof of the infinitude of the primes, *Amer. Math. Monthly*, **100**(1993) 388–392.

W. Sierpiński, *Elementary Number Theory*, (ed. A. Schinzel) PWN, Warszawa, 1987, p. 218.

W. Sierpiński, Sur une formule donnant tous les nombres premiers, *C.R. Acad. Sci. Paris*, **235**(1952) 1078–1079; *MR* **14**, 355.

W. Sierpiński, Les binômes $x^2 + n$ et les nombres premiers, *Bull. Soc. Royale Sciences Liège*, **33**(1964) 259–260.

B. R. Srinivasan, Formulae for the n-th prime, *J. Indian Math. Soc. (N.S.)*, **25**(1961) 33–39; *MR* **26** #1289.

B. R. Srinivasan, An arithmetical function and an associated formula for the n-th prime. I, *Norske Vid. Selsk. Forh. (Trondheim)*, **35**(1962) 68–71; *MR* **27** #101.

Garry J. Tee, Simple analytic expressions for primes, and for prime pairs, *New Zealand Math. Mag.*, **9**(1972) 32–44; *MR* **45** #8601.

E. Teuffel, Eine Rekursionsformel für Primzahlen, *Jber. Deutsch. Math. Verein.*, **57**(1954) 34–36; *MR* **15**, 685.

John Thompson, A method for finding primes, *Amer. Math. Monthly*, **60** (1953) 175; *MR* **14**, 621.

P. G. Tsangaris & James P. Jones, An old theorem on the GCD and its application to primes, *Fibonacci Quart.*, **30**(1992) 194–198; *MR* **93e**:11004.

Charles Vanden Eynden, A proof of Gandhi's formula for the n-th prime, *Amer. Math. Monthly*, **79**(1972) 625; *MR* **46** #3425.

E. Vantieghem, On a congruence only holding for primes, *Indag. Math.(N.S.)*, **2**(1991) 253–255; *MR* **92e**:11005.

C. P. Willans, On formulae for the Nth prime number, *Math. Gaz.*, **48**(1964) 413–415.

C. P. Wormell, Formulae for primes, *Math. Gaz.*, **51**(1967) 36–38.

E. M. Wright, A prime representing function, *Amer. Math. Monthly*, **58**(1951) 616–618; *MR* **13**, 321b.

E. M. Wright, A class of representing functions, *J. London Math. Soc.*, **29** (1954) 63–71; *MR* **15**, 288d.

A18 The Erdős-Selfridge classification of primes.

Erdős & Selfridge classify the primes as follows: p is in class 1 if the only prime divisors of $p+1$ are 2 or 3; and p is in class r if every prime factor of $p+1$ is in some class $\leq r-1$, with equality for at least one prime factor. For example:

class 1: 2 3 5 7 11 17 23 31 47 53 71 107 127 191 431 647 863 971 ...
class 2: 13 19 29 41 43 59 61 67 79 83 89 97 101 109 131 137 139 149 167 179 197 199 211 223 229 239 241 251 263 269 271 281 283 293 307 317 319 359 367 373 377 383 419 439 449 461 467 499 503 509 557 563 577 587 593 599 619 641 643 659 709 719 743 751 761 769 809 827 839 881 919 929 953 967 979 991 1019 ...
class 3: 37 103 113 151 157 163 173 181 193 227 233 257 277 311 331 337 347 353 379 389 397 401 409 421 457 463 467 487 491 521 523 541 547 571 601 607 613 631 653 683 701 727 733 773 787 811 821 829 853 857 859 877 883 911 937 947 983 997 1009 1013 1021 ...
class 4: 73 313 443 617 661 673 677 691 739 757 823 887 907 941 977 ...
class 5: 1321 1381 ...

It's easy to prove that the number of primes in class r, not exceeding n, is $o(n^\epsilon)$ for every $\epsilon > 0$ and all r. Prove that there are infinitely many primes in each class. If $p_1^{(r)}$ denotes the least prime in class r, so that $p_1^{(1)} = 2$, $p_1^{(2)} = 13$, $p_1^{(3)} = 37$, $p_1^{(4)} = 73$ and $p_1^{(5)} = 1021$, then Erdős thought that $(p_1^{(r)})^{1/r} \to \infty$, while Selfridge thought it quite likely to be bounded.

A similar classification arises if $p+1$ is replaced by $p-1$:

class 1: 2 3 5 7 13 17 19 37 73 97 109 163 193 433 487 577 769 1153 ...
class 2: 11 29 31 41 43 53 61 71 79 101 103 113 127 131 137 149 151 157 181 191 197 211 223 229 239 241 251 257 271 281 293 307 313 337 379 389 401 409 421 439 443 449 459 491 521 541 547 571 593 601 613 631 641 647 653 673 677 701 751 757 761 773 811 877 883 911 919 937 953 971 1009 1021 ...
class 3: 23 59 67 83 89 107 173 199 227 233 263 311 317 331 349 353 367 373 383 397 419 431 463 479 503 509 523 563 569 587 607 617 619 661 683 727 733 739 743 787 809 821 823 853 859 881 887 907 929 947 977 983 991 1031 1033 ...
class 4: 47 139 167 179 269 277 347 461 467 499 599 643 691 709 797 827 829 839 857 863 967 997 1013 1019 ...
class 5: 283 359 557 659 941 ...
class 6: 719 1319 ...
class 7: 1439 ...

for which similar answers are to be expected. Are corresponding classes equally dense? There is a connexion with Cunningham chains (**A7**).

P. Erdős, Problems in number theory and combinatorics, *Congr. Numer. XVIII*, Proc. 6th Conf. Numer. Math., Manitoba, 1976, 35–58 (esp. p. 53); *MR* **80e**:10005.

A19 Values of n making $n - 2^k$ prime. Odd numbers not of the form $\pm p^a \pm 2^b$.

Erdős conjectures that 4, 7, 15, 21, 45, 75 and 105 are the only values of n for which $n - 2^k$ is prime for all k such that $2 \le 2^k < n$. Mientka & Weitzenkamp have verified this for $n < 2^{44}$ and Uchiyama & Yorinaga have extended this to 2^{77}. Vaughan has proved that there are not too many such numbers, less than $x\exp\{-(\ln x)^c\}$ of them less than x, but he was unable to show that there were less than $x^{1-\epsilon}$.

Erdős also conjectures that for infinitely many n, all the integers $n-2^k$, $1 \le 2^k < n$ are squarefree (see also **F13**).

If we denote by $A(x)$ the number of $n \le x$ for which all $n-2^k$ are prime, $2 \le 2^k < n$, then Hooley showed that the extended Riemann hypothesis implies that $A(x) = O(x^c)$ with an explicit $c < 1$, and Narkiewicz improved this to $c < \frac{1}{2}$.

Cohen & Selfridge ask for the least positive odd number *not* of the form $\pm p^a \pm 2^b$, where p is prime, $a \ge 0$, $b \ge 1$ and any choice of signs may be made. They observe that the number is greater than 2^{18}, but at most

6120 6699060672 7677809211 5601756625 4819576161-
6319229817 3436854933 4512406741 7420946855 8999326569.

Crocker proved that there are infinitely many odd integers *not* of the form $2^k + 2^l + p$, where p is prime. Erdős suggests that there may be cx

of them less than x, but can $> x^\epsilon$ be proved? Can we show that covering congruences (**F13**) do not help here? I.e., does $p+2^u+2^v$ (or $p+2^u+2^v+2^w$) meet every arithmetic progression? More generally, Erdős asks if, for each r, there are infinitely many odd integers not the sum of a prime and r or fewer powers of 2. Is their density positive? Do they contain an infinite arithmetic progression? In the opposite direction, Gallagher has proved that for every $\epsilon > 0$ there is a sufficiently large r so that the lower density of sums of primes with r powers of 2 is greater than $1 - \epsilon$.

Erdős also asks if there is an odd integer *not* of the form $2^k + s$ where s is squarefree.

Let $f(n)$ be the number of representations of n as a sum $2^k + p$, and let $\{a_i\}$ be the sequence of values of n for which $f(n) > 0$. Does the density of $\{a_i\}$ exist? Erdős showed that $f(n) > c \ln\ln n$ infinitely often, but could not decide if $f(n) = o(\ln n)$. He conjectures that $\limsup(a_{i+1} - a_i) = \infty$. This would follow if there are covering systems with arbitrarily large least moduli.

Carl Pomerance notes that for $n = 210$, $n - p$ is prime for all p, $n/2 < p < n$, and asked if there is any other such n. With help from Deshouillers, Granville & Narkiewicz he later answered this negatively.

Fred Cohen & J. L. Selfridge, Not every number is the sum or difference of two prime powers, *Math. Comput.*, **29**(1975) 79–81; *MR* **51** #12758.

R. Crocker, On the sum of a prime and of two powers of two, *Pacific J. Math.*, **36**(1971) 103–107; *MR* **43** #3200.

Jean-Marc Deshouillers, Andrew Granville, Władysław Narkiewicz & Carl Pomerance, An upper bound in Goldbach's problem, *Math. Comput.*, **61**(1993) 209–213.

P. Erdős, On integers of the form $2^r + p$ and some related problems, *Summa Brasil. Math.*, **2**(1947-51) 113–123; *MR* **13**, 437.

Patrick X. Gallagher, Primes and powers of 2, *Inventiones Math.*, **29**(1975) 125–142; *MR* **52** #315.

C. Hooley, *Applications of Sieve Methods*, Academic Press, 1974, Chap. VIII.

Donald E. G. Malm, A graph of primes, *Math. Mag.*, **66**(1993) 317–320.

Walter E. Mientka & Roger C. Weitzenkamp, On f-plentiful numbers, *J. Combin. Theory*, **7**(1969) 374–377; *MR* **42** #3015.

W. Narkiewicz, On a conjecture of Erdős, *Colloq. Math.*, **37**(1977) 313–315; *MR* **58** #21971.

A. de Polignac, Recherches nouvelles sur les nombres premiers, *C. R. Acad. Sci. Paris*, **29**(1849) 397–401, 738–739.

Saburô Uchiyama & Masataka Yorinaga, Notes on a conjecture of P. Erdős, I, II, *Math. J. Okayama Univ.*, **19**(1977) 129–140; **20**(1978) 41–49; *MR* **56** #11929; **58** #570.

R. C. Vaughan, Some applications of Montgomery's sieve, *J. Number Theory*, **5**(1973) 64–79; *MR* **49** #7222.

B. Divisibility

We will denote by $d(n)$ the number of positive divisors of n, by $\sigma(n)$ the sum of those divisors, and by $\sigma_k(n)$ the sum of their kth powers, so that $\sigma_0(n) = d(n)$ and $\sigma_1(n) = \sigma(n)$. We use $s(n)$ for the sum of the **aliquot parts** of n, i.e., the positive divisors of n other than n itself, so that $s(n) = \sigma(n) - n$. The number of distinct prime factors of n will be denoted by $\omega(n)$ and the total number, counting repetitions, by $\Omega(n)$.

Iteration of various arithmetic functions will be denoted, for example, by $s^k(n)$, which is defined by $s^0(n) = n$ and $s^{k+1}(n) = s(s^k(n))$ for $k \geq 0$.

We use the notation $d|n$ to mean that d divides n, and $e \nmid n$ to mean that e does not divide n. The notation $p^k \| n$ is used to imply that $p^k | n$ but $p^{k+1} \nmid n$. By $[m, n]$ we will mean the consecutive integers $m, m+1, \ldots, n$.

B1 Perfect numbers.

A **perfect number** is one with $s(n) = n$. Euclid knew that $2^{p-1}(2^p - 1)$ was perfect if $2^p - 1$ is prime. For example, 6, 28, 496, ... ; see the list of Mersenne primes in **A3**. Euler showed that these were the only even perfect numbers.

The existence or otherwise of odd perfect numbers is one of the more notorious unsolved problems of number theory. The lower bound for an odd perfect number has now been pushed to 10^{300} by Brent, Cohen & te Riele. Brandstein has shown that the largest prime factor is > 500000 and Hagis that the second largest is > 1000. Cohen has shown that it contains a component (prime power divisor) $> 10^{20}$, and Sayers that there are at least 29 prime factors (not necessarily distinct).

Pomerance has shown that an odd perfect number with at most k distinct factors is less than

$$(4k)^{(4k)^{2^{k^2}}}$$

but Heath-Brown has much improved this by showing that if n is an odd number with $\sigma(n) = an$, then $n < (4d)^{4^k}$, where d is the denominator of a and k is the number of distinct prime factors of n. In particular, if n is an odd perfect number with k distinct prime factors, then $n < 4^{4^k}$.

John Leech asks for examples of spoof odd perfect numbers, like Descartes's

$$3^2 7^2 11^2 13^2 22021$$

which is perfect if you pretend that 22021 is prime.

For many earlier references, see the first edition of this book.

Michael S. Brandstein, New lower bound for a factor of an odd perfect number, #82T-10-240, *Abstracts Amer. Math. Soc.*, **3**(1982) 257.

Richard P. Brent & Graeme L. Cohen, A new lower bound for odd perfect numbers, *Math. Comput.*, **53**(1989) 431–437.

R. P. Brent, G. L. Cohen & H. J. J. te Riele, Improved techniques for lower bounds for odd perfect numbers, *Math. Comput.*, **57**(1991) 857–868; *MR* **92c**:11004.

Graeme L. Cohen, On the largest component of an odd perfect number, *J. Austral. Math. Soc. Ser. A*, **42**(1987) 280–286.

P. Hagis, Sketch of a proof that an odd perfect number relatively prime to 3 has at least eleven prime factors, *Math. Comput.*, **40**(1983) 399–404.

P. Hagis, On the second largest prime divisor of an odd perfect number, *Lecture Notes in Math.*, **899**, Springer-Verlag, New York, 1971, pp. 254–263.

D. R. Heath-Brown, Odd perfect numbers, (submitted).

Masao Kishore, Odd perfect numbers not divisible by 3 are divisible by at least ten distinct primes, *Math. Comput.*, **31**(1977) 274–279; *MR* **55** #2727.

Masao Kishore, Odd perfect numbers not divisible by 3. II, *Math. Comput.*, **40**(1983) 405–411.

M. D. Sayers, An improved lower bound for the total number of prime factors of an odd perfect number, M.App.Sc. Thesis, NSW Inst. Tech., 1986.

B2 Almost perfect, quasi-perfect, pseudoperfect, harmonic, weird, multiperfect and hyperperfect numbers.

Perhaps because they were frustrated by their failure to disprove the existence of odd perfect numbers, numerous authors have defined a number of closely related concepts and produced a raft of problems, many of which seem no more tractable than the original.

For a perfect number, $\sigma(n) = 2n$. If $\sigma(n) < 2n$, n is called **deficient**. A problem in *Abacus* was to prove that every number $n > 3$ is the sum of two deficient numbers, or to find a number that was not. If $\sigma(n) > 2n$, then n is called **abundant**. If $\sigma(n) = 2n - 1$, n has been called **almost perfect**. Powers of 2 are almost perfect; it is not known if any other numbers are. If $\sigma(n) = 2n + 1$, n has been called **quasi-perfect**. Quasi-perfect numbers must be odd squares, but no one knows if there are any. Masao Kishore shows that $n > 10^{30}$ and that $\omega(n) \geq 6$. Hagis & Cohen have improved these results to $n > 10^{35}$ and $\omega(n) \geq 7$. Cattaneo originally claimed to have proved that $3 \nmid n$, but Sierpiński and others have observed that his proof is fallacious. Kravitz, in a letter, makes a more general conjecture, that there are no numbers whose **abundance**, $\sigma(n) - 2n$, is an odd square. In this connexion Graeme Cohen writes that it is interesting that

$$\sigma(2^2 3^2 5^2) = 3(2^2 3^2 5^2) + 11^2$$

and that if $\sigma(n) = 2n + k^2$ with $n \perp k$, then $\omega(n) \geq 4$ and $n > 10^{20}$. He has also shown that if $k < 10^{10}$ then $\omega(n) \geq 6$, and that if $k < 44366047$ then n is primitive abundant (see below). Later, relaxing the condition $n \perp k$, he finds the solution

$$n = 2 \cdot 3^2 \cdot 238897^2, \quad k = 3^2 \cdot 23 \cdot 1999$$

and five solutions $n = 2^2 \cdot 7^2 \cdot p^2$, with

$p =$	53	277	541	153941	358276277
$k =$	$7 \cdot 29$	$5 \cdot 7 \cdot 23$	$5 \cdot 7 \cdot 43$	$5 \cdot 7 \cdot 103 \cdot 113$	$5 \cdot 7 \cdot 227 \cdot 229 \cdot 521$

He verifies that the first of these last five is the smallest integer with odd square abundance. Sidney Kravitz has since sent two more solutions,

$$n = 2^3 \cdot 3^2 \cdot 1657^2, \quad k = 3 \cdot 11 \cdot 359,$$

$$n = 2^4 \cdot 31^2 \cdot 79922201791288893^2, \quad k = 44498798693247589.$$

In the latter, 31 divides k. Erdős asks for a characterization of the large numbers for which $|\sigma(n) - 2n| < C$ for some constant C. For example, $n = 2^m$: for other infinite families, see Mąkowski's two papers.

Wall, Crews & Johnson showed that the density of abundant numbers lies between 0.2441 and 0.2909. In an 83-08-17 letter Wall claimed to have narrowed these bounds to 0.24750 and 0.24893. Erdős asks if the density is irrational.

Sierpiński called a number **pseudoperfect** if it was the sum of *some* of its divisors; e.g., $20 = 1 + 4 + 5 + 10$. Erdős has shown that their density exists and says that presumably there are integers n which are not pseudoperfect, but for which $n = ab$ with a abundant and b having many prime factors: can b in fact have many factors $< a$?

For $n \geq 3$ Abbott lets $l = l(n)$ be the least integer for which there are n integers $1 \leq a_1 < a_2 < \ldots < a_n = l$ such that $a_i | s = \sum a_i$ for each i (so that s is pseudoperfect). He can show that $l(n) > n^{c_1 \ln \ln n}$ for some $c_1 > 0$ and all $n \geq 3$ and that $l(n) < n^{c_2 \ln \ln n}$ for some $c_2 > 0$ and infinitely many n.

Call a number **primitive abundant** if it is abundant, but all its proper divisors are deficient, and **primitive pseudoperfect** if it is pseudoperfect, but none of its proper divisors are. If the harmonic mean of all the divisors of n is an integer, Pomerance called n a **harmonic number**.

A. & E. Zachariou call these "Ore numbers" and they call primitive pseudoperfect numbers "irreducible semiperfect". They note that every multiple of a pseudoperfect number is pseudoperfect and that the pseudoperfect numbers and the harmonic numbers both include the perfect numbers as a proper subset. The last result is due to Ore. All numbers $2^m p$ with $m \geq 1$ and p a prime between 2^m and 2^{m+1} are primitive pseudoperfect, but there are such numbers not of this form, e.g., 770. There are infinitely many primitive pseudoperfect numbers that are not harmonic numbers. The smallest odd primitive pseudoperfect number is 945. Erdős can show that the number of odd primitive pseudoperfect numbers is infinite.

García extended the list of harmonic numbers to include all 45 which are $< 10^7$, and he found more than 200 larger ones. The least one, apart from 1 and the perfect numbers, is 140. Are any of them squares, apart from 1? Are there infinitely many of them? If so, find upper and lower bounds on the number of them that are $< x$. Kanold has shown that their density is zero, and Pomerance that a harmonic number of the form $p^a q^b$ (p and q primes) is an even perfect number. If $n = p^a q^b r^c$ is harmonic, is it even?

Which values does the harmonic mean take? Presumably not 4, 12, 16, 18, 20, 22, ... ; does it take the value 23? Ore's own conjecture, that every harmonic number is even, implies that there are no odd perfect numbers!

Bateman, Erdős, Pomerance & Straus show that the set of n for which $\sigma(n)/d(n)$ is an integer has density 1, that the set for which $\sigma(n)/d(n)^2$ is an integer has density $\frac{1}{2}$, and that the number of rationals $r \leq x$ of the form $\sigma(n)/d(n)$ is $o(x)$. They ask for an asymptotic formula for

$$\frac{1}{x}\sum 1$$

where the sum is taken over those $n \leq x$ for which $d(n)$ does *not* divide $\sigma(n)$. They also note that the integers n for which $d(n)$ divides $s(n) = \sigma(n) - n$, have zero density, because for almost all n, $d(n)$ and $\sigma(n)$ are divisible by a high power of 2, while n is divisible only by a low power of 2.

Benkoski has called a number **weird** if it is abundant but not pseudoperfect. For example, 70 is not the sum of any subset of

$$1 + 2 + 5 + 7 + 10 + 14 + 35 = 74$$

There are 24 primitive weird numbers less than a million: 70, 836, 4030, 5830, 7192, ... Nonprimitive weird numbers include $70p$ with p prime and $p > \sigma(70) = 144$; $836p$ with $p = 421$, 487, 491, or p prime and ≥ 557; also $7192 \cdot 31$. Some large weird numbers were found by Kravitz, and Benkoski & Erdős showed that their density is positive. Here the open questions are: are there infinitely many primitive abundant numbers which are weird? Is every odd abundant number pseudoperfect (i.e., not weird)? Can $\sigma(n)/n$

be arbitrarily large for weird n? Benkoski & Erdős conjecture "no" in answer to the last question and Erdős offers \$10 and \$25 respectively for solutions to the last two questions.

He also asks if there are extra-weird numbers n for which $\sigma(n) > 3n$, but n is not the sum of distinct divisors of n in two ways without repetitions. For example, 180 does not qualify, because although $\sigma(180) = 546$, 180=30+60+90 and is the sum of all its other divisors except 6.

Numbers have been called **multiply perfect**, **multiperfect** or **k-fold perfect** if $\sigma(n) = kn$ with k an integer. For example, ordinary perfect numbers are 2-fold perfect and 120 is 3-fold perfect. Dickson's *History* records a long interest in such numbers. Lehmer has remarked that if n is odd, then n is perfect just if $2n$ is triperfect.

Selfridge and others have observed that there are just six known 3-perfect numbers and they come from $2^h - 1$ for $h = 4, 6, 9, 10, 14, 15$. For example, the third one is illustrated by

$$\sigma(2^8 \cdot 7 \cdot 73 \cdot 37 \cdot 19 \cdot 5) = (2^9 - 1)(2^3)(37 \cdot 2)(19 \cdot 2)(5 \cdot 2^2)(2 \cdot 3).$$

It appears that there may be a similar explanation for the 36 known 4-perfect numbers, the last of which was published by Poulet as long ago as 1929.

For many years the largest known value of k was 8, for which Alan L. Brown gave three examples and Franqui & García two others. Stephen Gretton found numerous multiperfect numbers, including many fivefold, sixfold and sevenfold perfect numbers, and an eightfold, namely
$2^{62} \cdot 3^{15} \cdot 5^9 \cdot 7^7 \cdot 11^3 \cdot 13^3 \cdot 17^2 \cdot 19 \cdot 23 \cdot 29 \cdot 31^2 \cdot 37 \cdot 41 \cdot 43 \cdot 53 \cdot 61^2 \cdot 71^2 \cdot$ $\cdot 73 \cdot 83 \cdot 89 \cdot 97^2 \cdot 127 \cdot 193 \cdot 283 \cdot 307 \cdot 317 \cdot 331 \cdot 337 \cdot 487 \cdot 521^2 \cdot 601 \cdot 1201 \cdot$ $\cdot 1279 \cdot 2557 \cdot 3169 \cdot 5113 \cdot 92737 \cdot 649657$. This is believed to be the record for the smallest such.

In late 1992 and early 1993, half a dozen examples with $k = 9$ have already been found by Fred Helenius. The smallest is
$2^{114} \cdot 3^{35} \cdot 5^{17} \cdot 7^{12} \cdot 11^4 \cdot 13^5 \cdot 17^3 \cdot 19^8 \cdot 23^2 \cdot 29^2 \cdot 31^2 \cdot 37^4 \cdot 41 \cdot 43 \cdot 47^2 \cdot 53 \cdot$ $\cdot 61^2 \cdot 67 \cdot 71 \cdot 73 \cdot 79^2 \cdot 83^2 \cdot 89^2 \cdot 97 \cdot 103 \cdot 109 \cdot 127 \cdot 131^2 \cdot 151 \cdot 157 \cdot 167 \cdot 179^2 \cdot$ $\cdot 197 \cdot 211 \cdot 227 \cdot 331 \cdot 347 \cdot 367 \cdot 379 \cdot 443 \cdot 523 \cdot 599 \cdot 709 \cdot 757 \cdot 829 \cdot 1151 \cdot 1699 \cdot$ $\cdot 1789 \cdot 2003 \cdot 2179 \cdot 2999 \cdot 3221 \cdot 4271 \cdot 4357 \cdot 4603 \cdot 5167 \cdot 8011 \cdot 8647 \cdot 8713 \cdot$ $\cdot 14951 \cdot 17293 \cdot 21467 \cdot 29989 \cdot 110563 \cdot 178481 \cdot 530713 \cdot 672827 \cdot 4036961 \cdot$ $\cdot 218834597 \cdot 16148168401 \cdot 151871210317 \cdot 2646507710984041$

Can k can be as large as we wish? Erdős conjectures that $k = o(\ln \ln n)$. It has even been suggested that there may be only finitely many k-perfect numbers with $k \geq 3$.

Rich Schroeppel is compiling as complete a list as possible of multiperfect numbers, and so would be a good checking point for those who believe that they have discovered new specimens. Since the previous three paragraphs were written, Shigeru Nakamura has drawn my attention to the work of Motoji Yoshitake, who lists 3 5-perfect, 30 6-perfect, 35 7-perfect and 8 8-perfect numbers. 2+20+8+0 of these are attributed to Carmichael, Mason or Cunningham. One of the 8-perfect numbers was given by Brown, and another can be derived from it by replacing $19^2 \cdot 127$ by $19^4 \cdot 151 \cdot 911$. This substitution, found by Cunningham in 1902, can be applied to the Carmichael-Mason table to give 50 multiperfect numbers. He also notes that Carmichael & Mason mistook $137561 = 151 \cdot 911$ and $485581 = 277 \cdot 1753$ for primes. In 1992 we knew of 700 k-perfect numbers with $k \geq 3$. In January, 1993, this number leapt to about 1150 from the discoveries of Fred Helenius which include 114 7-perfect, 327 8-perfect and two 9-perfect numbers. He continues to find dozens of new ones each month, so it is even more impossible to keep this section of the book up-to-date than it is to keep the rest; in March the total neared 1300; a postcript of a 93-09-08 letter from Schroepel gave 1526; by the time he mailed it next day it was 1605.

If n is an odd triperfect number, then McDaniel, Cohen, Kishore, Bugulov, Kishore, Cohen & Hagis, Reidlinger, and Kishore have respectively shown that $\omega(n) \geq 9$, 9, 10, 11, 11, 11, 12, and 12. Beck & Najar, Alexander, and Cohen & Hagis have shown that $n > 10^{50}, 10^{60}, 10^{70}$. Cohen & Hagis have shown that the largest prime factor of n is at least 100129 and that the second largest is at least 1009.

Shigeru Nakamura writes that Bugulov showed, in 1966, that odd k-perfect numbers contain at least ω distinct prime factors, where $(k,\omega) =$ (3,11), (4,21), (5,54) [incorrectly stated in *MR* **37** #5139 & rNT A32-96]. Nakamura claims to prove that for an even k-perfect number,

$$\omega > \max\{k^3/81 + \tfrac{5}{3},\ k^5/2500 + 2.9,\ k^{10}/(14 \cdot 10^8) + 2.9999\}$$

and for an odd k-perfect number,

$$\omega > \max\{k^5/60 + \tfrac{47}{12},\ k^5/50 - 20.8,\ 737k^{10}/10^9 + 11.5\}.$$

These improve the results of Cohen & Hendy and of Reidlinger; he also gives the improvements $(k,\omega) = (4,23)$, (5,56), (6,142), (7,373) to those of Bugulov.

Minoli & Bear say that n is k-**hyperperfect** if $n = 1 + k\sum d_i$, where the summation is taken over all proper divisors, $1 < d_i < n$, so that $k\sigma(n) = (k+1)n + k - 1$. For example, 21, 2133 and 19521 are 2-hyperperfect and 325 is 3-hyperperfect. They conjecture that there are k-hyperperfect numbers for every k.

Ron Graham asks if $s(n) = \lfloor n/2 \rfloor$ implies that n is 2 or a power of 3.

Erdős lets $f(n)$ be the smallest integer for which $n = \sum_{i=1}^{k} d_i$ for some k, where $1 = d_1 < d_2 < \ldots\ d_l = f(n)$ is the increasing sequence of divisors of $f(n)$. Is $f(n) = o(n)$? Or is this true only for almost all n, with $\limsup f(n)/n = \infty$?

n	1	2	3	4	5	6	7	8	9	10	11	12	13	14
$f(n)$	1	-	2	3	-	5	4	7	15	12	21	6	9	13

n	15	16	17	18	19	20	21	22	23	24	25	26	27	28
$f(n)$	8	12	30	10	42	19	18	20	57	14	36	46	30	12

Erdős defines n_k to be the smallest integer for which if you partition the proper divisors of n_k into k classes, n_k will always be the sum of distinct divisors from the same class. Clearly $n_1 = 6$, but he is not even able to prove the existence of n_2.

H. Abbott, C. E. Aull, Ezra Brown & D. Suryanarayana, Quasiperfect numbers, *Acta Arith.*, **22**(1973) 439–447; *MR* **47** #4915; corrections, **29**(1976) 427–428.

Leon Alaoglu & Paul Erdős, On highly composite and similar numbers, *Trans. Amer. Math. Soc.*, **56**(1944) 448–469; *MR* **6**, 117b.

L. B. Alexander, Odd triperfect numbers are bounded below by 10^{60}, M.A. thesis, East Carolina University, 1984.

M. M. Artuhov, On the problem of odd h-fold perfect numbers, *Acta Arith.*, **23**(1973) 249–255.

Paul T. Bateman, Paul Erdős, Carl Pomerance & E.G. Straus, The arithmetic mean of the divisors of an integer, in *Analytic Number Theory* (*Philadelphia, 1980*) 197–220, *Lecture Notes in Math.*, **899**, Springer, Berlin - New York, 1981; *MR* **84b**:10066.

Walter E. Beck & Rudolph M. Najar, A lower bound for odd triperfects, *Math. Comput.*, **38**(1982) 249–251.

S. J. Benkoski, Problem E2308, *Amer. Math. Monthly*, **79**(1972) 774.

S. J. Benkoski & P. Erdős, On weird and pseudoperfect numbers, *Math. Comput.*, **28**(1974) 617–623; *MR* **50** #228; corrigendum, S. Kravitz, **29**(1975) 673.

Alan L. Brown, Multiperfect numbers, *Scripta Math.*, **20**(1954) 103–106; *MR* **16**, 12.

E. A. Bugulov, On the question of the existence of odd multiperfect numbers (Russian), *Kabardino-Balkarsk. Gos. Univ. Ucen. Zap.*, **30**(1966) 9–19.

David Callan, Solution to Problem 6616, *Amer. Math. Monthly*, **99**(1992) 783–789.

R. D. Carmichael & T. E. Mason, Note on multiply perfect numbers, including a table of 204 new ones and the 47 others previously published, *Proc. Indiana Acad. Sci.*, **1911** 257–270.

Paolo Cattaneo, Sui numeri quasiperfetti, *Boll. Un. Mat. Ital.*(3), **6**(1951) 59–62; *Zbl.* **42**, 268.

Graeme L. Cohen, On odd perfect numbers II, multiperfect numbers and quasiperfect numbers, *J. Austral. Math. Soc. Ser. A*, **29**(1980) 369–384; *MR* **81m**:10009.

Graeme L. Cohen, The non-existence of quasiperfect numbers of certain forms, *Fibonacci Quart.*, **20**(1982) 81–84.

Graeme L. Cohen, On primitive abundant numbers, *J. Austral. Math. Soc. Ser. A*, **34**(1983) 123–137.

Graeme L. Cohen, Primitive α-abundant numbers, *Math. Comput.*, **43**(1984) 263–270.

Graeme L. Cohen, Stephen Gretton and his multiperfect numbers, Internal Report No. 28, School of Math. Sciences, Univ. of Technology, Sydney, Australia, Oct 1991.

G. L. Cohen & P. Hagis, Results concerning odd multiperfect numbers, *Bull. Malaysian Math. Soc.*, **8**(1985) 23–26.

G. L. Cohen & M. D. Hendy, On odd multiperfect numbers, *Math. Chronicle*, **9**(1980) 120–136; **10**(1981) 57–61.

Philip L. Crews, Donald B. Johnson & Charles R. Wall, Density bounds for the sum of divisors function, *Math. Comput.*, **26**(1972) 773–777; *MR* **48** #6042; Errata **31**(1977) 616; *MR* **55** #286.

J. T. Cross, A note on almost perfect numbers, *Math. Mag.*, **47**(1974) 230–231.

P. Erdős, On the density of the abundant numbers, *J. London Math. Soc.*, **9**(1934) 278–282.

P. Erdős, Problems in number theory and combinatorics, *Congressus Numerantium XVIII, Proc. 6th Conf. Numerical Math. Manitoba*, 1976, 35–58 (esp. pp. 53–54); *MR* **80e**:10005.

Benito Franqui & Mariano García, Some new multiply perfect numbers, *Amer. Math. Monthly*, **60**(1953) 459–462; *MR* **15**, 101.

Benito Franqui & Mariano García, 57 new multiply perfect numbers, *Scripta Math.*, **20**(1954) 169–171 (1955); *MR* **16**, 447.

Mariano García, A generalization of multiply perfect numbers, *Scripta Math.*, **19**(1953) 209–210; *MR* **15**, 199.

Mariano García, On numbers with integral harmonic mean, *Amer. Math. Monthly*, **61** (1954) 89–96; *MR* **15**, 506, 1140.

Peter Hagis, The third largest prime factor of an odd multiperfect number exceeds 100, *Bull. Malaysian Math. Soc.*, **9**(1986) 43–49.

Peter Hagis, A new proof that every odd triperfect number has at least twelve prime factors, *A tribute to Emil Grosswald: number theory and related analysis*, 445–450 *Contemp. Math.*, **143** Amer. Math. Soc., 1993. 43–49.

Peter Hagis & Graeme L. Cohen, Some results concerning quasiperfect numbers, *J. Austral. Math. Soc. Ser. A*, **33**(1982) 275–286.

B. E. Hardy & M. V. Subbarao, On hyperperfect numbers, Proc. 13th Manitoba Conf. Numer. Math. Comput., *Congressus Numerantium*, **42**(1984) 183–198; *MR* **86c**:11006.

B. Hornfeck & E. Wirsing, Über die Häufigkeit vollkommener Zahlen, *Math. Ann.*, **133**(1957) 431–438; *MR* **19**, 837; see also **137**(1959) 316–318; *MR* **21** #3389.

R. P. Jerrard & Nicholas Temperley, Almost perfect numbers, *Math. Mag.*, **46** (1973) 84–87.

H.-J. Kanold, Über mehrfach vollkommene Zahlen, *J. reine angew. Math.*, **194** (1955) 218–220; II **197**(1957) 82–96; *MR* **17**, 238; **18**, 873.

H.-J. Kanold, Über das harmonische Mittel der Teiler einer natürlichen Zahl, *Math. Ann.*, **133**(1957) 371–374.

H.-J. Kanold, Einige Bemerkungen über vollkommene und mehrfach vollkommene Zahlen, *Abh. Braunschweig. Wiss. Ges.*, **42**(1990/91) 49–55; *MR* **93c**: 11002.

David G. Kendall, The scale of perfection, *J. Appl. Probability*, **19A**(1982) 125–138; *MR* **83d**:10007.

Masao Kishore, Odd triperfect numbers, *Math. Comput.*, **42**(1984) 231–233; *MR* **85d**:11009.

Masao Kishore, Odd triperfect numbers are divisible by eleven distinct prime factors, *Math. Comput.*, **44** (1985) 261–263; *MR* **86k**:11007.

Masao Kishore, Odd triperfect numbers are divisible by twelve distinct prime factors, *J. Autral. Math. Soc. Ser. A*, **42**(1987) 173–182.

Masao Kishore, Odd integers N with 5 distinct prime factors for which $2 - 10^{-12} < \sigma(N)/N < 2 + 10^{-12}$, *Math. Comput.*, **32**(1978) 303–309.

M. S. Klamkin, Problem E1445*, *Amer. Math. Monthly*, **67**(1960) 1028; see also **82**(1975) 73.

Sidney Kravitz, A search for large weird numbers, *J. Recreational Math.*, **9**(1976-77) 82–85.

Richard Laatsch, Measuring the abundancy of integers, *Math. Mag.*, **59** (1986) 84–92.

A. Mąkowski, Remarques sur les fonctions $\theta(n)$, $\phi(n)$ et $\sigma(n)$, *Mathesis*, **69**(1960) 302–303.

A. Mąkowski, Some equations involving the sum of divisors, *Elem. Math.*, **34**(1979) 82; *MR* **81b**:10004.

Wayne L. McDaniel, On odd multiply perfect numbers, *Boll. Un. Mat. Ital.* (4), **3**(1970) 185–190; *MR* **41** #6764.

W. H. Mills, On a conjecture of Ore, *Proc. Number Theory Conf.*, Boulder CO, 1972, 142–146.

D. Minoli, Issues in non-linear hyperperfect numbers, *Math. Comput.*, **34** (1980) 639–645; *MR* **82c**:10005.

Daniel Minoli & Robert Bear, Hyperperfect numbers, *Pi Mu Epsilon J.*, **6**#3(1974-75) 153–157.

Shigeru Nakamura, On k-perfect numbers (Japanese), *J. Tokyo Univ. Merc. Marine(Nat. Sci.)*, **33**(1982) 43–50.

Shigeru Nakamura, On some properties of $\sigma(n)$, *J. Tokyo Univ. Merc. Marine(Nat. Sci.)*, **35**(1984) 85–93.

Shigeru Nakamura, On multiperfect numbers, (unpublished typescript).

Oystein Ore, On the averages of the divisors of a number, *Amer. Math. Monthly*, **55**(1948) 615–619.

Seppo Pajunen, On primitive weird numbers, *A collection of manuscripts related to the Fibonacci sequence*, 18th anniv. vol., Fibonacci Assoc., 162–166.

Carl Pomerance, On a problem of Ore: Harmonic numbers (unpublished typescript); see Abstract *709-A5, *Notices Amer. Math. Soc.*, **20**(1973) A-648.

Carl Pomerance, On multiply perfect numbers with a special property, *Pacific J. Math.*, **57**(1975) 511–517.

Carl Pomerance, On the congruences $\sigma(n) \equiv a \bmod n$ and $n \equiv a \bmod \phi(n)$, *Acta Arith.*, **26**(1975) 265–272.

Paul Poulet, *La Chasse aux Nombres*, Fascicule I, Bruxelles, 1929, 9–27.

Problem B-6, William Lowell Putnam Mathematical Competition, 1976-12-04.

Problem 14, *Abacus*, **1**(1984) 93.

Herwig Reidlinger, Über ungerade mehrfach vollkommene Zahlen [On odd multiperfect numbers], *Österreich. Akad. Wiss. Math.-Natur. Kl. Sitzungsber. II*, **192**(1983) 237–266; *MR* **86d**:11018.

Herman J. J. te Riele, Hyperperfect numbers with three different prime factors, *Math. Comput.*, **36**(1981) 297–298.

Neville Robbins, A class of solutions of the equation $\sigma(n) = 2n + t$, *Fibonacci Quart.*, **18**(1980) 137–147 (misprints in solutions for $t = 31$, 84, 86).

M. Satyanarayana, Bounds of $\sigma(N)$, *Math. Student*, **28**(1960) 79–81.

H. N. Shapiro, Note on a theorem of Dickson, *Bull. Amer. Math. Soc.*, **55**(1949) 450–452.

H. N. Shapiro, On primitive abundant numbers, *Comm. Pure Appl. Math.*, **21** (1968) 111–118.

W. Sierpiński, Sur les nombres pseudoparfaits, *Mat. Vesnik*, **2**(17)(1965) 212–213; *MR* **33** #7296.

W. Sierpiński, *Elementary Theory of Numbers* (ed. A. Schinzel), PWN–Polish Scientific Publishers, Warszawa, 1987, pp. 184–186.

D. Suryanarayana, Quasi-perfect numbers II, *Bull. Calcutta Math. Soc.*, **69** (1977) 421–426; *MR* **80m**:10003.

Charles R. Wall, The density of abundant numbers, Abstract 73T–A184, *Notices Amer. Math. Soc.*, **20**(1973) A-472.

Charles R. Wall, A Fibonacci-like sequence of abundant numbers, *Fibonacci Quart.*, **22**(1984) 349; *MR* **86d**:11018.

Charles R. Wall, Phillip L. Crews & Donald B. Johnson, Density bounds for the sum of divisors function, *Math. Comput.*, **26**(1972) 773–777.

Motoji Yoshitake, Abundant numbers, sum of whose divisors is equal to an integer times the number itself (Japanese), *Sūgaku Seminar*, **18**(1979) no. 3, 50–55.

Andreas & Eleni Zachariou, Perfect, semi-perfect and Ore numbers, *Bull. Soc. Math. Grèce*(N.S.), **13**(1972) 12–22; *MR* **50** #12905.

B3 Unitary perfect numbers.

If d divides n and $d \perp n/d$, call d a **unitary divisor** of n. A number n which is the sum of its unitary divisors, apart from n itself, is a **unitary perfect number**. There are no odd unitary perfect numbers, and Subbarao conjectures that there are only a finite number of even ones. He, Carlitz & Erdős each offer \$10.00 for settling this question and Subbarao offers 10¢ for each new example. If $n = 2^a m$, where m is odd and has r distinct prime factors, then Subbarao and others have shown that, apart from $2 \cdot 3$, $2^2 \cdot 3 \cdot 5$, $2 \cdot 3^2 \cdot 5$ and $2^6 \cdot 3 \cdot 5 \cdot 7 \cdot 13$, there are no unitary perfect numbers with $a \le 10$, or with $r \le 6$. S. W. Graham has shown that the first and third are the only unitary perfect numbers of shape $2^a m$ with m odd and squarefree, and Jennifer DeBoer that the second is the only one of shape $2^a 3^2 m$ with $m \perp 6$ and squarefree.

Wall has found the unitary perfect number

$$2^{18} \cdot 3 \cdot 5^4 \cdot 7 \cdot 11 \cdot 13 \cdot 19 \cdot 37 \cdot 79 \cdot 109 \cdot 157 \cdot 313$$

and shown that it is the fifth such. He can prove that any other unitary perfect number has an odd component greater than 2^{15}. Frey has shown that if $N = 2^m p_1^{a_1} \dots p_r^{a_r}$ is unitary perfect with $N \perp 3$, then $m > 144$, $r > 144$ and $N > 10^{440}$.

Peter Hagis investigates **unitary multiperfect numbers**: there are no odd ones. Write $\sigma^*(n)$ for the sum of the unitary divisors of n. If

$\sigma^*(n) = kn$ and n contains t distinct odd prime factors, then $k = 4$ or 6 implies $n > 10^{110}$, $t \geq 51$ and $2^{49}|n$; $k \geq 8$ implies $n > 10^{663}$ and $t \geq 247$; while k odd and $k \geq 5$ imply $n > 10^{461}$, $t \geq 166$ and $2^{166}|n$.

Cohen calls a divisor d of an integer n a 1-**ary divisor** of n if $d \perp n/d$, and he calls d a k-**ary divisor** of n (for $k > 1$), and writes $d|_k n$, if the greatest common $(k-1)$-ary divisor of d and n/d is 1 (written $(d, n/d)_{k-1} = 1$). In this notation $d|n$ and $d \parallel n$ are written $d|_0 n$ and $d|_1 n$. He also calls p^x an **infinitary divisor** of $p^y (y > 0)$ if $p^x|_{y-1} p^y$. This gives rise to infinitary analogs of earlier concepts. Write $\sigma_\infty(n)$ for the sum of the infinitary divisors of n. He found 14 infinitary perfect numbers, i.e., with $\sigma_\infty(n) = kn$ and $k = 2$; 13 numbers with $k = 3$; 7 with $k = 4$; and two with $k = 5$. There are no odd ones, and he conjectures that there are no infinitary multiperfect numbers not divisible by 3.

Note that Suryanarayana (who also uses the term 'k-ary divisor') and Alladi give *different* generalizations of unitary divisors.

K. Alladi, On arithmetic functions and divisors of higher order, *J. Austral. Math. Soc. Ser. A*, **23**(1977) 9–27.

Graeme L. Cohen, On an integer's infinitary divisors, *Math. Comput.*, **54** (1990) 395–411.

Graeme Cohen & Peter Hagis, Arithmetic functions associated with the infinitary divisors of an integer, *Internat. J. Math. Math. Sci.*, (to appear).

J. L. DeBoer, On the non-existence of unitary perfect numbers of certain type, *Pi Mu Epsilon J.* (submitted).

H. A. M. Frey, Über unitär perfekte Zahlen, *Elem. Math.*, **33**(1978) 95–96; *MR* **81a**:10007.

S. W. Graham, Unitary perfect numbers with squarefree odd part, *Fibonacci Quart.*, **27**(1989) 317–322; *MR* **90i**:11003.

Peter Hagis, Lower bounds for unitary multiperfect numbers, *Fibonacci Quart.*, **22**(1984) 140–143; *MR* **85j**:11010.

Peter Hagis, Odd nonunitary perfect numbers, *Fibonacci Quart.*, **28** (1990) 11–15; *MR* **90k**:11006.

Peter Hagis & Graeme Cohen, Infinitary harmonic numbers, *Bull. Austral. Math. Soc.*, **41**(1990) 151–158; *MR* **91d**:11001.

József Sándor, On Euler's arithmetical function, *Proc. Alg. Conf. Braşov 1988*, 121–125.

V. Siva Rama Prasad & D. Ram Reddy, On unitary abundant numbers, *Math. Student*, **52**(1984) 141–144 (1990) *MR* **91m**:11002.

V. Siva Rama Prasad & D. Ram Reddy, On primitive unitary abundant numbers, *Indian J. Pure Appl. Math.*, **21**(1990) 40–44; *MR* **91f**:11004.

M. V. Subbarao, Are there an infinity of unitary perfect numbers? *Amer. Math. Monthly*, **77**(1970) 389–390.

M. V. Subbarao & D. Suryanarayana, Sums of the divisor and unitary divisor functions, *J. reine angew. Math.*, **302**(1978) 1–15; *MR* **80d**:10069.

M. V. Subbarao & L. J. Warren, Unitary perfect numbers, *Canad. Math. Bull.*, **9**(1966) 147–153; *MR* **33** #3994.

M. V. Subbarao, T. J. Cook, R. S. Newberry & J. M. Weber, On unitary perfect numbers, *Delta*, **3**#1(Spring 1972) 22–26.

D. Suryanarayana, The number of k-ary divisors of an integer, *Monatsh. Math.*, **72**(1968) 445–450.

Charles R. Wall, The fifth unitary perfect number, *Canad. Math. Bull.*, **18**(1975) 115–122. See also *Notices Amer. Math. Soc.*, **16**(1969) 825.

Charles R. Wall, Unitary harmonic numbers, *Fibonacci Quart.*, **21**(1983) 18–25.

Charles R. Wall, On the largest odd component of a unitary perfect number, *Fibonacci Quart.*, **25**(1987) 312–316; *MR* **88m**:11005.

B4 Amicable numbers.

Unequal numbers m, n are called **amicable** if each is the sum of the aliquot parts of the other, i.e., $\sigma(m) = \sigma(n) = m + n$. Several thousand such pairs are known. The smaller member, 220, of the smallest pair, occurs in *Genesis*, xxxii, 14, and amicable numbers intrigued the Greeks and Arabs and many others since. For their history see the articles of Lee & Madachy. The *Genesis* reference, from the King James Bible, is achieved by amalgamating 200 females and 20 males. Aviezri Fraenkel writes that in his Pentateuch, they occur at xxxii, 15, and gives the more convincing occurrences of 220 in *Ezra* viii, 20 and in *1 Chronicles* xv, 6; and of 284 in *Nehemiah* xi, 18. He notes that the three places are amicably related: all are connected to the tribe of Levi, whose name derives from the wish of Levi's mother to be amicably related to his father (*Genesis* xxix, 34).

It is not known if there are infinitely many, but it is believed that there are. In fact Erdős conjectures that the number, $A(x)$, of such pairs with $m < n < x$ is at least $x^{1-\epsilon}$. He improved a result of Kanold to show that $A(x) = o(x)$ and his method can be used to obtain $A(x) \le cx/\ln\ln\ln x$. Pomerance obtained the further improvement

$$A(x) \le x \exp\{-c(\ln\ln\ln x \ln\ln\ln\ln x)^{1/2}\}.$$

Erdős conjectured that $A(x) = o(x/(\ln x)^k)$ for every k whereupon Pomerance proved the stronger result

$$A(x) \le x \exp\{-(\ln x)^{1/3}\}.$$

This implies that the sum of the reciprocals of the amicable numbers is finite, a fact not earlier known. He also notes that his proof can be modified to give the slightly stronger result

$$A(x) \ll x \exp\{-c(\ln x \ln\ln x)^{1/3}\}.$$

Herman te Riele has found all 1427 amicable pairs whose lesser members are less than 10^{10}. He remarks that the quantity $A(x)(\ln x)^3/x^{1/2}$ "remains very close to 174.6", but I suspect that a much more powerful telescope would require the exponent 1/2 to be increased much nearer to 1. Moews & Moews have continued the complete search to beyond $2 \cdot 10^{11}$. Through the efforts of Battiato and others, more than 40,000 pairs are known.

Some very large amicable pairs, with 32, 40, 81 and 152 decimal digits, discovered by te Riele, are mentioned by Kaplansky under "Mathematics" in the 1975 *Encyclopedia Britannica Yearbook.* The largest previously known had 25 decimal digits. More recently te Riele has constructed, from a "mother" list of 92 known amicable pairs, more than 2000 new pairs of sizes up to 38 decimal digits, and five pairs with from 239 to 282 digits. The largest amicable pair known in mid-1993 has 1041 decimal digits:

$$(2^9 p^{20} q_1 rstu, 2^9 p^{20} q_2 v)$$

with $p = 5661346302015448219060051$; q_1, q_2 of shape $bc^{20} - 1$ with $b_1 = 5797874220719830725124352$, $b_2 = 5531348900141215019827200$, $c = 5661346302015448219060051$; and $r = 569$, $s = 5039$, $t = 1479911$, $u = 30636732851$; and $v = 1365279187043825060064301$. It was found by Holger Wiethaus, a student at Dortmund in July 1988.

Elvin J. Lee has given half a dozen rules for amicable pairs of type $(2^n pq, 2^n rs)$ where p, q, r, s are primes of appropriate shape. E.g.,

$$p = 3 \cdot 2^{n-1} - 1, \quad q = 35 \cdot 2^{n+1} - 29, \quad r = 7 \cdot 2^{n-1} - 1, \quad s = 15 \cdot 2^{n+1} - 13,$$

but the simultaneous discovery of four such primes is a rare event.

Borho, Hoffman & te Riele have made considerable advances, both with proliferation of generalized Thabit rules, and with actual computation. Of the 1427 amicable pairs mentioned above, all but 17 have $m+n \equiv 0 \bmod 9$. The smallest exception is Poulet's pair

$$2^4 \cdot 331 \cdot \begin{cases} 19 \cdot 6619 \\ 199 \cdot 661 \end{cases}$$

with $m + n \equiv 5 \bmod 9$: te Riele gives the first examples

$$2^4 \cdot \begin{cases} 19^2 \cdot 103 \cdot 1627 \\ 3847 \cdot 16763 \end{cases} \quad \text{and} \quad 2^2 \cdot 19 \cdot \begin{cases} 13^2 \cdot 37 \cdot 43 \cdot 139 \\ 41 \cdot 151 \cdot 6709 \end{cases}$$

with m, n even, $m + n \equiv 3 \bmod 9$.

It is not known if an amicable pair exists with m and n of opposite parity, or with $m \perp n$. Bratley & McKay conjectured that both members of all odd amicable pairs are divisible by 3, but Battiato & Borho produced 15 counterexamples with from 36 to 73 decimal digits. In an 87-05-15 letter te Riele announced a 33-digit specimen

$$5 \cdot 7^2 \cdot 11^2 \cdot 13 \cdot 17 \cdot 19^3 \cdot 23 \cdot 37 \cdot 181 \begin{cases} 101 \cdot 8643 \cdot 1947938229 \\ 365147 \cdot 47303071129 \end{cases}$$

Is this the smallest such pair? Is there an odd amicable pair with one member, but not both, divisible by 3?

An old conjecture of Charles Wall is that odd amicable pairs must be incongruent modulo 4. Its truth, he says, implies that there are no odd perfect numbers, so it may be more prudent to look for a counterexample than to try to prove it.

On p. 169 of *Mathematical Magic Show*, Vintage Books, 1978, Martin Gardner makes a conjecture about the digital roots of amicable numbers. Lee confirms this in part by showing that if $(2^n pqr, 2^n stu)$ is an amicable pair whose sum is not divisible by 9, then each number is congruent to 7, modulo 9.

Unitary amicable numbers have been studied by Peter Hagis and by Mariano García, who list 82 pairs.

J. Alanen, O. Ore & J. G. Stemple, Systematic computations on amicable numbers, *Math. Comput.*, **21**(1967) 242–245; *MR* **36** #5058.

M. M. Artuhov, On some problems in the theory of amicable numbers (Russian), *Acta Arith.*, **27**(1975) 281–291.

S. Battiato, *Über die Produktion von 37803 neuen befreundeten Zahlenpaaren mit der Brütermethode*, Master's thesis, Wuppertal, June 1988.

S. Battiato & W. Borho, Are there odd amicable numbers not divisible by three? *Math. Comput.*, **50**(1988) 633–636; *MR* **89c**:11015.

W. Borho, On Thabit ibn Kurrah's formula for amicable numbers, *Math. Comput.*, **26**(1972) 571–578.

W. Borho, Befreundete Zahlen mit gegebener Primteileranzahl, *Math. Ann.*, **209**(1974) 183–193.

W. Borho, Eine Schranke für befreundete Zahlen mit gegebener Teileranzahl, *Math. Nachr.*, **63**(1974) 297–301.

W. Borho, Some large primes and amicable numbers, *Math. Comput.*, **36** (1981) 303–304.

W. Borho & H. Hoffmann, Breeding amicable numbers in abundance, *Math. Comput.*, **46**(1986) 281–293.

P. Bratley & J. McKay, More amicable numbers, *Math. Comput.*, **22**(1968) 677–678; *MR* **37** #1299.

P. Bratley, F. Lunnon & J. McKay, Amicable numbers and their distribution, *Math. Comput.*, **24**(1970) 431–432.

B. H. Brown, A new pair of amicable numbers, *Amer. Math. Monthly*, **46** (1939) 345.

Patrick Costello, Four new amicable pairs, *Notices Amer. Math. Soc.*, **21** (1974) A-483.

Patrick Costello, Amicable pairs of Euler's first form, *Notices Amer. Math. Soc.*, **22**(1975) A-440.

Patrick Costello, Amicable pairs of the form $(i, 1)$, *Math. Comput.*, **56**(1991) 859–865; *MR* **91k**:11009.

P. Erdős, On amicable numbers, *Publ. Math. Debrecen*, **4**(1955) 108–111; *MR* **16**, 998.

P. Erdős & G. J. Rieger, Ein Nachtrag über befreundete Zahlen, *J. reine angew. Math.*, **273**(1975) 220.

E. B. Escott, Amicable numbers, *Scripta Math.*, **12**(1946) 61–72; *MR* **8**, 135.

M. García, New amicable pairs, *Scripta Math.*, **23**(1957) 167–171; *MR* **20** #5158.

Mariano García, New unitary amicable couples, *J. Recreational Math.*, **17** (1984-5) 32–35.

Mariano García, *K*-fold isotopic amicable numbers, *J. Recreational Math.*, **19** (1987) 12–14

Mariano García, Some useful substitutions for finding amicable numbers (preprint March 1987).

Mariano García, Favorable conditions for amicability, *Hostos Community Coll. Math. J.*, New York, Spring 1989, 20–25.

A. A. Gioia & A. M. Vaidya, Amicable numbers with opposite parity, *Amer. Math. Monthly*, **74**(1967) 969–973; correction **75**(1968) 386; *MR* **36** #3711, **37** #1306.

Peter Hagis, On relatively prime odd amicable numbers, *Math. Comput.*, **23**(1969) 539–543; *MR* **40** #85.

Peter Hagis, Lower bounds for relatively prime amicable numbers of opposite parity, *Math. Comput.*, **24**(1970) 963–968.

Peter Hagis, Relatively prime amicable numbers of opposite parity, *Math. Mag.*, **43**(1970) 14–20.

Peter Hagis, Unitary amicable numbers, *Math. Comput.*, **25**(1971) 915–918.

H.-J. Kanold, Über die Dichten der Mengen der vollkommenen und der befreundeten Zahlen, *Math. Z.*, **61**(1954) 180–185; *MR* **16**, 337.

H.-J. Kanold, Über befreundete Zahlen I, *Math. Nachr.*, **9**(1953) 243–248; II *ibid.*, **10** (1953) 99–111; *MR* **15**, 506.

H.-J. Kanold, Über befreundete Zahlen III, *J. reine angew. Math.*, **234**(1969) 207–215; *MR* **39** #122.

E. J. Lee, Amicable numbers and the bilinear diophantine equation, *Math. Comput.*, **22**(1968) 181–187; *MR* **37** #142.

E. J. Lee, On divisibility by nine of the sums of even amicable pairs, *Math. Comput.*, **23**(1969) 545–548; *MR* **40** #1328.

E. J. Lee & J. S. Madachy, The history and discovery of amicable numbers, part 1, *J. Recreational Math.*, **5**(1972) 77–93; part 2, 153–173; part 3, 231–249.

O. Ore, *Number Theory and its History*, McGraw-Hill, New York, 1948, p. 89.

Carl Pomerance, On the distribution of amicable numbers, *J. reine angew. Math.*, **293/294**(1977) 217–222; II **325**(1981) 183–188; *MR* **56** #5402, **82m**: 10012.

P. Poulet, 43 new couples of amicable numbers, *Scripta Math.*, **14**(1948) 77.

H. J. J. te Riele, Four large amicable pairs, *Math. Comput.*, **28**(1974) 309–312.

H. J. J. te Riele, On generating new amicable pairs from given amicable pairs, *Math. Comput.*, **42**(1984) 219–223.

Herman J. J. te Riele, New very large amicable pairs, in *Number Theory* Noordwijkerhout 1983, *Springer Lecture Notes in Math.*, **1068**(1984) 210–215.

H. J. J. te Riele, Computation of all the amicable pairs below 10^{10}, *Math. Comput.*, **47**(1986) 361–368 & S9–S40.

H. J. J. te Riele, A new method for finding amicable pairs, in *Mathematics of Computation 1943–1993* (Vancouver, 1993), *Proc. Sympos. Appl. Math.* **43**, Amer. Math. Soc., Providence RI, 1994.

H. J. J. te Riele, W. Borho, S. Battiato, H. Hoffmann & E.J. Lee, *Table of Amicable Pairs between* 10^{10} *and* 10^{52}, Centrum voor Wiskunde en Informatica, Note NM-N8603, Stichting Math. Centrum, Amsterdam, 1986.

Dale Woods, Construction of amicable pairs, #789-10-21, *Abstracts Amer. Math. Soc.*, **3**(1982) 223.

B5 Quasi-amicable or betrothed numbers.

García has called a pair of numbers (m,n), $m < n$, **quasi-amicable** if

$$\sigma(m) = \sigma(n) = m + n + 1.$$

For example, (48,75), (140,195), (1575,1648), (1050,1925) and (2024,2295). Rufus Isaacs, noting that each of m and n is the sum of the *proper* divisors of the other (i.e., omitting 1 as well as the number itself) has much more appropriately named them **betrothed numbers**.

Mąkowski gave examples of betrothed numbers and also of **amicable triples**

$$\sigma(a) = \sigma(b) = \sigma(c) = a + b + c,$$

e.g., $2^2 3^2 5 \cdot 11$, $2^5 3^2 7$, $2^2 3^2 71$. Similarly, in a 92-07-20 letter, Yasutoshi Kohmoto calls the set $\{a,b,c,d\}$ **quadri-amicable** if

$$\sigma(a) = \sigma(b) = \sigma(c) = \sigma(d) = a + b + c + d.$$

As examples which are not multiples of 3 he gives

$$a = x \cdot 173 \cdot 1933058921 \cdot 149 \cdot 103540742849 \quad b = x \cdot 173 \cdot 1933058921 \cdot 15531111427499$$

$$c = x \cdot 336352252427 \cdot 149 \cdot 103540742849 \quad d = x \cdot 336352252427 \cdot 15531111427499$$

where x is the product of

$$5^9 \cdot 7^2 \cdot 11^4 \cdot 17^2 \cdot 19 \cdot 29^2 \cdot 67 \cdot 71^2 \cdot 109 \cdot 131 \cdot 139 \cdot 179 \cdot 307 \cdot 431 \cdot 521 \cdot 653$$
$$\cdot 1019 \cdot 1279 \cdot 2557 \cdot 3221 \cdot 5113 \cdot 5171 \cdot 6949$$

with a perfect number $2^{p-1}M_p$, $M_p = 2^p - 1$ being a Mersenne prime (see **A3**) with $p > 3$.

Hagis & Lord have found all 46 pairs of betrothed numbers with $m < 10^7$. All of them are of opposite parity. No pairs are known with m, n having the same parity. If there are such, then $m > 10^{10}$. If $m \perp n$, then mn contains at least four distinct prime factors, and if mn is odd, then mn contains at least 21 distinct prime factors.

Beck & Najar call such pairs *reduced* amicable pairs, and call numbers m, n such that

$$\sigma(m) = \sigma(n) = m + n - 1$$

augmented amicable pairs. They found 11 augmented amicable pairs. They found no reduced or augmented *unitary* amicable or sociable numbers (see **B8**) with $n < 10^5$.

Walter E. Beck & Rudolph M. Najar, More reduced amicable pairs, *Fibonacci Quart.*, **15**(1977) 331–332; *Zbl.* **389**.10004.

Walter E. Beck & Rudolph M. Najar, Fixed points of certain arithmetic functions, *Fibonacci Quart.*, **15**(1977) 337–342; *Zbl.* **389**.10005.

Peter Hagis & Graham Lord, Quasi-amicable numbers, *Math. Comput.*, **31** (1977) 608–611; *MR* **55** #7902; *Zbl.* **355**.10010.

M. Lal & A. Forbes, A note on Chowla's function, *Math. Comput.*, **25**(1971) 923–925; *MR* **45** #6737; *Zbl.* **245**.10004.

Andrzej Mąkowski, On some equations involving functions $\phi(n)$ and $\sigma(n)$, *Amer. Math. Monthly*, **67**(1960) 668–670; correction **68**(1961) 650; *MR* **24** #A76.

B6 Aliquot sequences.

Since some numbers are abundant and some deficient, it is natural to ask what happens when you iterate the function $s(n) = \sigma(n) - n$ and produce an **aliquot sequence**, $\{s^k(n)\}$, $k = 0, 1, 2, \ldots$. Catalan and Dickson conjectured that all such sequences were bounded, but we now have heuristic arguments and experimental evidence that some sequences, perhaps almost all of those with n even, go to infinity. The smallest n for which there was ever doubt was 138, but D. H. Lehmer eventually showed that after reaching a maximum

$$s^{117}(138) = 179931\,895322 = 2 \cdot 61 \cdot 929 \cdot 1587569$$

the sequence terminated at $s^{177}(138) = 1$. The next value for which there continues to be real doubt is 276. A good deal of computation by Lehmer, subsequently assisted by Godwin, Selfridge, Wunderlich and others, pushed the calculation as far as $s^{469}(276)$, which was quoted in the first edition.

Thomas Struppeck factored this term and computed two more iterates. Andy Guy wrote a PARI program which started from scratch and overnight verified all the earlier calculations and reached $s^{487}(276)$.

The first few sequences whose fate was unknown are the "Lehmer six" starting from 276, 552, 564, 660, 840 and 966. Our program has shown that the 840 sequence hit the prime $s^{746}(840) = 601$ having established a new record

$$s^{287}(840) = 3\,463982\,260143\,725017\,429794\,136098\,072146\,586526\,240388$$

$$= 2^2 \cdot 64970467217 \cdot 6237379309797547 \cdot 21369655584781129900003$$

for the maximum of a terminating sequence. This has recently been beaten by Mitchell Dickerman who found that the 1248 sequence has length 1075 after reaching a maximum $s^{583}(1248) =$

$$1231\,636691\,923602\,991963\,829388\,638861\,714770\,651073\,275257\,065104 = 2^4 p$$

of 58 digits. He has pursued the 276 sequence to its 628th term, which has

65 decimal digits. Godwin investigated the fourteen main sequences starting between 1000 and 2000 whose outcome was unknown and discovered that the sequence 1848 terminated. We have found that those for 2580, 2850, 4488, 4830, 6792, 7752, 8862 and 9540 also terminate.

H. W. Lenstra has proved that it is possible to construct arbitrarily long monotonic increasing aliquot sequences. See the quadruple paper cited under **B41**. The last of the following references has a bibliography of 60 items concerning the iteration of number-theoretic functions.

Jack Alanen, Empirical study of aliquot series, *Math. Rep.*, **133** Stichting Math. Centrum Amsterdam, 1972; see *Math. Comput.*, **28**(1974) 878–880.

E. Catalan, Propositions et questions diverses, *Bull. Soc. Math. France*, **16** (1887–88) 128–129.

John Stanley Devitt, Aliquot Sequences, MSc thesis, The Univ. of Calgary, 1976; see *Math. Comput.*, **32**(1978) 942–943.

J. S. Devitt, R. K. Guy & J. L. Selfridge, Third report on aliquot sequences, *Congr. Numer.* XVIII, Proc. 6th Manitoba Conf. Numer. Math., 1976, 177–204; *MR* **80d**:10001.

L. E. Dickson, Theorems and tables on the sum of the divisors of a number, *Quart. J. Math.*, **44**(1913) 264–296.

Paul Erdős, On asymptotic properties of aliquot sequences, *Math. Comput.*, **30**(1976) 641–645.

Andrew W. P. Guy & Richard K. Guy, A record aliquot sequence, in *Mathematics of Computation 1943–1993* (Vancouver, 1993), *Proc. Sympos. Appl. Math.*, (1994) Amer. Math. Soc., Providence RI, 1984.

Richard K. Guy, Aliquot sequences, in *Number Theory and Algebra*, Academic Press, 1977, 111–118; *MR* **57** #223; *Zbl.* **367**.10007.

Richard K. Guy & J. L. Selfridge, Interim report on aliquot sequences, *Congr. Numer.* V, Proc. Conf. Numer. Math., Winnipeg, 1971, 557–580; *MR* **49** #194; *Zbl.* **266**.10006.

Richard K. Guy & J. L. Selfridge, Combined report on aliquot sequences, The Univ. of Calgary Math. Res. Rep. **225**(May, 1974).

Richard K. Guy & J. L. Selfridge, What drives an aliquot sequence? *Math. Comput.*, **29**(1975) 101–107; *MR* **52** #5542; *Zbl.* **296**.10007. Corrigendum, *ibid.*, **34**(1980) 319–321; *MR* **81f**:10008; *Zbl.* **423**.10005.

Richard K. Guy & M. R. Williams, Aliquot sequences near 10^{12}, *Congr. Numer.* XII, Proc. 4th Manitoba Conf. Numer. Math., 1974, 387–406; *MR* **52** #242; *Zbl.* **359**.10007.

Richard K. Guy, D. H. Lehmer, J. L. Selfridge & M. C. Wunderlich, Second report on aliquot sequences, *Congr. Numer.* IX, Proc. 3rd Manitoba Conf. Numer. Math., 1973, 357–368; *MR* **50** #4455; *Zbl.* **325**.10007.

H. W. Lenstra, Problem 6064, *Amer. Math. Monthly*, **82**(1975) 1016; solution **84** (1977) 580.

G. Aaron Paxson, Aliquot sequences (preliminary report), *Amer. Math. Monthly*, **63**(1956) 614. See also *Math. Comput.*, **26** (1972) 807–809.

P. Poulet, La chasse aux nombres, Fascicule I, Bruxelles, 1929.

P. Poulet, Nouvelles suites arithmétiques, *Sphinx*, Deuxième Année (1932) 53–54.

H. J. J. te Riele, A note on the Catalan-Dickson conjecture, *Math. Comput.*, **27**(1973) 189–192; *MR* **48** #3869; *Zbl.* **255**.10008.
H. J. J. te Riele, Iteration of number theoretic functions, Report NN 30/83, Math. Centrum, Amsterdam, 1983.

B7 Aliquot cycles or sociable numbers.

Poulet discovered two cycles of numbers, showing that $s^k(n)$ can have the periods 5 and 28, in addition to 1 and 2. For $k \equiv 0, 1, 2, 3, 4 \bmod 5$, $s^k(12496)$ takes the values

$$12496 = 2^4 \cdot 11 \cdot 71, \quad 14288 = 2^4 \cdot 19 \cdot 47, \quad 15472 = 2^4 \cdot 967,$$
$$14536 = 2^3 \cdot 23 \cdot 79, \quad 14264 = 2^3 \cdot 1783.$$

For $k \equiv 0, 1, \ldots, 27 \bmod 28$, $s^k(14316)$ takes the values

14316	19116	31704	47616	83328	177792	295488
629072	589786	294896	358336	418904	366556	274924
275444	243760	376736	381028	285778	152990	122410
97946	48976	45946	22976	22744	19916	17716

After a gap of over 50 years, and the advent of high-speed computing, Henri Cohen discovered nine cycles of period 4, and Borho, David and Root also discovered some. Recently Moews & Moews have made an exhaustive search for such cycles with greatest member less than 10^{10}. There are twenty-four: their smallest members are

1264460	7169104	46722700	330003580	2387776550	4424606020
2115324	18048976	81128632	498215416	2717495235	4823923384
2784580	18656380	174277820	1236402232	2879697304	5373457070
4938136	28158165	209524210	1799281330	3705771825	8653956136

Moews & Moews give five larger 4-cycles, and, in a 90-09-01 letter, another whose least member is:

$$2^6 \cdot 79 \cdot 1913 \cdot 226691 \cdot 207722852483$$

They also found an 8-cycle:

1095447416	1259477224	1156962296	1330251784
1221976136	1127671864	1245926216	1213138984

Ren Yuanhua had already found three of the 4-cycles and Achim Flammenkamp had also found many of them, as well as a second 8-cycle:

1276254780	2299401444	3071310364	2303482780
2629903076	2209210588	2223459332	1697298124

and a 9-cycle:

805984760 1268997640 1803863720 2308845400 3059220620
3367978564 2525983930 2301481286 1611969514

Moews & Moews have continued their exhaustive search to uncover all cycles, of any length, whose member preceding the largest member is less than $3.6 \cdot 10^{10}$. There are three more 4-cycles, with least members

15837081520, 17616303220, 21669628904,

and a 6-cycle, all of whose members are odd:

$21548919483 = 3^5 \cdot 7^2 \cdot 13 \cdot 17 \cdot 19 \cdot 431,$ $23625285957 = 3^5 \cdot 7^2 \cdot 13 \cdot 19 \cdot 29 \cdot 277,$
$24825443643 = 3^2 \cdot 7^2 \cdot 11 \cdot 13 \cdot 19 \cdot 20719,$ $26762383557 = 3^4 \cdot 7^2 \cdot 13 \cdot 19 \cdot 27299,$
$25958284443 = 3^2 \cdot 7^2 \cdot 13 \cdot 19 \cdot 167 \cdot 1427,$ $23816997477 = 3^2 \cdot 7^2 \cdot 13 \cdot 19 \cdot 218651.$

It has been conjectured that there are no 3-cycles. On the other hand it has been conjectured that for each k there are infinitely many k-cycles.

Walter Borho, Über die Fixpunkte der k-fach iterierten Teilersummenfunktion, *Mitt. Math. Gesellsch. Hamburg*, **9**(1969) 34–48; *MR* **40** #7189.
Achim Flammenkamp, New sociable numbers, *Math. Comput.*, **56**(1991) 871–873.
David Moews & Paul C. Moews, A search for aliquot cycles below 10^{10}, *Math. Comput.*, **57**(1991) 849–855; *MR* **92e**:11151.
David Moews & Paul C. Moews, A search for aliquot cycles and amicable pairs, *Math. Comput.*, **61**(1993) 935–938.

B8 Unitary aliquot sequences.

The ideas of aliquot sequence and aliquot cycle can be adapted to the case where only the *unitary* divisors are summed, leading to **unitary aliquot sequences** and **unitary sociable numbers**. We use $\sigma^*(n)$ and $s^*(n)$ for the analogs of $\sigma(n)$ and $s(n)$ when just the unitary divisors are summed (compare **B3**).

Are there unbounded unitary aliquot sequences? Here the balance is more delicate than in the ordinary aliquot sequence case. The only sequences which deserve serious consideration are those involving odd multiples of 6, which is a unitary perfect number as well as an ordinary one. Now the sequences tend to increase if $3 \| n$, but decrease when a higher power of 3 is present, and it is a moot point as to which situation will dominate. Once a term of a sequence is $6m$, with m odd, then $\sigma^*(6m)$ is an even multiple of 6, making $s^*(6m)$ an odd multiple of 6 again, except in the extremely rare case that m is 4 raised to an odd power.

te Riele pursued all unitary aliquot sequences for $n < 10^5$. The only one which did not terminate or become periodic was 89610. Later calculations showed that this reached a maximum,

$$645\,856907\,610421\,353834 = 2 \cdot 3^2 \cdot 13 \cdot 19 \cdot 73 \cdot 653 \cdot 3047409443791$$

at its 568th term, and terminated at its 1129th.

One can hardly expect typical behavior until the expected number of prime factors is large. Since this number is $\ln\ln n$, such sequences are well beyond computer range. Of 80 sequences examined near 10^{12}, all have terminated or become periodic. One sequence exceeded 10^{23}.

Unitary amicable pairs and unitary sociable numbers may occur rather more frequently than their ordinary counterparts. Lal, Tiller & Summers found cycles of periods 1, 2, 3, 4, 5, 6, 14, 25, 39 and 65. Examples of unitary amicable pairs are (56430,64530) and (1080150,1291050), while (30,42,54) is a 3-cycle and

$$(1482, 1878, 1890, 2142, 2178)$$

is a 5-cycle.

Cohen (see **B3** for definitions and a reference) finds 62 infinitary amicable pairs with smaller member less than a million, eight infinitary aliquot cycles of order 4 and three of order 6. The only other such cycle of order less than 17 and least member less than a million is of order 11:

448800, 696864, 1124448, 1651584, 3636096, 6608784,
5729136, 3736464, 2187696, 1572432, 895152.

A type of aliquot sequence which can be unbounded has been suggested by David Penney & Carl Pomerance and is based on Dedekind's function: see **B41**.

Erdős, looking for a number-theoretic function whose iterates might be bounded, suggested defining $w(n) = n\sum 1/p_i^{\alpha_i}$ where $n = \prod p_i^{\alpha_i}$, and $W^k(n) = w(w^{k-1}(n))$. Note that $w(n) \perp n$. Can it be proved that $w^k(n)$, $k = 1, 2, \ldots$, is bounded? Is $|\{w(n) : 1 \leq n \leq x\}| = o(x)$?

Erdős & Selfridge called n a **barrier** for a number-theoretic function $f(m)$ if, for all $m < n$, $m + f(m) \leq n$. Euler's ϕ-function (see **B36**) and $\sigma(m)$ increase too fast to have barriers, but does $\omega(m)$ have infinitely many barriers? The numbers 2, 3, 4, 5, 6, 8, 9, 10, 12, 14, 17, 18, 20, 24, 26, 28, 30, ..., are barriers for $\omega(m)$. Does $\Omega(m)$ have infinitely many barriers? Selfridge observes that 99840 is the largest barrier for $\Omega(m)$ that is $< 10^5$. Mąkowski observes that $n = 1$ is a barrier for every function, and that 2 is a barrier for every function $f(n)$ with $f(1) = 1$; in particular for $d(m)$, the number of divisors of m. The inequality

$$\max\{d(n-1) + n - 1, d(n-2) + n - 2\} \geq n + 2$$

holds for $n \geq 7$, but not for $n = 6$. But $d(n-1) + n - 1 \geq n + 1$ for $n \geq 3$, so $d(m)$ has no barriers ≥ 3. Does

$$\max_{m<n}(m + d(m)) = n + 2$$

have infinitely many solutions? It is very doubtful. One solution is $n = 24$; the next larger is probably beyond computer range.

Paul Erdős, A mélange of simply posed conjectures with frustratingly elusive solutions, *Math. Mag.*, **52**(1979) 67–70.

P. Erdős, Problems and results in number theory and graph theory, *Congressus Numerantium* **27**, Proc. 9th Manitoba Conf. Numerical Math. Comput., 1979, 3–21.

Richard K. Guy & Marvin C. Wunderlich, Computing unitary aliquot sequences – a preliminary report, *Congressus Numerantium* **27**, Proc. 9th Manitoba Conf. Numerical Math. Comput., 1979, 257–270.

P. Hagis, Unitary amicable numbers, *Math. Comput.*, **25**(1971) 915–918; *MR* **45** #8599.

Peter Hagis, Unitary hyperperfect numbers, *Math. Comput.*, **36**(1981) 299–301.

M. Lal, G. Tiller & T. Summers, Unitary sociable numbers, *Congressus Numerantium* **7**, Proc. 2nd Manitoba Conf. Numerical Math., 1972, 211–216: *MR* **50** #4471.

H. J. J. te Riele, *Unitary Aliquot Sequences*, MR139/72, Mathematisch Centrum, Amsterdam, 1972; reviewed *Math. Comput.*, **32**(1978) 944–945; *Zbl.* 251. 10008.

H. J. J. te Riele, *Further Results on Unitary Aliquot Sequences*, NW12/73, Mathematisch Centrum, Amsterdam, 1973; reviewed *Math. Comput.*, **32**(1978) 945.

H. J. J. te Riele, *A Theoretical and Computational Study of Generalized Aliquot Sequences*, MCT72, Mathematisch Centrum, Amsterdam, 1976; reviewed *Math. Comput.*, **32**(1978) 945–946; *MR* **58** #27716.

C. R. Wall, Topics related to the sum of unitary divisors of an integer, PhD thesis, Univ. of Tennessee, 1970.

B9 Superperfect numbers.

Suryanarayana defines **superperfect numbers** n by $\sigma^2(n) = 2n$, i.e., $\sigma(\sigma(n)) = 2n$. He and Kanold show that the even ones are just the numbers 2^{p-1} where $2^p - 1$ is a Mersenne prime. Are there any odd superperfect numbers? If so, Kanold shows that they are perfect squares, and Dandepat and others that n or $\sigma(n)$ is divisible by at least three distinct primes.

More generally, Bode defines m-**superperfect numbers** as numbers n for which $\sigma^m(n) = 2n$, and shows that for $m \geq 3$ there are no even m-superperfect numbers. He also shows that for $m = 2$ there is no superperfect number $< 10^{10}$. Hunsucker & Pomerance have raised this bound to 7×10^{24} and have unpublished results on the numbers of distinct prime factors of n and of $\sigma(n)$ if n is superperfect.

If $\sigma^2(n) = 2n+1$, it would be consistent with earlier terminology to call n quasi-superperfect. The Mersenne primes are such. Are there others? Are there "almost superperfect numbers" for which $\sigma^2(n) = 2n - 1$?

Erdős asks if $(\sigma^k(n))^{1/k}$ has a limit as $k \to \infty$. He conjectures that it is infinite for each $n > 1$.

Schinzel asks if $\liminf \sigma^k(n)/n < \infty$ for each k, as $n \to \infty$, and observes that it follows for $k = 2$ from a deep theorem of Rényi. Mąkowski & Schinzel give an elementary proof for $k = 2$ that the limit is 1. Helmut Maier has used sieve methods to prove the result for $k = 3$.

Dieter Bode, Über eine Verallgemeinerung der volkommenen Zahlen, Dissertation, Braunschweig, 1971.

P. Erdős, Some remarks on the iterates of the ϕ and σ functions, *Colloq. Math.*, **17**(1967) 195–202.

J. L. Hunsucker & C. Pomerance, There are no odd super perfect numbers less than $7 \cdot 10^{24}$, *Indian J. Math.*, **17**(1975) 107–120; *MR* **82b**:10010.

H.-J. Kanold, Über "Super perfect numbers," *Elem. Math.*, **24**(1969) 61–62; *MR* **39** #5463.

Graham Lord, Even perfect and superperfect numbers, *Elem. Math.*, **30** (1975) 87–88.

Helmut Maier, On the third iterates of the ϕ- and σ-functions, *Colloq. Math.*, **49**(1984) 123–130.

Andrzej Mąkowski, On two conjectures of Schinzel, *Elem. Math.*, **31**(1976) 140–141.

A. Mąkowski & A. Schinzel, On the functions $\phi(n)$ and $\sigma(n)$, *Colloq. Math.*, **13**(1964-65) 95–99.

A. Schinzel, Ungelöste Probleme Nr. 30, *Elem. Math.*, **14**(1959) 60–61.

D. Suryanarayana, Super perfect numbers, *Elem. Math.*, **24**(1969) 16–17; *MR* **39** #5706.

D. Suryanarayana, There is no superperfect number of the form $p^{2\alpha}$, *Elem. Math.*, **28**(1973) 148–150; *MR* **48** #8374.

B10 Untouchable numbers.

Erdős has proved that there are infinitely many n such that $s(x) = n$ has no solution. Alanen calls such n **untouchable**. In fact Erdős shows that the untouchable numbers have positive lower density. Here are the untouchable numbers less than 1000:

2	5	52	88	96	120	124	146	162	178	188	206	210	216	238	246
248	262	268	276	288	290	292	304	306	322	324	326	336	342	372	406
408	426	430	448	472	474	498	516	518	520	530	540	552	556	562	576
584	612	624	626	628	658	668	670	714	718	726	732	738	748	750	756
766	768	782	784	792	802	804	818	836	848	852	872	892	894	896	898
902	916	926	936	964	966	976	982	996							

In view of the plausibility of the Goldbach conjecture (**C1**), it seems likely that 5 is the only odd untouchable number since if $2n+1 = p+q+1$ with p and q prime, then $s(pq) = 2n + 1$. Can this be proved independently?

Are there arbitrarily long sequences of consecutive even numbers which are untouchable? How large can the gaps between untouchable numbers be?

P. Erdős, Über die Zahlen der Form $\sigma(n)-n$ und $n-\phi(n)$, *Elem. Math.*, **28**(1973) 83–86; *MR* **49** #2502.

Paul Erdős, Some unconventional problems in number theory, *Astérisque*, **61** (1979) 73–82; *MR* **81h**:10001.

B11 Solutions of $m\sigma(m) = n\sigma(n)$.

Leo Moser has observed that while $n\phi(n)$ determines n uniquely, $n\sigma(n)$ does not. [$\phi(n)$ is Euler's totient function; see **B36**.] For example, $m\sigma(m) = n\sigma(n)$ for $m = 12$, $n = 14$. The multiplicativity of $\sigma(n)$ now ensures an infinity of solutions, $m = 12q$, $n = 14q$, where $q \perp 42$. So Moser asked if there is an infinity of *primitive* solutions, in the sense that (m^*, n^*) is *not* a solution for any $m^* = m/d$, $n^* = n/d$, $d > 1$. The example we've given is the least of the set $m = 2^{p-1}(2^q - 1)$, $n = 2^{q-1}(2^p - 1)$, where $2^p - 1$, $2^q - 1$ are distinct Mersenne primes, so that only a finite number of such solutions is known. Another set of solutions is $m = 2^7 \cdot 3^2 \cdot 5^2 \cdot (2^p - 1)$, $n = 2^{p-1} \cdot 5^3 \cdot 17 \cdot 31$, where 2^{p-1} is a Mersenne prime other than 3 or 31; also $p = 5$ gives a primitive solution on deletion of the common factor 31. There are other solutions, such as $m = 2^4 \cdot 3 \cdot 5^3 \cdot 7$, $n = 2^{11} \cdot 5^2$ and $m = 2^9 \cdot 5$, $n = 2^3 \cdot 11 \cdot 31$. An example with $m \perp n$ is $m = 2^5 \cdot 5$, $n = 3^3 \cdot 7$. If $m\sigma(m) = n\sigma(n)$, is m/n bounded?

Erdős observes that if n is squarefree, then integers of the form $n\sigma(n)$ are distinct. He can also prove that the number of solutions of $m\sigma(m) = n\sigma(n)$ with $m < n < x$ is $cx + o(x)$. In answer to the question, are there three distinct numbers l, m, n such that $l\sigma(l) = m\sigma(m) = n\sigma(n)$, Mąkowski observes that for distinct Mersenne primes M_{p_i}, $1 \le i \le s$, we have $n_i\sigma(n_i)$ is constant for $n_i = A/M_{p_i}$, where $A = \prod_{j=1}^{s} M_{p_j}$. Is there an infinity of primitive solutions of the equation $\sigma(a)/a = \sigma(b)/b$? Without restricting the solutions to being primitive, Erdős can show that their number with $a < b < x$ is at least $cx + o(x)$; with the restriction $a \perp b$, no solution is known at all.

Erdős believes that the number of solutions of $x\sigma(x) = n$ is less than $n^{\epsilon/\ln\ln n}$ for every $\epsilon > 0$, and says that the number may be less than $(\ln n)^c$.

P. Erdős, Remarks on number theory II: some problems on the σ function, *Acta Arith.*, **5**(1959) 171–177; *MR* **21** #6348.

B12 Analogs with $d(n)$, $\sigma_k(n)$.

Analogous questions may be asked with $\sigma_k(n)$ in place of $\sigma(n)$, where $\sigma_k(n)$ is the sum of the k-th powers of the divisors of n. For example, are there distinct numbers m and n such that $m\sigma_2(m) = n\sigma_2(n)$? For $k = 0$ we have $md(m) = nd(n)$ for $(m,n) = (18,27)$, $(24,32)$, $(56,64)$ and $(192,224)$. The last pair can be supplemented by 168 to give three distinct numbers such that $ld(l) = md(m) = nd(n)$. There are primitive solutions (m,n) of shape

$$m = 2^{qt-1}p, \qquad n = 2^{pt\cdot 2^{tu}-1}q$$

where p and $q = u + p \cdot 2^{tu}$ are primes, but it does not immediately follow that these are infinitely numerous. Many other solutions can be constructed; for example $(2^{70}, 2^{63} \cdot 71)$, $(3^{19}, 3^{17} \cdot 5)$ and $(5^{51}, 5^{49} \cdot 13)$.

Bencze proves the inequalities

$$\frac{n^k+1}{2} \geq \frac{\sigma_k(n)}{\sigma_{k-l}(n)} \geq \sqrt{n^l}$$

for $0 \leq l \leq k$ and gives no fewer than 60 applications.

Mihály Bencze, A contest problem and its application (Hungarian), *Mat. Lapok Ifjúsági Folyóirat (Románia)*, **91**(1986) 179–186.

B13 Solutions of $\sigma(n) = \sigma(n+1)$.

Sierpiński has asked if $\sigma(n) = \sigma(n+1)$ infinitely often. Hunsucker, Nebb & Stearns extended the tabulations of Mąkowski and of Mientka & Vogt and have found just 113 solutions

14, 206, 957, 1334, 1364, 1634, 2685, 2974, 4364, ...

less than 10^7. They also obtain statistics concerning the equation $\sigma(n) = \sigma(n+l)$, of which Mientka & Vogt had asked: for what l (if any) is there an infinity of solutions? They found many solutions if l is a factorial, but only two solutions for $l = 15$ and $l = 69$. They also ask whether, for each l and m, there is an n such that $\sigma(n) + m = \sigma(n+l)$.

One can ask corresponding questions for $\sigma_k(n)$, the sum of the k-th powers of the divisors of n. [For $k = 0$, see **B15**.] The only solution of $\sigma_2(n) = \sigma_2(n+1)$ is $n = 6$, since $\sigma_2(2n) > \sigma_2(2n+1)$ for $n > 7$ and $\sigma_2(2n) > 5n^2 > (\pi^2/8)(2n-1)^2 > \sigma_2(2n-1)$. Note that $\sigma_2(24) = \sigma_2(26)$; Erdős doubts that $\sigma_2(n) = \sigma_2(n+2)$ has infinitely many solutions, and thinks that $\sigma_3(n) = \sigma_3(n+2)$ has no solutions at all.

Richard K. Guy & Daniel Shanks, A constructed solution of $\sigma(n) = \sigma(n+1)$, *Fibonacci Quart.*, **12**(1974) 299; *MR* **50** #219.

John L. Hunsucker, Jack Nebb & Robert E. Stearns, Computational results concerning some equations involving $\sigma(n)$, *Math. Student*, **41**(1973) 285–289.

W. E. Mientka & R. L. Vogt, Computational results relating to problems concerning $\sigma(n)$, *Mat. Vesnik*, **7**(1970) 35–36.

B14 Some irrational series.

Is $\sum_{n=1}^{\infty}(\sigma_k(n)/n!)$ irrational? It is for $k=1$ and 2.

Erdős established the irrationality of the series

$$\sum_{n=1}^{\infty}\frac{1}{2^n-1}=\sum_{n=1}^{\infty}\frac{d(n)}{2^n}$$

and Peter Borwein showed that

$$\sum_{n=1}^{\infty}\frac{1}{q^n+r}\quad\text{and}\quad\sum_{n=1}^{\infty}\frac{(-1)^n}{q^n+r}$$

are irrational if q is an integer other than 0, ±1 and r is a rational other than 0 or $-q^n$.

Peter B. Borwein, On the irrationality of $\sum 1/(q^n+r)$, *J. Number Theory*, **37**(1991) 253–259.

Peter B. Borwein, On the irrationality of certain series, *Math. Proc. Cambridge Philos. Soc.*, **112**(1992) 141–146; *MR* **93g**:11074.

P. Erdős, On arithmetical properties of Lambert series, *J. Indian Math. Soc.(N.S.)* **12**(1948) 63–66.

P. Erdős, On the irrationality of certain series: problems and results, in *New Advances in Transcendence Theory*, Cambridge Univ. Press, 1988, pp. 102–109.

P. Erdős & M. Kac, Problem 4518, *Amer. Math. Monthly*, **60**(1953) 47. Solution R. Breusch, **61**(1954) 264–265.

B15 Solutions of $\sigma(q)+\sigma(r)=\sigma(q+r)$.

Max Rumney (*Eureka*, **26**(1963) 12) asked if the equation $\sigma(q)+\sigma(r)=\sigma(q+r)$ has infinitely many solutions which are primitive in a sense similar to that used in **B11**. If $q+r$ is prime, the only solution is $(q,r)=(1,2)$. If $q+r=p^2$ where p is prime, then one of q and r, say q, is prime, and $r=2^nk^2$ where $n\geq1$ and k is odd. If $k=1$, there is a solution if $p=2^n-1$ is a Mersenne prime and $q=p^2-2^n$ is prime; this is so for $n=2$, 3, 5, 7, 13 and 19. For $k=3$ there are no solutions, and none for $k=5$ with $n<189$. For $k=7$, $n=1$ and 3 give $(q,r,q+r)=(5231,2\cdot7^2,73^2)$ and $(213977,2^3\cdot7^2,463^2)$. Other solutions are $(k,n)=(11,1)$ (11,3), (19,5), (25,1), (25,9), (49,9), (53,1), (97,5), (107,5), (131,5), (137,1), (149,5), (257,5), (277,1), (313,3) and (421,3). Solutions with $q+r=p^3$ and p prime are $\sigma(2)+\sigma(6)=\sigma(8)$ and

$$\sigma(11638687)+\sigma(2^2\cdot13\cdot1123)=\sigma(227^3)$$

Erdős asks how many solutions (not necessarily primitive) are there with $q+r<x$; is it $cx+o(x)$ or is it of higher order? If $s_1<s_2<\cdots$ are the numbers for which $\sigma(s_i) = \sigma(q) + \sigma(s_i - q)$ has a solution with $q < s_i$, what is the density of the sequence $\{s_i\}$?

M. Sugunamma, PhD thesis, Sri Venkataswara Univ., 1969.

B16 Powerful numbers.

Erdős & Szekeres studied numbers n such that if a prime p divides n, then p^i divides n where i is a given number greater than one. Golomb named these numbers **powerful** and exhibited infinitely many pairs of consecutive ones. In answer to his conjecture that 6 was not representable as the difference of two powerful numbers, Władysław Narkiewicz noted that $6 = 5^4 7^3 - 463^2$, and that there were infinitely many such representations. In fact in 1971 Richard P. Stanley (unpublished) used the theory of the Pell equation to show that every non-zero integer is the difference between two powerful numbers and that 1 is the difference between two non-square powerful numbers, each in infinitely many ways.

Erdős denotes by $u_1^{(k)} < u_2^{(k)} < \ldots$ the integers all of whose prime factors have exponents $\geq k$; sometimes called **k-ful numbers**. He asks if the equation $u_{i+1}^{(2)} - u_i^{(2)} = 1$ has infinitely many solutions which do not come from Pell equations $x^2 - dy^2 = \pm 1$. Is there a constant c, such that the number of solutions with $u_i < x$ is less than $(\ln x)^c$? Does $u_{i+1}^{(3)} - u_i^{(3)} = 1$ have no solutions? Do the equations $u_{i+2}^{(2)} - u_{i+1}^{(2)} = 1$, $u_{i+1}^{(2)} - u_i^{(2)} = 1$ have no simultaneous solutions? And several other questions, some of which have been answered by Mąkowski.

For example, Mąkowski notes that $7^3x^2 - 3^3y^2 = 1$ has infinitely many solutions, and that this is not usually counted as a Pell equation. He also notes that

$$(2^{k+1} - 1)^k, \quad 2^k(2k + 1 - 1)^k \quad \text{and} \quad (2^{k+1} - 1)^{k+1}$$

are k-ful numbers in A.P., and that if $a_1, a_2, \ldots, a_s$ are k-ful and in A.P. with common difference d then

$$a_1(a_s + d)^k, \quad a_2(a_s + d)^k, \quad \ldots, \quad a_s(a_s + d)^k, \quad (a_s + d)^{k+1}$$

are $s + 1$ such numbers. As

$$a^k(a^l + \ldots + 1)^k + a^{k+1}(a^l + \ldots + 1)^k + \ldots + a^{k+l}(a^l + \ldots + 1)^k = a^k(a^l + \ldots + 1)^{k+1},$$

the sum of $l + 1$ k-ful numbers can be k-ful. He says that these last two questions become difficult when we require that the numbers be relatively prime. However, Nitaj constructs three infinite families of solutions of $x + y = z$ in relatively prime 3-ful numbers. A specific example is

$$17^3 \cdot 106219^3 + 2^7 \cdot 3^4 \cdot 5^3 \cdot 7^3 \cdot 2287^3 = 37^3 \cdot 197^3 \cdot 307^3$$

Heath-Brown has shown that every sufficiently large number is the sum of three powerful numbers; his proof would be much shortened if his conjecture could be proved that the quadratic form $x^2 + y^2 + 125z^2$ represents every sufficiently large $n \equiv 7 \bmod 8$. Erdős suggested that this may follow from work of Duke and Iwaniec: in fact see the forthcoming paper by Moroz.

Are there only finitely many powerful numbers n such that $n^2 - 1$ is also powerful? (See the Granville reference at **D2**.)

Gerry Myerson notes that the following conjecture is still open. If p is an odd prime and u, v are the smallest positive integers such that $u^2 - pv^2 = 1$, then

$$¿ \qquad p \nmid v \qquad ?$$

This has been verified for $p \equiv 1 \bmod 4$, $p < 6270713$ and for $p \equiv -1 \bmod 4$, $p < 7679299$. The conjecture is false if p is not prime; Myerson believes that 46 and 430 are the two smallest counterexamples.

N. C. Ankeny, E. Artin & S. Chowla, The class-number of real quadratic number fields, *Ann. of Math.*(2), **56**(1952) 479–493; *MR* **14**, 251.

B. D. Beach, H. C. Williams & C. R. Zarnke, Some computer results on units in quadratic and cubic fields, *Proc. 25th Summer Meet. Canad. Math. Congress*, Lakehead, 1971, 609–648; *MR* **49** #2656.

David Drazin & Robert Gilmer, Complements and comments, *Amer. Math. Monthly*, **78**(1971) 1104–1106 (esp. p. 1106).

W. Duke, Hyperbolic distribution problems and half-integral weight Maass forms, *Invent. Math.*, **92**(1988) 73–90; *MR* **89d**:11033.

P. Erdős, Problems and results on consecutive integers, *Eureka*, **38**(1975–76) 3–8.

P. Erdős & G. Szekeres, Über die Anzahl der Abelschen Gruppen gegebener Ordnung und über ein verwandtes zahlentheoretisches Problem, *Acta Litt. Sci. Szeged*, **7**(1934) 95–102; *Zbl.* **10**, 294.

S. W. Golomb, Powerful numbers, *Amer. Math. Monthly*, **77**(1970) 848–852; *MR* **42** #1780.

D. R. Heath-Brown, Ternary quadratic forms and sums of three square-full numbers, *Séminaire de Théorie des Nombres, Paris, 1986-87*, Birkhäuser, Boston, 1988; *MR* **91b**:11031.

D. R. Heath-Brown, Sums of three square-full numbers, in Number Theory, I (Budapest, 1987), *Colloq. Math. Soc. János Bolyai*, **51**(1990) 163–171; *MR* **91i**:11036.

D. R. Heath-Brown, Square-full numbers in short intervals, *Math. Proc. Cambridge Philos. Soc.*, **110**(1991) 1–3; *MR* **92c**:11090.

Aleksander Ivić, On the asymptotic formulas for powerful numbers, *Publ. Math. Inst. Beograd (N.S.)*, **23(37)**(1978) 85–94; *MR* **58** #21977.

A. Ivić & P. Shiu, The distribution of powerful integers, *Illinois J. Math.*, **26**(1982) 576–590; *MR* **84a**:10047.

H. Iwaniec, Fourier coefficients of modular forms of half-integral weight, *Invent. Math.*, **87**(1987) 385–401; *MR* **88b**:11024.

C.-H. Jia, On square-full numbers in short intervals, *Acta Math. Sinica (N.S.)* **5**(1987) 614–621.

Liu Hong-Quan, On square-full numbers in short intervals, *Acta Math. Sinica* (*N.S.*), **6**(1990) 148–164; *MR* **91g**:11105.

Andrzej Mąkowski, On a problem of Golomb on powerful numbers, *Amer. Math. Monthly*, **79**(1972) 761.

Andrzej Mąkowski, Remarks on some problems in the elementary theory of numbers, *Acta Math. Univ. Comenian.*, **50/51**(1987) 277–281; *MR* **90e**:11022.

Wayne L. McDaniel, Representations of every integer as the difference of powerful numbers, *Fibonacci Quart.*, **20**(1982) 85–87.

Richard A. Mollin, The power of powerful numbers, *Internat. J. Math. Math. Sci.*, **10**(1987) 125–130; *MR* **88e**:11008.

Richard A. Mollin & P. Gary Walsh, On non-square powerful numbers, *Fibonacci Quart.*, **25**(1987) 34–37; *MR* **88f**:11006.

Richard A. Mollin & P. Gary Walsh, On powerful numbers, *Internat. J. Math. Math. Sci.*, **9**(1986) 801–806; *MR* **88f**:11005.

Richard A. Mollin & P. Gary Walsh, A note on powerful numbers, quadratic fields and the Pellian, *CR Math. Rep. Acad. Sci. Canada*, **8**(1986) 109–114; *MR* **87g**:11020.

Richard A. Mollin & P. Gary Walsh, Proper differences of non-square powerful numbers, *CR Math. Rep. Acad. Sci. Canada*, **10**(1988) 71–76; *MR* **89e**:11003.

L. J. Mordell, On a pellian equation conjecture, *Acta Arith.*, **6**(1960) 137–144; *MR* **22** #9470.

B. Z. Moroz, On representation of large integers by integral ternary positive definite quadratic forms, *Journées Arithmetiques*, Geneva.

Abderrahmane Nitaj, On a conjecture of Erdős on 3-powerful numbers, *London Math. Soc.*, (submitted).

Peter Georg Schmidt, On the number of square-full integers in short intervals, *Acta Arith.*, **50**(1988) 195–201; corrigendum, **54**(1990) 251–254; *MR* **89f**:11131.

W. A. Sentance, Occurrences of consecutive odd powerful numbers, *Amer. Math. Monthly*, **88**(1981) 272–274.

P. Shiu, On square-full integers in a short interval, *Glasgow Math. J.*, **25** (1984) 127–134.

P. Shiu, The distribution of cube-full numbers, *Glasgow Math. J.*, **33**(1991) 287–295. *MR* **92g**:11091.

P. Shiu, Cube-full numbers in short intervals, *Math. Proc. Cambridge Philos. Soc.*, **112** (1992) 1–5; *MR* **93d**:11097.

A. J. Stephens & H. C. Williams, Some computational results on a problem concerning powerful numbers, *Math. Comput.*, **50**(1988) 619–632.

D. Suryanarayana, On the distribution of some generalized square-full integers, *Pacific J. Math.*, **72**(1977) 547–555; *MR* **56** #11933.

D. Suryanarayana & R. Sitaramachandra Rao, The distribution of square-full integers, *Ark. Mat.*, **11**(1973) 195–201; *MR* **49** #8948.

Charles Vanden Eynden, Differences between squares and powerful numbers, *816-11-305, *Abstracts Amer. Math. Soc.*, **6**(1985) 20.

David T. Walker, Consecutive integer pairs of powerful numbers and related Diophantine equations, *Fibonacci Quart.*, **14**(1976) 111–116; *MR* **53** #13107.

Yuan Ping-Zhi, On a conjecture of Golomb on powerful numbers (Chinese. English summary), *J. Math. Res. Exposition*, **9**(1989) 453–456; *MR* **91c**:11009.

B17 Exponential-perfect numbers.

If $n = p_1^{a_1} p_2^{a_2} \cdots p_r^{a_r}$, then Straus & Subbarao call d an **exponential divisor** (e-divisor) of n if $d|n$ and $d = p_1^{b_1} p_2^{b_2} \cdots p_r^{b_r}$ where $b_j | a_j$ $(1 \le j \le r)$, and they call n **e-perfect** if $\sigma_e(n) = 2n$, where $\sigma_e(n)$ is the sum of the e-divisors of n. Some examples of e-perfect numbers are

$$2^2 \cdot 3^2, \quad 2^2 \cdot 3^3 \cdot 5^2, \quad 2^3 \cdot 3^2 \cdot 5^2, \quad 2^4 \cdot 3^2 \cdot 11^2, \quad 2^4 \cdot 3^3 \cdot 5^2 \cdot 11^2,$$

$$2^6 \cdot 3^2 \cdot 7^2 \cdot 13^2, \quad 2^6 \cdot 3^3 \cdot 5^2 \cdot 7^2 \cdot 13^2, \quad 2^7 \cdot 3^2 \cdot 5^2 \cdot 7^2 \cdot 13^2, \quad 2^8 \cdot 3^2 \cdot 5^2 \cdot 7^2 \cdot 139^2$$

and

$$2^{19} \cdot 3^2 \cdot 5^2 \cdot 7^2 \cdot 11^2 \cdot 13^2 \cdot 19^2 \cdot 37^2 \cdot 79^2 \cdot 109^2 \cdot 157^2 \cdot 313^2.$$

If m is squarefree, $\sigma_e(m) = m$, so if n is e-perfect and m is squarefree with $m \perp n$, then mn is e-perfect. So it suffices to consider only powerful (**B16**) e-perfect numbers.

Straus & Subbarao show that there are no odd e-perfect numbers, in fact no odd n which satisfy $\sigma_e(n) = kn$ for any integer $k > 1$. They also show that for each r the number of (powerful) e-perfect numbers with r prime factors is finite, and that the same holds for e-**multiperfect numbers** $(k > 2)$.

Is there an e-perfect number which is *not* divisible by 3?

Straus & Subbarao conjecture that there is only a finite number of e-perfect numbers *not* divisible by any given prime p.

Are there any e-multiperfect numbers?

E. G. Straus & M. V. Subbarao, On exponential divisors, *Duke Math. J.*, **41**(1974) 465–471; *MR* **50** #2053.

M. V. Subbarao, On some arithmetic convolutions, *Proc. Conf. Kalamazoo MI, 1971, Springer Lecture Notes in Math.*, **251**(1972) 247–271; *MR* **49** #2510.

M. V. Subbarao & D. Suryanarayana, Exponentially perfect and unitary perfect numbers, *Notices Amer. Math. Soc.*, **18**(1971) 798.

B18 Solutions of $d(n) = d(n+1)$.

Claudia Spiro has proved that $d(n) = d(n+5040)$ has infinitely many solutions and Heath-Brown used her ideas to show that there are infinitely many numbers n such that $d(n) = d(n+1)$, and Pinner has extended this to $d(n) = d(n+a)$ for any integer a. Many examples arise from pairs of consecutive numbers which are products of just two distinct primes, and it has been conjectured that there is an infinity of *triples* of consecutive products of two primes, n, $n+1$, $n+2$. For example, $n = 33$, 85, 93, 141, 201, 213, 217, 301, 393, 445, 633, 697, 921, ... It is clearly not possible to have *four* such numbers, but it *is* possible to have longer sequences of consecutive numbers with the same number of divisors. For example,

$$d(242) = d(243) = d(244) = d(245) = 6 \quad \text{and}$$

$$d(40311) = d(40312) = d(40313) = d(40314) = d(40315) = 8.$$

How long can such sequences be? In an 87-07-16 letter Stephane Vandemergel sent the sequence of seven numbers: $171893 = 19 \cdot 83 \cdot 109$, $171894 = 2 \cdot 3 \cdot 28649$, $171895 = 5 \cdot 31 \cdot 1109$, $171896 = 2^3 \cdot 21487$, $171897 = 3 \cdot 11 \cdot 5209$, $171898 = 2 \cdot 61 \cdot 1409$, $171899 = 7 \cdot 13 \cdot 1889$, each with 8 divisors. In 1990, Ivo Düntsch & Roger Eggleton discovered several such sequences of 7 numbers, two of 8 and one of 9, each with 48 divisors; the last example starts at 17796126877482329126044, presumably not the smallest of its kind.

Erdős believes that there are sequences of length k for every k, but does not see how to give an upper bound for k in terms of n.

Erdős, Pomerance & Sárközy showed that the number of $n \le x$ with $d(n) = d(n+1)$ is $\ll x/(\ln\ln x)^{1/2}$, and Hildebrand showed that this number is $\gg x/(\ln\ln x)^3$. The former authors also showed that the number of $n \le x$ with the ratio $d(n)/d(n+1)$ in the set $\{2^{-3}, 2^{-2}, 2^{-1}, 1, 2, 2^2, 2^3\}$ is $\asymp x/(\ln\ln x)^{1/2}$.

Erdős showed that the density of numbers n with $d(n+1) > d(n)$ is $\frac{1}{2}$. This, with the above results, settles a conjecture of S. Chowla. Fabrykowski & Subbarao extend this to the case with $n+h$ in place of $n+1$.

Erdős also lets

$$1 = d_1 < d_2 < \cdots < d_\tau = n$$

be the set of all divisors of n, listed in order, defines

$$f(n) = \sum_{1}^{\tau-1} d_i/d_{i+1}$$

and asks us to prove that $\sum_{n=1}^{x} f(n) = (1 + o(1))x \ln x$.

Erdős & Mirsky ask for the largest k so that the numbers $d(n)$, $d(n+1)$, ..., $d(n+k)$ are all distinct. They only have trivial bounds; probably $k = (\ln n)^c$.

P. Erdős, Problem P. 307, *Canad. Math. Bull.*, **24**(1981) 252.

P. Erdős & L. Mirsky, The distribution of values of the divisor function $d(n)$, *Proc. London Math. Soc.*(3), **2**(1952) 257–271.

P. Erdős, C. Pomerance & A. Sárközy, On locally repeated values of certain arithmetic functions, II, *Acta Math. Hungarica*, **49**(1987) 251–259; *MR* **88c**:11008.

J. Fabrykowski & M. V. Subbarao, Extension of a result of Erdős concerning the divisor function, *Utilitas Math.*, **38**(1990) 175–181; *MR* **92d**:11101.

D. R. Heath-Brown, A parity problem from sieve theory, *Mathematika*, **29** (1982) 1–6 (esp. p. 6).

D. R. Heath-Brown, The divisor function at consecutive integers, *Mathematika*, **31**(1984) 141–149.

Adolf Hildebrand, The divisors function at consecutive integers, *Pacific J. Math.*, **129** (1987) 307–319; *MR* **88k**:11062.

M. Nair & P. Shiu, On some results of Erdős and Mirsky, *J. London Math. Soc.*(2), **22**(1980) 197–203; and see *ibid.*, **17**(1978) 228–230.

C. Pinner, M.Sc. thesis, Oxford, 1988.

A. Schinzel, Sur un problème concernant le nombre de diviseurs d'un nombre naturel, *Bull. Acad. Polon. Sci. Ser. sci. math. astr. phys.*, **6**(1958) 165–167.

A. Schinzel & W. Sierpiński, Sur certaines hypothèses concernant les nombres premiers, *Acta Arith.*, **4**(1958) 185–208.

W. Sierpiński, Sur une question concernant le nombre de diviseurs premiers d'un nombre naturel, *Colloq. Math.*, **6**(1958) 209–210.

B19 $(m, n+1)$ and $(m+1, n)$ with same set of prime factors.

Motzkin & Straus asked for all pairs of numbers m, n such that m and $n+1$ have the same set of distinct prime factors, and similarly for n and $m+1$. It was thought that such pairs were necessarily of the form $m = 2^k + 1$, $n = m^2 - 1$ ($k = 0, 1, 2, \ldots$) until Conway observed that if $m = 5 \cdot 7$, $n + 1 = 5^4 \cdot 7$, then $n = 2 \cdot 3^7$, $m + 1 = 2^2 \cdot 3^2$. Are there others?

Similarly, Erdős asks if there are numbers m, n ($m < n$) other than $m = 2^k - 2$, $n = 2^k(2^k - 2)$ such that m and n have the same prime factors and similarly for $m+1$, $n+1$. Mąkowski found the pair $m = 3 \cdot 5^2$, $n = 3^5 \cdot 5$ for which $m + 1 = 2^2 \cdot 19$, $n + 1 = 2^6 \cdot 19$. Compare problem **B29**.

Pomerance has asked if there are any odd numbers $n > 1$ such that n and $\sigma(n)$ have the same prime factors. He conjectures that there are not.

The example $1 + 2 \cdot 3^7 = 5^4 7$ in the first paragraph is of interest in connexion with the "**ABC conjecture**":

Many of the classical problems of number theory (Goldbach conjecture, twin primes, the Fermat problem, Waring's problem, the Catalan conjecture) owe their difficulty to a clash between multiplication and addition. Roughly, if there's an additive relation between three numbers, their prime factors can't all be small.

Suppose that $A + B = C$ with $\gcd(A, B, C) = 1$. Define the radical R to be the maximum squarefree integer dividing ABC and the power P by

$$P = \frac{\ln \max(|A|, |B|, |C|)}{\ln R}$$

then for a given η are there only finitely many triples $\{A, B, C\}$ with $P \geq \eta$? A stronger form of this conjecture is that $\limsup P = 1$; both forms of the conjecture seem to be hopelessly beyond reach. The example just given is the fifth in the list below.

Joe Kanapka, a student of Noam Elkies, has produced a list of all examples with $C < 2^{32}$ and $P > 1.2$. There are nearly 1000 of them. The "top ten" as far as I know are

P	A	B	C	author
1.629912	2	$3^{10} \cdot 109$	23^5	Reyssat
1.625991	11^2	$3^2 \cdot 5^6 \cdot 7^3$	$2^{21} \cdot 23$	de Weger(**D10**)
1.623490	$19 \cdot 1307$	$7 \cdot 29^2 \cdot 31^8$	$2^8 \cdot 3^{22} \cdot 5^4$	Browkin-Brzeziński
1.580756	283	$5^{11} \cdot 13^2$	$2^8 \cdot 3^8 \cdot 17^3$	Br-Br, Nitaj
1.567887	1	$2 \cdot 3^7$	$5^4 \cdot 7$	Lehmer(**B29**)
1.547075	7^3	3^{10}	$2^{11} \cdot 29$	de Weger
1.526999	$13 \cdot 19^6$	$2^{30} \cdot 5$	$3^{13} \cdot 11^2 \cdot 31$	Nitaj
1.502839	239	$5^8 \cdot 17^3$	$2^{10} \cdot 37^4$	Br-Br, Nitaj
1.497621	$5^2 \cdot 7937$	7^{13}	$2^{18} \cdot 3^7 \cdot 13^2$	de Weger
1.492432	$2^2 \cdot 11$	$3^2 \cdot 13^{10} \cdot 17 \cdot 151 \cdot 4423$	$5^9 \cdot 139^6$	Nitaj

Browkin & Brzeziński generalize the ABC-conjecture (which is their case $n = 3$) to an "n-conjecture" on $a_1 + \cdots + a_n = 0$ in coprime integers with non-vanishing subsums. With R and P defined analogously, they conjecture that $\limsup P = 2n - 5$. They prove that $\limsup P \geq 2n - 5$. They give a lot of examples for the ABC-conjecture with $P > 1.4$. Their method is to look for rational numbers approximating roots of integers (note that the best example above is connected to the good approximation $23/9$ for $109^{1/5}$). Abderrahmane Nitaj used a similar method. Some of these were found independently by Robert Styer (**D10**). The Catalan relation $1 + 2^3 = 3^2$ gives a comparatively poor $P \approx 1.22629$.

For connexions between the ABC conjecture and the Fermat problem, see the Granville references at **D2**. Indeed, if $A = a^p$, $B = b^p$, $C = c^p$ and the Fermat equation $A + B = C$ is satisfied, then the elliptic curve

$$y^2 = x(x - A)(x + B)$$

has discriminant $(4ABC)^2$.

Jerzy Browkin & Juliusz Brzeziński, Some remarks on the *abc*-conjecture, *Math. Comput.*, (to appear).

Noam D. Elkies, *ABC* implies Mordell, *Internat. Math. Res. Notices*, **1991** no. 7, 99–109; *MR* **93d**:11064.

Serge Lang, Old and new conjectured diophantine inequalities, *Bull. Amer. Math. Soc.*, **23**(1990) 37–75.

A. Mąkowski, On a problem of Erdős, *Enseignement Math.*(2), **14**(1968) 193.

Abderrahmane Nitaj, 1993 preprint.

András Sárközy, On sums $a+b$ and numbers of the form $ab+1$ with many prime factors, *Österreichisch-Ungarisch-Slowakisches Kolloquium über Zahlentheorie* (Maria Trost, 1992), 141–154, *Grazer Math. Ber.*, **318** Karl-Franzens-Univ. Graz, 1993.

C. L. Stewart & Yu Kun-Rui, On the *abc* conjecture, *Math. Ann.*, **291**(1991) 225–230; *MR* **92k**:11037.

R. Tijdeman, The number of solutions of Diophantine equations, in Number Theory, II (Budapest, 1987), *Colloq. Math. Soc. János Bolyai*, **51**(1990) 671–696.

B20 Cullen numbers.

Some interest has been shown in the **Cullen numbers**, $n \cdot 2^n + 1$, which are all composite for $2 \le n \le 1000$, except for $n = 141$. This is probably a good example of the Strong Law of Small Numbers, because for small n, where the density of primes is large, the Cullen numbers are very likely to be composite because Fermat's (little) theorem tells us that $(p-1)2^{p-1}+1$ and $(p-2)2^{p-2}+1$ are both divisible by p. Moreover, as John Conway observes, the Cullen numbers are divisible by $2n-1$ if that is a prime of shape $8k \pm 3$. He asks if p and $p \cdot 2^p + 1$ can both be prime. Wilfrid Keller notes that Conway's remark can be generalized as follows. Write $C_n = n \cdot 2^n + 1$, $W_n = n \cdot 2^n - 1$: then a prime p divides $C_{(p+1)/2}$ and $W_{(3p-1)/2}$ or it divides $C_{(3p-1)/2}$ and $W_{(p+1)/2}$ according as the Legendre symbol (see **F5**) $\left(\frac{2}{p}\right)$ is -1 or $+1$. Keller has found prime Cullen numbers with $n = 4713$, 5795, 6611 and 18496. There are no others with $n \le 30000$.

Riesel observes that the corresponding numbers $n \cdot 2^n - 1$ are prime for $n = 2$, 3, 6, 30, 75, 81 and 115; Jönsson for $n = 362$ and Keller for $n = 123$, 249, 384, 462, 512 (i.e. M_{521}), 751, 882, 5312, 7755, 9531, 12379, 15822 and 18885. Many of these were also found by Waldemar Gorzkowski. There are no others with $n \le 20000$. In parallel with Conway's question above, Keller notes that here 3, 751 and 12379 are primes.

Ingemar Jönsson, On certain primes of Mersenne-type, *Nordisk Tidskr. Informationsbehandling* (*BIT*), **12** (1972) 117–118; *MR* **47** #120.

Wilfrid Keller, New Cullen primes, (92-11-20 preprint).

Hans Riesel, *En Bok om Primtal* (Swedish), Lund, 1968; supplement Stockholm, 1977; *MR* **42** #4507, **58** #10681.

B21 $k \cdot 2^n + 1$ composite for all n.

Let $N(x)$ be the number of odd positive integers k, not exceeding x, such that $k \cdot 2^n + 1$ is prime for no positive integer n. Sierpiński used covering congruences (see **F13**) to show that $N(x)$ tends to infinity with x. For example, if

$$k \equiv 1 \bmod 641 \cdot (2^{32} - 1) \quad \text{and} \quad k \equiv -1 \bmod 6700417,$$

then every member of the sequence $k \cdot 2^n + 1$ ($n = 0, 1, 2, \ldots$) is divisible by just one of the primes 3, 5, 17, 257, 641, 65537 or 6700417. He also noted that at least one of 3, 5, 7, 13, 17, 241 will always divide $k \cdot 2^n + 1$ for certain other values of k.

Erdős & Odlyzko have shown that

$$(\frac{1}{2} - c_1)x \geq N(x) \geq c_2 x.$$

What is the least value of k such that $k \cdot 2^n + 1$ is composite for all values of n? Selfridge discovered that one of 3, 5, 7, 13, 19, 37, 73 always divides $78557 \cdot 2^n + 1$. He also noted that there is a prime of the form $k \cdot 2^n + 1$ for each $k < 383$ and Hugh Williams discovered the prime $383 \cdot 2^{6393} + 1$.

In the first edition we wrote that the determination of the least k may now be within computer reach, though Keller has expressed his doubts about this. Extensive calculations have been made by Baillie, Cormack & Williams, by Keller, and by Buell & Young. The answer seems almost certain to be $k = 78557$, but there remain the 35 possibilities

4847 5297 5359 7013 10223 13787 19249 21181 22699 24737
25819 27653 27923 28433 33661 34999 39781 44131 46157 46187
46471 47897 48833 50693 54767 55459 59569 60443 60541 63017
65567 67607 69109 74191 74269

for none of which is there a prime with $n \leq 50000$.

A very full bibliography is appended to the thorough survey of the subject in the second of Keller's papers.

Robert Baillie, New primes of the form $k \cdot 2^n + 1$, *Math. Comput.*, **33**(1979) 1333–1336; *MR* **80h**:10009.

Robert Baillie, G. V. Cormack & H. C. Williams, The problem of Sierpiński concerning $k \cdot 2^n + 1$, *Math. Comput.*, **37**(1981) 229–231; corrigendum, **39**(1982) 308.

Wieb Bosma, Explicit primality criteria for $h \cdot 2^k \pm 1$, *Math. Comput.*, **61**(1993) 97–109.

D. A. Buell & J. Young, Some large primes and the Sierpiński problem, SRC Technical Report 88-004, Supercomputing Research Center, Lanham MD, May 1988.

G. V. Cormack & H. C. Williams, Some very large primes of the form $k \cdot 2^n + 1$, *Math. Comput.*, **35**(1980) 1419–1421; *MR* **81i**:10011; corrigendum, Wilfrid Keller, **38**(1982) 335; *MR* **82k**:10011.

Paul Erdős & Andrew M. Odlyzko, On the density of odd integers of the form $(p-1)2^{-n}$ and related questions, *J. Number Theory*, **11**(1979) 257–263; *MR* **80i**:10077.

G. Jaeschke, On the smallest k such that all $k \cdot 2^N + 1$ are composite, *Math. Comput.*, **40**(1983) 381–384; *MR* **84k**:10006; corrigendum, **45**(1985) 637; *MR* **87b**:11009.

Wilfrid Keller, Factors of Fermat numbers and large primes of the form $k \cdot 2^n + 1$, *Math. Comput.*, **41**(1983) 661–673; *MR* **85b**:11119; II (incomplete draft, 92-02-19).

Wilfrid Keller, Woher kommen die größten derzeit bekannten Primzahlen? *Mitt. Math. Ges. Hamburg*, **12**(1991) 211–229;*MR* **92j**:11006.

N. S. Mendelsohn, The equation $\phi(x) = k$, *Math. Mag.*, **49**(1976) 37–39; *MR* **53** #252.

Raphael M. Robinson, A report on primes of the form $k \cdot 2^n + 1$ and on factors of Fermat numbers, *Proc. Amer. Math. Soc.*, **9**(1958) 673–681; *MR* **20** #3097.

J. L. Selfridge, Solution of problem 4995, *Amer. Math. Monthly*, **70**(1963) 101.

W. Sierpiński, Sur un problème concernant les nombres $k \cdot 2^n + 1$, *Elem. Math.*, **15**(1960) 73–74; *MR* **22** #7983; corrigendum, **17**(1962) 85.

W. Sierpiński, *250 Problems in Elementary Number Theory*, Elsevier, New York, 1970, Problem 118, pp. 10 & 64.

R. G. Stanton & H. C. Williams, Further results on covering of the integers $1 + k2^n$ by primes, *Combinatorial Math.* VIII, *Lecture Notes in Math.*, **884**, Springer-Verlag, Berlin–New York, 1980, 107–114.

B22 Factorial n as the product of n large factors.

Straus, Erdős & Selfridge have asked that $n!$ be expressed as the product of n factors, with the least one, l, as large as possible. For example, for $n = 56$, $l = 15$,

$$56! = 15 \cdot 16^3 \cdot 17^3 \cdot 18^8 \cdot 19^2 \cdot 20^{12} \cdot 21^9 \cdot 22^5 \cdot 23^2 \cdot 26^4 \cdot 29 \cdot 31 \cdot 37 \cdot 41 \cdot 43 \cdot 47 \cdot 53$$

Selfridge has two conjectures: (a) that, except for $n = 56$, $l \geq \lfloor 2n/7 \rfloor$; (b) that for $n \geq 300000$, $l \geq n/3$. If the latter is true, by how much can 300000 be reduced?

Straus was reputed to have shown that for $n > n_0 = n_0(\epsilon)$, $l > n/(e+\epsilon)$, but a proof was not found in his Nachlaß. It is clear from Stirling's formula that this is best possible. It is also clear that l is a monotonic, though not strictly monotonic, increasing function of n. On the other hand it does not take all integer values: for $n = 124$, 125, l is respectively 35 and 37. Erdős asks how large the gaps in the values of l can be, and can l be constant for arbitrarily long stretches?

Alladi & Grinstead write $n!$ as a product of prime powers each as large as $n^{\delta(n)}$ and let $\alpha(n) = \max \delta(n)$ and show that $\lim_{n\to\infty} \alpha(n) = e^{c-1} = \alpha$, say, where

$$c = \sum_2^\infty \frac{1}{k} \ln \frac{k}{k-1} \quad \text{so that} \quad \alpha = 0.809394020534\ldots .$$

K. Alladi & C. Grinstead, On the decomposition of $n!$ into prime powers, *J. Number Theory*, **9**(1977) 452–458; *MR* **56** #11934.

P. Erdős, Some problems in number theory, *Computers in Number Theory*, Academic Press, London & New York, 1971, 405–414.

B23 Equal products of factorials.

Suppose that $n! = a_1!a_2!\ldots a_r!$, $r \geq 2$, $a_1 \geq a_2 \geq \ldots \geq a_r \geq 2$. A trivial example is $a_1 = a_2!\ldots a_r!-1$, $n = a_2!\ldots a_r!$ Dean Hickerson notes that the only nontrivial examples with $n \leq 410$ are $9! = 7!3!3!2!$, $10! = 7!6! = 7!5!3!$ and $16! = 14!5!2!$ and asks if there are any others. Jeffrey Shallit & Michael Easter have extended the search to $n = 18160$.

Erdős observes that if $P(n)$ is the largest prime factor of n and if it were known that $P(n(n+1))/\ln n$ tends to infinity with n, then it would follow that there are only finitely many nontrivial examples.

He & Graham have studied the equation $y^2 = a_1!a_2!\ldots a_r!$ They define the set F_k to be those m for which there is a set of integers $m = a_1 > a_2 > \ldots > a_r$ with $r \leq k$ which satisfies this equation for some y, and write D_k for $F_k - F_{k-1}$. They have various results, for example: for almost all primes p, $13p$ does not belong to F_5; and the least element of D_6 is 527. If $D_4(n)$ is the number of elements of D_4 which are $\leq n$, they do not know the order of growth of $D_4(n)$. They conjecture that $D_6(n) > cn$ but cannot prove this.

Earl Ecklund & Roger Eggleton, Prime factors of consecutive integers, *Amer. Math. Monthly*, **79**(1972) 1082–1089.

E. Ecklund, R. Eggleton, P. Erdős & J. L. Selfridge, on the prime factorization of binomial coefficients, *J. Austral. Math. Soc. Ser. A*, **26**(1978) 257–269; *MR* **80e**:10009.

P. Erdős, Problems and results on number theoretic properties of consecutive integers and related questions, *Congressus Numerantium XVI* (Proc. 5th Manitoba Conf. Numer. Math. 1975), 25–44.

P. Erdős & R. L. Graham, On products of factorials, *Bull. Inst. Math. Acad. Sinica, Taiwan*, **4**(1976) 337–355.

B24 The largest set with no member dividing two others.

Let $f(n)$ be the size of the largest subset of $[1,n]$ no member of which divides two others. Erdős asks how large can $f(n)$ be? By taking $[m+1, 3m+2]$ it is clear that one can have $\lceil 2n/3 \rceil$. D.J. Kleitman shows that $f(29) = 21$ by taking [11,30] and omitting 18, 24 and 30, which then allows the inclusion of 6, 8, 9 and 10. However, this example does not seem to generalize. In fact Lebensold has shown that if n is large, then

$$0.6725n \leq f(n) \leq 0.6736n.$$

Erdős also asks if $\lim f(n)/n$ is irrational.

Dually, one can ask for the largest number of numbers $\leq n$, with no number a multiple of any two others. Kleitman's example serves this purpose also. More generally, Erdős asks for the largest number of numbers with no one divisible by k others, for $k > 2$. For $k = 1$, the answer is $\lceil n/2 \rceil$.

For some related problems, see **E2**.

Driss Abouabdillah & Jean M. Turgeon, On a 1937 problem of Paul Erdős concerning certain finite sequences of integers none divisible by another, Proc. 15th S.E. Conf. Combin. Graph Theory Comput., Baton Rouge, 1984, *Congr. Numer.*, **43**(1984) 19–22; *MR* **86h**:11020.

P. Erdős, On a problem in elementary number theory and a combinatorial problem, *Math. Comput.*, (1964) 644–646; *MR* **30** #1087.

Kenneth Lebensold, A divisibility problem, *Studies in Appl. Math.*, **56**(1976–77) 291–294; *MR* **58** #21639.

Emma Lehmer, Solution to Problem 3820, *Amer. Math. Monthly*, **46**(1939) 240–241.

B25 Equal sums of geometic progressions with prime ratios.

Bateman asks if $31 = (2^5 - 1)/(2 - 1) = (5^3 - 1)/(5 - 1)$ is the only prime which is expressible in more than one way in the form $(p^r - 1)/(p - 1)$ where p is prime and $r \geq 3$ and $d \geq 1$ are integers. Trivially one has $7 = (2^3 - 1)/(2 - 1) = ((-3)^3 - 1)/(-3 - 1)$, but there are no others $< 10^{10}$. If the condition that p be prime is relaxed, the problem goes back to Goormaghtigh and we have the solution

$$8191 = (2^{13} - 1)/(2 - 1) = (90^3 - 1)/(90 - 1)$$

E. T. Parker observed that the very long proof by Feit & Thompson that every group of odd order is solvable would be shortened if it could be proved that $(p^q - 1)/(p - 1)$ never divides $(q^p - 1)/(q - 1)$ where p, q are distinct odd primes. In fact it has been conjectured that that these two expressions are relatively prime, but Nelson Stephens noticed that when $p = 17$, $q = 3313$ they have a common factor $2pq + 1 = 112643$. McKay has established that $p^2 + p + 1 \nmid 3^p - 1$ for $p < 53 \cdot 10^6$.

P. T. Bateman & R. M. Stemmler, Waring's problem for algebraic number fields and primes of the form $(p^r - 1)/(p^d - 1)$, *Illinois J. Math.*, **6**(1962) 142–156; *MR* **25** #2059.

Ted Chinburg & Melvin Henriksen, Sums of kth powers in the ring of polynomials with integer coefficients, *Bull. Amer. Math. Soc.*, **81**(1975) 107–110; *MR* **51** #421; *Acta Arith.*, **29**(1976) 227–250; *MR* **53** #7942.

A. Mąkowski & A. Schinzel, Sur l'équation indéterminée de R. Goormaghtigh, *Mathesis*, **68**(1959) 128–142; *MR* **22** # 9472; **70**(1965) 94–96.

N. M. Stephens, On the Feit-Thompson conjecture, *Math. Comput.*, **25**(1971) 625; *MR* **45** #6738.

B26 Densest set with no l pairwise coprime.

Erdős asks what is the maximum k so that the integers a_i, $1 \le a_1 < a_2 < \cdots < a_k \le n$ have no l among them which are pairwise relatively prime. He conjectures that this is the number of integers $\le n$ which have one of the first $l-1$ primes as a divisor. He says that this is easy to prove for $l=2$ and not difficult for $l=3$; he offers \$10.00 for a general solution.

Dually one can ask for the largest subset of $[1,n]$ whose members have pairwise least common multiples not exceeding n. If $g(n)$ is the cardinality of such a maximal subset, then Erdős showed that

$$\frac{3}{2\sqrt{2}}n^{1/2} - 2 < g(n) \le 2n^{1/2}$$

where the first inequality follows by taking the integers from 1 to $(n/2)^{1/2}$ together with the even integers from $(n/2)^{1/2}$ to $(2n)^{1/2}$. Choi improved the upper bound to $1.638n^{1/2}$.

S. L. G. Choi, The largest subset in $[1,n]$ whose integers have pairwise l.c.m. not exceeding n, *Mathematika*, **19**(1972) 221–230; **47** #8461.

S. L. G. Choi, On sequences containing at most three pairwise coprime integers, *Trans. Amer. Math. Soc.*, **183**(1973) 437–440; **48** #6052.

P. Erdős, Extremal problems in number theory, *Proc. Sympos. Pure Math. Amer. Math. Soc.*, **8**(1965) 181–189; *MR* **30** #4740.

B27 The number of prime factors of $n+k$ which don't divide $n+i$, $0 \le i < k$.

Erdős & Selfridge define $v(n;k)$ as the number of prime factors of $n+k$ which do not divide $n+i$ for $0 \le i < k$, and $v_0(n)$ as the maximum of $v(n;k)$ taken over all $k \ge 0$. Does $v_0(n) \to \infty$ with n? They show that $v_0(n) > 1$ for all n except 1, 2, 3, 4, 7, 8 and 16. More generally, define $v_l(n)$ as the maximum of $v(n;k)$ taken over $k \ge l$. Does $v_l(n) \to \infty$ with n? They are unable to prove even that $v_1(n) = 1$ has only a finite number of solutions. Probably the greatest n for which $v_1(n) = 1$ is 330.

They also denote by $V(n;k)$ the number of primes p for which p^α is the highest power of p dividing $n+k$, but p^α does not divide $n+i$ for $0 \le i < k$, and by $V_l(n)$ the maximum of $V(n;k)$ taken over $k \ge l$. Does $V_1(n) = 1$ have only a finite number of solutions? Perhaps $n = 80$ is the largest solution. What is the largest n such that $V_0(n) = 2$?

Some further problems are given in their paper.

P. Erdős & J. L. Selfridge, Some problems on the prime factors of consecutive integers, *Illinois J. Math.*, **11**(1967) 428–430.

A. Schinzel, Unsolved problem 31, *Elem. Math.*, **14**(1959) 82–83.

B28 Consecutive numbers with distinct prime factors.

Selfridge asked: do there exist n consecutive integers, each having either two distinct prime factors less than n or a repeated prime factor less than n? He gives two examples:

1. the numbers $a+11+i$ $(1 \le i \le n = 115)$ where $a \equiv 0 \bmod 2^2 \cdot 3^2 \cdot 5^2 \cdot 7^2 \cdot 11^2$ and $a+p \equiv 0 \bmod p^2$ for each prime p, $13 \le p \le 113$;

2. the numbers $a+31+i$ $(1 \le i \le n = 1329)$ where $a+p \equiv 0 \bmod p^2$ for each prime p, $37 \le p \le 1327$ and $a \equiv 0 \bmod 2^2 \cdot 3^2 \cdot 5^2 \cdot 7^2 \cdot 11^2 \cdot 13^2 \cdot 17^2 \cdot 19^2 \cdot 23^2 \cdot 29^2 \cdot 31^2$.

It is harder to find examples of n consecutive numbers, each one divisible by two distinct primes less than n or by the square of a prime $< n/2$, though he believes that they could be found by computer.

This is related to the problem: find n consecutive integers, each having a composite common factor with the product of the other $n-1$. If the composite condition is relaxed, and one asks merely for a common factor greater than 1, then $2184+i$ $(1 \le i \le n = 17)$ is a famous example.

Alfred Brauer, On a property of k consecutive integers, *Bull. Amer. Math. Soc.*, **47**(1941) 328–331; *MR* **2**, 248.

Ronald J. Evans, On blocks of N consecutive integers, *Amer. Math. Monthly*, **76**(1969) 48–49.

Ronald J. Evans, On N consecutive integers in an arithmetic progression, *Acta Sci. Math. Univ. Szeged*, **33**(1972) 295–296; *MR* **47** #8408.

Heiko Harborth, Eine Eigenschaft aufeinanderfolgender Zahlen, *Arch. Math. (Basel)* **21**(1970) 50–51; *MR* **41** #6771.

Heiko Harborth, Sequenzen ganzer Zahlen, *Zahlentheorie* (*Tagung, Math. Forschungsinst. Oberwolfach*, 1970) 59–66; *MR* **51** #12775.

S. S. Pillai, On m consecutive integers I, *Proc. Indian Acad. Sci. Sect. A*, **11**(1940) 6–12; *MR* **1**, 199; II **11**(1940) 73–80; *MR* **1**, 291; III **13**(1941) 530–533; *MR* **3**, 66; IV *Bull. Calcutta Math. Soc.*, **36**(1944) 99–101; *MR* **6**, 170.

B29 Is x determined by the prime divisors of $x+1$, $x+2$, ..., $x+k$?

Alan R. Woods asks if there is a positive integer k such that every x is uniquely determined by the (sets of) prime divisors of $x+1$, $x+2$, ..., $x+k$. Perhaps $k=3$?

For primes less than 23 there are four ambiguous cases for $k=2$: $(x+1, x+2) = (2,3)$ or $(8,9)$; $(6,7)$ or $(48,49)$; $(14,15)$ or $(224,225)$; $(75,76)$ or $(1215,1216)$. The first three of these are members of the infinite family $(2^n-2, 2^n-1)$, $(2^n(2^n-2), (2^n-1)^2)$. Compare **B19**.

D. H. Lehmer, On a problem of Størmer, *Illinois J. Math.*, **8**(1964) 57–79; *MR* **28** #2072.

B30 A small set whose product is square.

Erdős, Graham & Selfridge want us to find the least value of t_n so that the integers $n+1$, $n+2$, ..., $n+t_n$ contain a subset the product of whose members with n is a square. The Thue–Siegel theorem implies that $t_n \to \infty$ with n, faster than a power of $\ln n$. Selfridge has shown that $t_n \le \max(P(n), 3\sqrt{n})$, where $P(n)$ is the largest prime factor of n.

Alternatively, is it true that for every c there is an n_0 so that for every $n > n_0$ the products $\prod a_i$, taken over $n < a_1 < \ldots < a_k < n + (\ln n)^c$ ($k = 1, 2, \ldots$) are all distinct? They proved this for $c < 2$.

Selfridge conjectures that if n is not a square, and t is the next larger number than n such that nt is a square, then, unless $n = 8$ or 392, it is always possible to find r and s, $n < r < s < t$ such that nrs is a square. E.g., if $n = 240 = 2^4 3 \cdot 5$ then $t = 375 = 3 \cdot 5^3$ and we can find $r = 243 = 3^5$ and $s = 245 = 5 \cdot 7^2$.

P. Erdős & Jan Turk, Products of integers in short intervals, *Acta Arith.*, **44**(1984) 147–174; *MR* **86d**:11073.

B31 Binomial coefficients.

Earl Ecklund, Roger Eggleton, Erdős & Selfridge (see **B23**) write the **binomial coefficient** $\binom{n}{k} = n!/k!(n-k)!$ as a product UV in which every prime factor of U is at most k and every prime factor of V is greater than k. There are only finitely many cases with $n \ge 2k$ for which $U > V$. They determine all such cases except when $k = 3$, 5 or 7.

S. P. Khare lists all cases with $n \le 551$: $k = 3$, $n = 8$, 9, 10, 18, 82, 162; $k = 5$, $n = 10$, 12, 28; and $k = 7$, $n = 21$, 30, 54.

Most binomial coefficients $\binom{n}{k}$ with $n \ge 2k$ have a prime factor $p \le n/k$. After some computing with Lacampagne & Erdős, Selfridge conjectured that this inequality is true whenever $n > 17.125k$. A slightly stronger conjecture is that any such binomial coefficient has least prime factor $p \le n/k$ or $p \le 17$ with just 4 exceptions: $\binom{62}{6}$, $\binom{959}{56}$, $\binom{474}{66}$, $\binom{284}{28}$ for which $p = 19$, 19, 23 and 29 respectively.

These authors define the **deficiency** of the binomial coefficient $\binom{n+k}{k}$, $k \le n$, as the number of i for which $b_i = 1$, where $n + i = a_i b_i$, $1 \le i \le k$, the prime factors of b_i are greater than k, and $\prod a_i = k!$ Then $\binom{44}{8}$, $\binom{74}{10}$, $\binom{174}{12}$, $\binom{239}{14}$, $\binom{5179}{27}$, $\binom{8413}{28}$, $\binom{8414}{28}$ and $\binom{96622}{42}$ each have deficiency 2; $\binom{46}{10}$, $\binom{47}{10}$, $\binom{241}{16}$, $\binom{2105}{25}$, $\binom{1119}{27}$ and $\binom{6459}{33}$ have deficiency 3; $\binom{47}{11}$ has deficiency 4; and $\binom{284}{28}$ has deficiency 9; and they conjecture that there are no others with deficiency greater than 1. Are there only finitely many binomial coefficients with deficiency 1?

Erdős & Selfridge noted that if $n \geq 2k \geq 4$, then there is at least one value of i, $0 \leq i \leq k-1$, such that $n-i$ does not divide $\binom{n}{k}$, and asked for the least n_k for which there was only one such i. For example, $n_2 = 4$, $n_3 = 6$, $n_4 = 9$, $n_5 = 12$. $n_k \leq k!$ for $k \geq 3$.

Harry Ruderman asks for a proof or disproof that for every pair (p, q) of nonnegative integers there is a positive integer n such that

$$\frac{(2n-p)!}{n!(n+q)!}$$

is an integer.

A problem which has briefly baffled good mathematicians is: is $\binom{n}{r}$ ever prime to $\binom{n}{s}$, $0 < r < s \leq n/2$? The negative answer follows from the identity

$$\binom{n}{s}\binom{s}{r} = \binom{n}{r}\binom{n-r}{s-r}.$$

Erdős & Szekeres ask if the greatest prime factor of the g.c.d. is always greater than r; the only counterexample with $r > 3$ that they noticed is

$$\gcd\left(\binom{28}{5}, \binom{28}{14}\right) = 2^3 \cdot 3^3 \cdot 5$$

Wolstenholme's theorem states that if n is a prime > 3, then

$$\binom{2n-1}{n} \equiv 1 \bmod n^3.$$

James P. Jones asks if the converse is true. For other problems and results on the divisors of binomial coefficients, see **B33**.

D. F. Bailey, Two p^3 variations of Lucas's theorem, *J. Number Theory*, **35**(1990) 208–215; *MR* **90f**:11008.

Paul Erdős, C. B. Lacampagne & J. L. Selfridge, Estimates of the least prime factor of a binomial coefficient, *Math. Comput.*, **61**(1993) 215–224; *MR* **93k**:11013.

P. Erdős & J. L. Selfridge, Problem 6447, *Amer. Math. Monthly*, **90**(1983) 710; **92**(1985) 435–436.

P. Erdős & G. Szekeres, Some number theoretic problems on binomial coefficients, *Austral. Math. Soc. Gaz.*, **5**(1978) 97–99; *MR* **80e**:10010 is uninformative.

Richard J. McIntosh, A generalization of a congruential property of Lucas, *Amer. Math. Monthly*, **99**(1992) 231–238.

Harry D. Ruderman, Problem 714, *Crux Math.*, **8**(1982) 48; **9**(1983) 58.

David Segal, Problem E435, partial solution by H.W. Brinkman, *Amer. Math. Monthly*, **48**(1941) 269–271.

B32 Grimm's conjecture.

Grimm has conjectured that if $n+1$, $n+2$, ..., $n+k$ are all composite, then there are distinct primes p_{i_j} such that $p_{i_j}|(n+j)$ for $1 \le j \le k$. For example

1802 1803 1804 1805 1806 1807 1808 1809 1810

are respectively divisible by

53 601 41 19 43 139 113 67 181

and

114 115 116 117 118 119 120 121 122 123 124 125 126

by

19 23 29 13 59 17 2 11 61 41 31 5 7

Ramachandra, Shorey & Tijdeman proved, under the hypothesis of Schinzel mentioned in **A2**, that there are only finitely many exceptions to Grimm's conjecture.

Erdős & Selfridge asked for an estimate of $f(n)$, the least number such that for each m there are distinct integers $a_1, a_2, \ldots, a_{\pi(n)}$ in the interval $[m+1, m+f(n)]$ with $p_i|a_i$ where p_i is the ith prime. They and Pomerance show that, for large n,

$$(3-\epsilon)n \le f(n) \ll n^{3/2}(\ln n)^{-1/2}$$

P. Erdős, Problems and results in combinatorial analysis and combinatorial number theory, in *Proc. 9th S.E. Conf. Combin. Graph Theory, Comput., Boca Raton, Congressus Numerantium* XXI, Utilitas Math. Winnipeg, 1978, 29–40.

P. Erdős & C. Pomerance, Matching the natural numbers up to n with distinct multiples in another interval, *Nederl. Akad. Wetensch. Proc. Ser. A*, **83**(= *Indag. Math.*, **42**)(1980) 147–161; *MR* **81i**:10053.

Paul Erdős & Carl Pomerance, An analogue of Grimm's problem of finding distinct prime factors of consecutive integers, *Utilitas Math.*, **24**(1983) 45–46; *MR* **85b**:11072.

P. Erdős & J. L. Selfridge, Some problems on the prime factors of consecutive integers II, in *Proc. Washington State Univ. Conf. Number Theory*, Pullman, 1971, 13–21.

C. A. Grimm, A conjecture on consecutive composite numbers, *Amer. Math. Monthly*, **76**(1969) 1126–1128.

Michel Langevin, Plus grand facteur premier d'entiers en progression arithmétique, *Sém. Delange-Pisot-Poitou*, **18**(1976/77) *Théorie des nombres: Fasc.* 1, *Exp. No.* 3, Paris, 1977; *MR* **81a**:10011.

Carl Pomerance, Some number theoretic matching problems, in *Proc. Number Theory Conf.*, Queen's Univ., Kingston, 1979, 237–247.

Carl Pomerance & J. L. Selfridge, Proof of D.J. Newman's coprime mapping conjecture, *Mathematika*, **27**(1980) 69–83; *MR* **81i**:10008.

K. Ramachandra, T. N. Shorey & R. Tijdeman, On Grimm's problem relating to factorization of a block of consecutive integers, *J. reine angew. Math.*, **273**(1975) 109–124.

B33 Largest divisor of a binomial coefficient.

What can one say about the largest divisor, less than n, of the binomial coefficient $\binom{n}{k} = n!/k!(n-k)!$? Erdős points out that it is easy to show that it is at least n/k and conjectures that there may be one between cn and n for any $c < 1$ and n sufficiently large. Marilyn Faulkner showed that if p is the least prime $> 2k$ and $n \geq p$, then $\binom{n}{k}$ has a prime divisor $\geq p$, except for $\binom{9}{2}$ and $\binom{10}{3}$. Earl Ecklund showed that if $n \geq 2k > 2$ then $\binom{n}{k}$ has a prime divisor $p \leq n/2$, except for $\binom{7}{3}$.

John Selfridge conjectures that if $n \geq k^2 - 1$, then, apart from the exception $\binom{62}{6}$, there is a prime divisor $\leq n/k$ of $\binom{n}{k}$. Among those binomial coefficients whose least prime factor p is $\geq n/k$ there may be only a finite number with $p \geq 13$, but there could be infinitely many with $p = 7$. That there are infinitely many with $p = 5$ was proved by Erdős, Lacampagne & Selfridge (**B31**).

A classical theorem, discovered independently by Sylvester and Schur, stated that the product of k consecutive integers, each greater than k, has a prime divisor greater than k. Leo Moser conjectured that the Sylvester-Schur theorem holds for primes $\equiv 1 \bmod 4$, in the sense that for n sufficiently large (and $\geq 2k$), $\binom{n}{k}$ has a prime divisor $\equiv 1 \bmod 4$ which is greater than k. However, Erdős does not think that this is true, but it may not be at all easy to settle. In this connexion John Leech notices that the fourteen integers 280213, ..., 280226 have no prime factor of the form $4m + 1 > 13$.

Thanks to Ira Gessel and John Conway, we can say that the generalization of the **Catalan numbers** $\frac{1}{n+1}\binom{2n}{n}$, requested in the first edition by Neil Sloane, is $\frac{(n,r)}{n}\binom{n}{r}$, which is always an integer (multiply by n and by r and Euclid knew that (n,r) is a linear combination of n and r). These are also known as generalized ballot numbers and they occur when enumerating certain lattice paths.

If $f(n)$ is the sum of the reciprocals of those primes $< n$ which do not divide $\binom{2n}{n}$, then Erdős, Graham, Ruzsa & Straus conjectured that there is an absolute constant c so that $f(n) < c$ for all n. Erdős also conjectured that $\binom{2n}{n}$ is never squarefree for $n > 4$. Since $4 \mid \binom{2n}{n}$ unless $n = 2^k$, it suffices to consider

$$\binom{2^{k+1}}{2^k}.$$

Sárközy proved this for n sufficiently large and Sander has shown, in a precise sense, that binomial coefficients near the centre of the Pascal triangle are not squarefree. Granville & Ramaré completed Sárközy's proof by showing that $k > 300000$ was sufficiently large, and checking it computationally for $2 \leq k \leq 300000$. They also improved Sander's result by showing that there is a constant δ, $0 < \delta < 1$, such that if $\binom{n}{k}$ is squarefree then k or $n-k$ must be $< n^\delta$ for sufficiently large n. They conjecture that k or $n-k$ must in fact be $< (\ln n)^{2-\delta}$, and that this is best possible in the sense that there are infinitely many squarefree $\binom{n}{k}$ with $\frac{1}{2}n > k > c(\ln n)^2$ for some $c > 0$. They prove such a result for $\frac{1}{2}n > k > \frac{1}{5}\ln n$. They show that there is a constant $\rho_k > 0$ such that the number of $n \leq N$ with $\binom{n}{k}$ squarefree is $\sim \rho_k N$. Since $\rho_k < c/k^2$ for some $c > 0$, they conjecture that there is a constant $\gamma > 0$ such that the number of squarefree entries in the first N rows of Pascal's triangle is $\sim \gamma N$.

Erdős has also conjectured that for $k > 8$, 2^k is not the sum of distinct powers of 3 [$2^8 = 3^5 + 3^2 + 3 + 1$]. If that's true, then for $k \geq 9$,

$$3 \,\Big|\, \binom{2^{k+1}}{2^k}.$$

In answer to the question, is $\binom{342}{171}$ the largest $\binom{2n}{n}$ which is not divisible by the square of an odd prime, Eugene Levine gave the examples $n = 784$ and 786. Erdős feels sure that there are no larger such n.

Denote by $e = e(n)$ the largest exponent such that, for some prime p, p^e divides $\binom{2n}{n}$. It is not known whether $e \to \infty$ with n. On the other hand Erdős cannot disprove $e > c\ln n$.

Ron Graham offers \$100.00 for deciding if $(\binom{2n}{n}, 105) = 1$ infinitely often. Kummer knew that n, when written in base 3 or 5 or 7, would have to have only the digits 0, 1 or 0, 1, 2 or 0, 1, 2, 3 respectively. H. Gupta & S. P. Khare found the 14 values 1, 10, 756, 757, 3160, 3186, 3187, 3250, 7560, 7561, 7651, 20007, 59548377, 59548401 of n less than 7^{10}, while Peter Montgomery, Khare and others found many larger values.

Erdős, Graham, Ruzsa & Straus showed that for any *two* primes p, q there are infinitely many n for which $(\binom{2n}{n}, pq) = 1$. If $g(n)$ is the smallest odd prime factor of $\binom{2n}{n}$, then $g(3160) = 13$ and $g(n) \leq 11$ for $3160 < n < 10^{10000}$.

E. F. Ecklund, On prime divisors of the binomial coefficient, *Pacific J. Math.*, **29**(1969) 267–270.

P. Erdős, A theorem of Sylvester and Schur, *J. London Math. Soc.*, **9**(1934) 282–288.

Paul Erdős, A mélange of simply posed conjectures with frustratingly elusive solutions, *Math. Mag.*, **52**(1979) 67–70.

P. Erdős & R. L. Graham, On the prime factors of $\binom{n}{k}$, *Fibonacci Quart.*, **14**(1976) 348–352.

P. Erdős, R. L. Graham, I. Z. Ruzsa & E. Straus, On the prime factors of $\binom{2n}{n}$, *Math. Comput.*, **29**(1975) 83–92.

M. Faulkner, On a theorem of Sylvester and Schur, *J. London Math. Soc.*, **41**(1966) 107–110.

Andrew Granville & Olivier Ramaré, Explicit bounds on exponential sums and the scarcity of squarefree binomial coefficients (see Abstract 882-11-124, *Abstracts Amer. Math. Soc.*, **14**(1993) 419.)

Hansraj Gupta, On the parity of $(n+m-1)!(n,m)/n!m!$, *Res. Bull. Panjab Univ. (N.S.)*, **20**(1969) 571–575; *MR* **43** #3201.
L. Moser, Insolvability of $\binom{2n}{n} = \binom{2a}{a}\binom{2b}{b}$, *Canad. Math. Bull.*, **6**(1963)167–169.
J. W. Sander, Prime power divisors of $\binom{2n}{n}$, *J. Number Theory*, **39**(1991) 65–74; *MR* **92i**:11097.
J. W. Sander, On prime divisors of binomial coefficients, *Bull. London Math. Soc.*, **24**(1992) 140–142; *MR* **93g**:11019.
J. W. Sander, Prime power divisors of binomial coefficients, *J. reine angew. Math.*, **430**(1992) 1–20; *MR* **93h**:11021; reprise **437**(1993) 217–220.
J. W. Sander, On primes not dividing binomial coefficients, *Math. Proc. Cambridge Philos. Soc.*, **113**(1993) 225–232; *MR* **93m**:11099.
J. W. Sander, An asymptotic formula for ath powers dividing binomial coefficients, *Mathematika*, **39**(1992) 25–36; *MR* **93i**:11110.
J. W. Sander, On primes not dividing binomial coefficients, *Math. Proc. Cambridge Philos. Soc.*, **113**(1993) 225–232.
A. Sárközy, On divisors of binomial coefficients I, *J. Number Theory*, **20**(1985) 70–80; *MR* **86c**:11002.
Renate Scheidler & Hugh C. Williams, A method of tabulating the number-theoretic function $g(k)$, *Math. Comput.*, **59**(1992) 251–257; *MR* **92k**:11146.
I. Schur, Einige Sätze über Primzahlen mit Anwendungen und Irreduzibilitätsfragen I, *S.-B. Preuss, Akad. Wiss. Phys.-Math. Kl.*, **14**(1929) 125–136.
J. Sylvester, On arithmetical series, *Messenger of Math.*, **21**(1892) 1–19, 87–120.
W. Utz, A conjecture of Erdős concerning consecutive integers, *Amer. Math. Monthly*, **68**(1961) 896–897.

B34 If there's an i such that $n-i$ divides $\binom{n}{k}$.

If $H_{k,n}$ is the proposition: there is an i, $0 \le i < k$ such that $n-i$ divides $\binom{n}{k}$, then Erdős asked if $H_{k,n}$ is true for all k when $n \ge 2k$. Schinzel gave the counterexample $n = 99215$, $k = 15$. If H_k is the proposition: $H_{k,n}$ is true for all n, then Schinzel showed that H_k is false for $k = 15$, 21, 22, 33, 35 and thirteen other values of k. He showed that H_k is true for all other $k \le 32$ and asked if there are infinitely many k, other than prime-powers, for which H_k is true: he conjectures not and later reported that it is true for $k = 34$, but for no other non-prime-powers between 34 and 201.

E. Burbacka & J. Piekarczyk, P. 217, R. 1, *Colloq. Math.*, **10**(1963) 365.
A. Schinzel, Sur un problème de P. Erdős, *Colloq. Math.*, **5**(1957–58) 198–204.

B35 Products of consecutive numbers with the same prime factors.

Let $f(n)$ be the least integer such that at least one of the numbers n, $n+1, \ldots, n+f(n)$ divides the product of the others. It is easy to see that

$f(k!) = k$ and $f(n) > k$ for $n > k!$ Erdős has also shown that

$$f(n) > \exp((\ln n)^{1/2-\epsilon})$$

for an infinity of values of n, but it seems difficult to find a good upper bound for $f(n)$.

Erdős asks if $(m+1)(m+2)\cdots(m+k)$ and $(n+1)(n+2)\cdots(n+l)$ with $k \geq l \geq 3$ can contain the same prime factors infinitely often. For example $(2\cdot 3\cdot 4\cdot 5\cdot 6)\cdot 7\cdot 8\cdot 9\cdot 10$ and $14\cdot 15\cdot 16$ and $48\cdot 49\cdot 50$; also $(2\cdot 3\cdot 4\cdot 5\cdot 6)\cdot 7\cdot 8\cdot 9\cdot 10\cdot 11\cdot 12$ and $98\cdot 99\cdot 100$. For $k = l \geq 3$ he conjectures that this happens only finitely many times.

If $L(n;k)$ is the l.c.m. of $n+1$, $n+2$, ..., $n+k$, then Erdős conjectures that for $l > 1$, $n \geq m+k$, $L(m;k) = L(n;l)$ has only a finite number of solutions. Examples are $L(4;3) = L(13;2)$ and $L(3;4) = L(19;2)$. He asks if there are infinitely many n such that for all k $(1 \leq k < n)$ we have $L(n;k) > L(n-k;k)$. What is the largest $k = k(n)$ for which this inequality can be reversed? He notes that it is easy to see that $k(n) = o(n)$, but he believes that much more is true. He expects that for every $\epsilon > 0$ and $n > n_0(\epsilon)$, $k(n) < n^{1/2+\epsilon}$ but cannot prove this.

P. Erdős, How many pairs of products of consecutive integers have the same prime factors? *Amer. Math. Monthly*, **87**(1980) 391–392.

B36 Euler's totient function.

Euler's **totient function**, $\phi(n)$, is the number of numbers not greater than n and prime to n. For example $\phi(1) = \phi(2) = 1$, $\phi(3) = \phi(4) = \phi(6) = 2$, $\phi(5) = \phi(8) = \phi(10) = \phi(12) = 4$, $\phi(7) = \phi(9) = 6$. Are there infinitely many pairs of consecutive numbers, n, $n+1$, such that $\phi(n) = \phi(n+1)$? For example, $n = 1$, 3, 15, 104, 164, 194, 255, 495, 584, 975. It is not even known if $|\phi(n+1) - \phi(n)| < n^\epsilon$ has an infinity of solutions for each $\epsilon > 0$. Baillie extended the work of others to find 306 solutions of $\phi(n) = \phi(n+1)$ below 10^8 and 85 between 10^8 and $2\cdot 10^8$.

Schinzel conjectures that for every even k the equation $\phi(n+k) = \phi(n)$ has an infinity of solutions. He observes that the corresponding conjecture with k odd is implausible. For $k = 1$, the problem is that of the previous paragraph. For $k = 3$ he found only the solutions $n = 3$ and $n = 5$ in the range $n < 10^4$, and D.H. Lehmer extended this to 10^6. Sierpiński has shown that $\phi(n+k) = \phi(n)$ has at least one solution for each value of k and Schinzel & Wakulicz have shown that there are at least two solutions for each $k < 2\cdot 10^{58}$. Mąkowski has shown that $\phi(n+k) = 2\phi(n)$ has at least one solution for every k. For the equation $\phi(n+k) = 3\phi(n)$ see the solution to problem E 3215 in *Amer. Math. Monthly*, **96**(1989) 63–64.

Mąkowski (see reference at **B5**) also discusses the equation $\phi(x+k) = \phi(x)+\phi(k)$. J. Browkin showed that if $k=3$, then there was no solution with $x < 37182142$.

Three curiosities are $\phi(5186) = \phi(5187) = \phi(5188) = 2^5 3^4$, $\phi(25930) = \phi(25935) = \phi(25940) = \phi(25942) = 2^7 3^4$ and $\phi(404471) = \phi(404473) = \phi(404477) = 2^8 3^2 5^2 7$.

Nontotients are positive *even* values of n for which $\phi(x) = n$ has no solution; for example, $n = 14$, 26, 34, 38, 50, 62, 68, 74, 76, 86, 90, 94, 98. The number, $\#(y)$, of these less than y has been calculated by the Lehmers.

y	10^3	10^4	$2\cdot10^4$	$3\cdot10^4$	$4\cdot10^4$	$5\cdot10^4$	$6\cdot10^4$	$7\cdot10^4$	$8\cdot10^4$	$9\cdot10^4$
$\#(y)$	210	2627	5515	8458	11438	14439	17486	20536	23606	26663

Erdős & Hall have shown that the number, $\Phi(y) = y - \#(n)$, of n for which $\phi(x) = n$ *has* a solution is $ye^{f(y)}/\ln y$, where $f(y)$ lies between $c(\ln\ln\ln y)^2$ and $c(\ln y)^{1/2}$. Maier & Pomerance more recently showed that the lower bound was correct, with $c \approx 0.8178$. Erdős conjectures that $\Phi(cy)/\Phi(y) \to c$, and that this, if true, may be the best substitute that one can find for an asymptotic formula for $\Phi(y)$.

Noncototients are positive values of n for which $x - \phi(x) = n$ has no solution; for example, $n = 10$, 26, 34, 50, 52, 58, 86, 100. Sierpiński and Erdős conjecture that there are infinitely many noncototients.

Erdős once asked if it was true that for every ϵ there is an n with $\phi(n) = m$, $m < \epsilon n$ and for no $t < n$ is $\phi(t) = m$; perhaps there are many such n.

Michael Ecker has asked for which values of x do each of the series $\sum_{n=1}^{\infty} \phi(n)/n^x$ and $\sum_{n=1}^{\infty}(-1)^{n+1}\phi(n)/n^x$ converge.

Robert Baillie, Table of $\phi(n) = \phi(n+1)$, *Math. Comput.*, **30**(1976) 189–190.

David Ballew, Janell Case & Robert N. Higgins, Table of $\phi(n) = \phi(n+1)$, *Math. Comput.*, **29**(1975) 329–330.

Michael W. Ecker, Problem E-1, *The AMATYC Review*, **5**(1983) 55; comment **6**(1984)55.

P. Erdős, Über die Zahlen der Form $\sigma(n)-n$ und $n-\phi(n)$, *Elem. Math.*, **28**(1973) 83–86.

P. Erdős & R. R. Hall, Distinct values of Euler's ϕ-function, *Mathematika*, **23**(1976) 1–3.

Patricia Jones, On the equation $\phi(x)+\phi(k) = \phi(x+k)$, *Fibonacci Quart.*, **28**(1990) 162–165; *MR* **91e**:11008.

M. Lal & P. Gillard, On the equation $\phi(n) = \phi(n+k)$, *Math. Comput.*, **26**(1972) 579–582.

Helmut Maier & Carl Pomerance, On the number of distinct values of Euler's ϕ-function, *Acta Arith.*, **49**(1988) 263–275.

Andrzej Mąkowski, On the equation $\phi(n+k) = 2\phi(n)$, *Elem. Math.*, **29**(1974) 13.

Kathryn Miller, UMT **25**, *Math. Comput.*, **27**(1973) 447-448.

A. Schinzel, Sur l'équation $\phi(x+k) = \phi(x)$, *Acta Arith.*, **4**(1958) 181–184; *MR* **21** #5597.

A. Schinzel & A. Wakulicz, Sur l'équation $\phi(x+k) = \phi(x)$ II, *Acta Arith.*, **5**(1959) 425–426; *MR* **23** #A831.

W. Sierpiński, Sur un propriété de la fonction $\phi(n)$, *Publ. Math. Debrecen*, **4**(1956) 184–185.

Charles R. Wall, Density bounds for Euler's function, *Math. Comput.*, **26** (1972) 779-783 with microfiche supplement; *MR* **48** #6043.

Masataka Yorinaga, Numerical investigation of some equations involving Euler's ϕ-function, *Math. J. Okayama Univ.*, **20**(1978) 51–58.

B37 Does $\phi(n)$ properly divide $n-1$?

D. H. Lehmer has conjectured that there is no composite value of n such that $\phi(n)$ is a divisor of $n-1$, i.e., that for no value of n is $\phi(n)$ a *proper* divisor of $n-1$. Such an n must be a Carmichael number (**A13**). He showed that it would have to be the product of at least seven distinct primes, and Lieuwens has shown that if $3|n$, then $n > 5.5 \cdot 10^{571}$ and $\omega(n) \geq 212$; if the smallest prime factor of n is 5, then $\omega(n) \geq 11$; if the smallest prime factor of n is at least 7, then $\omega(n) \geq 13$. This supersedes and corrects the work of Schuh. Masao Kishore has shown that at least 13 primes are needed in any case, and Cohen & Hagis have improved this to 14. Siva Rama Prasad & Subbarao improve Lieuwens's 212 result to $\omega(n) \geq 1850$ and Hagis to $\omega(n) \geq 298848$. Siva Rama Prasad & Rangamma show that if $3|n$, n composite, $M\phi(n) = n-1$, $M \neq 4$, then $\omega(n) \geq 5334$.

Pomerance has proved that the number of composite n less than x for which $\phi(n)|n-1$ is

$$O(x^{1/2}(\ln x)^{3/4}(\ln\ln x)^{-1/2})$$

and Shan Zun improved the exponent $\frac{3}{4}$ to $\frac{1}{2}$.

Schinzel notes that if $n = p$ or $2p$, where p is prime, then $\phi(n)+1$ divides n and asks if the converse is always true. Segal (see paper with Cohen) observes that Schinzel's question reduces to that of Lehmer, that it arises in group theory, and may have been raised by G. Hajós (see Miech's paper, though there it is attributed to Gordon).

If n is prime, it divides $\phi(n)d(n)+2$. Is this true for any composite n other than $n = 4$? Subbarao also notes that if n is prime, then $n\sigma(n) \equiv 2 \bmod \phi(n)$, and also if $n = 4$, 6 or 22; is it true for infinitely many composite n?

Subbarao has an analogous conjecture to Lehmer's, based on the function $\phi^*(n) = \prod(p^\alpha - 1)$, where the product is taken over the maximal prime power divisors of n, $p^\alpha \| n$. He conjectures that $\phi^*(n)|(n-1)$ if and only if n is a power of a prime. He also has a 'dual' of Lehmer's conjecture, namely that $\psi(n) \equiv 1 \bmod n$ only when n is a prime, where $\psi(n)$ is Dedekind's function (see **B41**).

Ron Graham makes the following conjecture

¿ For all k there are infinitely many n such that $\phi(n)|(n-k)$?

He observes that it is true for $k=0$, $k=2^a$ $(a \geq 0)$ and $k=2^a3^b$ $(a, b>0)$ for example. Pomerance (see *Acta Arith.* paper quoted in **B2**) has treated Graham's problem. Victor Meally notes that $\phi(n)$ sometimes divides $n+1$, e.g., for $n=n_1=3\cdot5\cdot17\cdot353\cdot929$ and $n=n_1\cdot83623937$. [Note that $353=11\cdot2^5+1$, $929=29\cdot2^5+1$, $83623937=11\cdot29\cdot2^{18}+1$ and $(353-2^8)(929-2^8)=2^{16}-2^8+1$.]

Ronald Alter, Can $\phi(n)$ properly divide $n-1$? *Amer. Math. Monthly*, **80** (1973) 192–193.

G. L. Cohen & P. Hagis, On the number of prime factors of n if $\phi(n)|n-1$, *Nieuw Arch. Wisk.* (3), **28**(1980) 177–185.

G. L. Cohen & S. L. Segal, A note concerning those n for which $\phi(n)+1$ divides n, *Fibonacci Quart.*, **27**(1989)285–286.

Masao Kishore, On the equation $k\phi(M)=M-1$, *Nieuw Arch. Wisk.* (3), **25**(1977) 48–53; see also *Notices Amer. Math. Soc.*, **22**(1975) A501–502.

D. H. Lehmer, On Euler's totient function, *Bull. Amer. Math. Soc.*, **38**(1932) 745–751.

E. Lieuwens, Do there exist composite numbers for which $k\phi(M)=M-1$ holds? *Nieuw Arch. Wisk.* (3), **18**(1970) 165–169; *MR* **42** #1750.

R. J. Miech, An asymptotic property of the Euler function, *Pacific J. Math.*, **19**(1966) 95–107; *MR* **34** #2541.

Carl Pomerance, On composite n for which $\phi(n)|n-1$, *Acta Arith.*, **28**(1976) 387–389; II, *Pacific J. Math.*, **69**(1977) 177–186; *MR* **55** #7901; see also *Notices Amer. Math. Soc.*, **22**(1975) A542.

József Sándor, On the arithmetical functions $\sigma_k(n)$ and $\phi_k(n)$, *Math. Student*, **58**(1990) 49–54; *MR* **91h**:11005.

Fred. Schuh, Can $n-1$ be divisible by $\phi(n)$ when n is composite? *Mathematica*, Zutphen B, **12**(1944) 102–107.

V. Siva Rama Prasad & M. Rangamma, On composite n satisfying a problem of Lehmer, *Indian J. Pure Appl. Math.*, **16**(1985) 1244–1248; *MR* **87g**:11017.

V. Siva Rama Prasad & M. Rangamma, On composite n for which $\phi(n)|n-1$, *Nieuw Arch. Wisk.* (4), **5**(1987) 77–81; *MR* **88k**:11008.

M. V. Subbarao, On two congruences for primality, *Pacific J. Math.*, **52**(1974) 261–268; *MR* **50** #2049.

M. V. Subbarao, On composite n satisfying $\psi(n) \equiv 1 \bmod n$, Abstract 882-11-60 *Abstracts Amer. Math. Soc.*, **14**(1993) 418.

David W. Wall, Conditions for $\phi(N)$ to properly divide $N-1$, *A Collection of Manuscripts Related to the Fibonacci Sequence*, 18th Anniv. Vol., Fibonacci Assoc., 205–208.

Shan Zun, On composite n for which $\phi(n)|n-1$, *J. China Univ. Sci. Tech.*, **15**(1985) 109–112; *MR* **87h**:11007.

B38 Solutions of $\phi(m) = \sigma(n)$.

Are there infinitely many pairs of numbers m, n such that $\phi(m) = \sigma(n)$? Since for p prime $\phi(p) = p - 1$ and $\sigma(p) = p + 1$ this question would be answered affirmatively if there were infinitely many twin primes (**A7**). Also if there were infinitely many Mersenne primes (**A3**) $M_p = 2^p - 1$, since $\sigma(M_p) = 2^p = \phi(2^{p+1})$. However there are many solutions other than these, sometimes displaying little noticeable pattern, e.g., $\phi(780) = 192 = \sigma(105)$.

Erdős remarks that the equation $\phi(x) = n!$ is solvable, and (apart from $n = 2$) $\sigma(y) = n!$ is probably solvable also. Charles R. Wall can show that $\psi(n) = n!$ is solvable for $n \neq 2$, where ψ is Dedekind's function (see **B41**).

Le Mao-Hua, A note on primes p with $\sigma(p^m) = z^n$, *Colloq. Math.*, **62**(1991) 193–196.

B39 Carmichael's conjecture.

Carmichael's conjecture. For every n it appears to be possible to find an m, not equal to n, such that $\phi(m) = \phi(n)$ and for a few years early in this century it was thought that Carmichael had proved this. Klee verified the conjecture for $\phi(n) < 10^{400}$, and for all n not divisible by $2^{42} \cdot 3^{47}$. Masai & Valette have raised the bound to 10^{10000}, and Schlafly & Wagon to $10^{1360000}$ and are proceeding to $10^{10000000}$. Pomerance has shown that if n is such that for every prime p for which $p - 1$ divides $\phi(n)$ we have p^2 divides n, then n is a counterexample. He can also show (unpublished) that if the first k primes $p \equiv 1 \pmod q$ (where q is prime) are all less than q^{k+1}, then there are no numbers n which satisfy his theorem. This also implies the truth of his conjecture that $p_k - 1 | \prod_{i<k} p_i(p_i - 1)$. The truth of this last conjecture for all k also implies that there are no numbers n which satisfy his theorem.

Define the **multiplicity** of an integer as the number of times it occurs as a value of $\phi(n)$. For example, 6 has multiplicity 4 because $\phi(n) = 6$ for $n = 7, 9, 14, 18$ and no other values of n. The multiplicity may be zero (for any odd $n > 1$, and $n = 14, 26, 34, \ldots$), but not, according to the Carmichael conjecture, equal to one. Sierpiński conjectured that all integers greater than 1 occur as multiplicities and Erdős has shown that if a multiplicity occurs once it occurs infinitely often. Schlafly & Wagon have found examples of all multiplicities from 2 through 65.

There are examples of *even* numbers n such that there is no *odd* number m such that $\phi(m) = \phi(n)$. Lorraine Foster has given $n = 33817088 = 2^9 \cdot 257^2$ as the least such.

Erdős proved that if $\phi(x) = k$ has exactly s solutions, then there are infinitely many other k for which there are exactly s solutions, and that $s > k^c$ for infinitely many k. If C is the least upper bound of those c for which this is true, then Wooldridge showed that $C \geq 3 - 2\sqrt{2} > 0.17157$.

Pomerance used Hooley's improvement on the Brun–Titchmarsh theorem to improve this to $C \geq 1 - 625/512e > 0.55092$ and notes that further improvements by Iwaniec enable him to get $C > 0.55655$ so that $s > k^{5/9}$ for infinitely many k. Erdős conjectures that $C = 1$. In the other direction Pomerance also shows that

$$s < k \exp\{-(1 + o(1)) \ln k \ln\ln\ln k \,/\, \ln\ln k\}$$

and gives a heuristic argument to support the belief that this is best possible.

R. D. Carmichael, Note on Euler's ϕ-function, *Bull. Amer. Math. Soc.*, **28** (1922) 109–110; and see **13**(1907) 241–243.

P. Erdős, On the normal number of prime factors of $p - 1$ and some other related problems concerning Euler's ϕ-function, *Quart. J. Math. Oxford Ser.*, **6**(1935) 205–213.

P. Erdős, Some remarks on Euler's ϕ-function and some related problems, *Bull. Amer. Math. Soc.*, **51**(1945) 540–544.

P. Erdős, Some remarks on Euler's ϕ-function, *Acta Arith.*, **4**(1958) 10–19; *MR* **22**#1539.

Lorraine L. Foster, Solution to problem E3361, *Amer. Math. Monthly*, **98** (1991) 443.

Peter Hagis, On Carmichael's conjecture concerning the Euler phi function (Italian summary), *Boll. Un. Mat. Ital.* (6), **A5**(1986) 409–412.

C. Hooley, On the greatest prime factor of $p+a$, *Mathematika*, **20**(1973) 135–143.

Henryk Iwaniec, On the Brun-Tichmarsh theorem and related questions, *Proc. Queen's Number Theory Conf., Kingston, Ont. 1979*, Queen's Papers Pure Appl. Math., **54**(1980) 67–78; *Zbl.* **446**.10036.

V. L. Klee, On a conjecture of Carmichael, *Bull. Amer. Math. Soc.*, **53**(1947) 1183–1186; *MR* **9**, 269.

P. Masai & A. Valette, A lower bound for a counterexample to Carmichael's conjecture, *Boll. Un. Mat. Ital. A* (6), **1** (1982) 313–316; *MR* **84b**:10008.

Carl Pomerance, On Carmichael's conjecture, *Proc. Amer. Math. Soc.*, **43** (1974) 297–298.

Carl Pomerance, Popular values of Euler's function, *Mathematika*, **27** (1980) 84–89; *MR* **81k**:10076.

Aaron Schlafly & Stan Wagon, Carmichael's conjecture is valid below $10^{2,000,000}$, *Math. Comput.*,

M. V. Subbarao & L.-W. Yip, Carmichael's conjecture and some analogues, in *Théorie des Nombres* (Québec, 1987), de Gruyter, Berlin–New York, 1989, 928–941 (and see *Canad. Math. Bull.*, **34**(1991) 401–404.

Alain Valette, Fonction d'Euler et conjecture de Carmichael, *Math. et Pédag.*, Bruxelles, **32**(1981) 13–18.

Stan Wagon, Carmichael's 'Empirical Theorem', *Math. Intelligencer*, **8**(1986) 61-63; *MR* **87d**:11012.

K. R. Wooldridge, Values taken many times by Euler's phi-function, *Proc. Amer. Math. Soc.*, **76**(1979) 229–234; *MR* **80g**:10008.

B40 Gaps between totatives.

If $a_1 < a_2 < \ldots < a_{\phi(n)}$ are the integers less than n and prime to it, then Erdős conjectured that $\sum(a_{i+1}-a_i)^2 < cn^2/\phi(n)$ and offered \$500.00 for a proof. Hooley showed that, for $1 \le \alpha < 2$, $\sum(a_{i+1}-a_i)^\alpha \ll n(n/\phi(n))^{\alpha-1}$ and that $\sum(a_{i+1}-a_i)^2 \ll n(\ln\ln n)^2$, Vaughan established the conjecture "on the average" and he & Montgomery finally won the prize.

Jacobsthal asked what bounds can be placed on $J(n) = \max(a_{i+1}-a_i)$. Erdős asks if, for infinitely many x, there are two integers n_1, n_2, $n_1 < n_2 < x$, $n_1 \perp n_2$, $J(n_1) > \ln x$, $J(n_2) > \ln x$.

P. Erdős, On the integers relatively prime to n and on a number-theoretic function considered by Jacobsthal, *Math. Scand.*, **10**(1962) 163–170; *MR* **26** #3651.

C. Hooley, On the difference of consecutive numbers prime to n, *Acta Arith.*, **8**(1962/63) 343–347; *MR* **27** #5741.

H. L. Montgomery & R. C. Vaughan, On the distribution of reduced residues, *Ann. of Math.* (2), **123**(1986) 311–333; *MR* **87g**:11119.

R. C. Vaughan, Some applications of Montgomery's sieve, *J. Number Theory*, **5**(1973) 64–79.

B41 Iterations of ϕ and σ.

There is a close relative to the sum of divisors and the sum of the unitary divisors function, which complements Euler's totient function and which is often named for Dedekind. If $n = p_1^{a_1}p_2^{a_2}\ldots p_k^{a_k}$, denote by $\psi(n)$ the product $\prod p_i^{a_i-1}(p_i+1)$, i.e., $\psi(n) = n\prod(1+p^{-1})$, where the product is taken over the distinct prime divisors of n. It is easy to see that iteration of the function leads eventually to terms of the form 2^a3^b where b is fixed and a increases by one in successive terms. Given any value of b there are infinitely many values of n which lead to such terms, for example, $\psi^k(2^a3^b7^c) = 2^{a+4k}3^b7^{c-k}$ $(0 \le k \le c)$ and $\psi^k(2^a3^b7^c) = 2^{a+5k-c}3^b$ $(k > c)$.

David E. Penney & Pomerance, in an unpublished paper, show that there are values of n for which the iterates of the function $\psi(n) - n$ are unbounded as the number of iterations tends to infinity; the least such is $n = 318$.

If we average ψ with the ϕ-function, $\frac{1}{2}(\phi+\psi)$, and iterate, we produce sequences whose terms become constant whenever they are prime powers; for example 24, $\frac{1}{2}(8+48) = 28$, $\frac{1}{2}(12+48) = 30$, $\frac{1}{2}(8+72) = 40$, $\frac{1}{2}(16+72) = 44$, $\frac{1}{2}(20+72) = 46$, $\frac{1}{2}(22+72) = 47$, $\frac{1}{2}(46+48) = 47, \ldots$. Charles R. Wall gives examples where iteration leads to an unbounded sequence: start with 45, 48, ... or 50, 55, ... and continue 56, 60, 80, 88, 92, 94, 95, 96, ... ; each term after the 35th is the double of the last but seven!

We can also average the σ- and ϕ-functions, and iterate. Since $\phi(n)$ is always even for $n > 2$ and $\sigma(n)$ is odd when n is a square or twice a square, we will sometimes get a noninteger value. For example, 54, 69, 70, 84, 124, 142, 143 ,144, $225\frac{1}{2}$; in this case we say that the sequence **fractures**. It is easy to show that $(\sigma(n)+\phi(n))/2 = n$ just if $n = 1$ or a prime, so sequences can become constant, for example, 60, 92, 106, 107, 107, ... Are there sequences which increase indefinitely without fracturing?

Of course, if we iterate the ϕ-function, it eventually arrives at 1. Call the least integer k for which $\phi^k(n) = 1$ the **class** of n.

k	n
1	2
2	3 4 6
3	5 7 8 9 10 12 14 18
4	11 13 15 16 19 20 21 22 24 26 ...
5	17 23 25 29 31 32 33 34 35 37 39 40 43...
6	41 47 51 53 55 59 61 64 65 67 68 69 71 73 ...
7	83 85 89 97 101 103 107 113 115 119 121 122 123 125 128 ...

The set of least values of the classes is $M = \{2, 3, 5, 11, 17, 41, 83, \ldots\}$. Shapiro conjectured that M contained only prime values, but Mills found several composite members. If S is the union, for all k, of the members of class k which are $< 2^k$, then

$$S = \{3; 5, 7; 11, 13, 15; 17, 23, 25, 29, 31; 41, 47, 51, 53, 55, 59, 61; 83, 85, \ldots\}$$

and Shapiro showed that the factors of an element of S is also in S. Catlin showed that if m is an odd element of M, then the factors of M are in M, and that there are finitely many primes in M just if there are finitely many odd numbers in M. Does S contain infinitely many odd numbers? Does M contain infinitely many odd numbers?

Pillai showed that the class, $k = k(n)$, of n satisfies

$$\left\lfloor \frac{\ln n}{\ln 3} \right\rfloor \le k(n) \le \left\lfloor \frac{\ln n}{\ln 2} \right\rfloor$$

and it's easy to see (look at 2^a3^b) that $k(n)/\ln n$ is dense in the interval $[1/\ln 3, 1/\ln 2]$. What is the average and normal behavior of $k(n)$? Erdős, Granville, Pomerance & Spiro conjecture that there is a constant α such that the normal order of $k(n)$ is $\alpha \ln n$ and prove this under the assumption of the Elliott-Halberstam conjecture. They also showed that the normal order of $\phi^h(n)/\phi^{h+1}(n)$ is $he^\gamma \ln\ln\ln n$ for each positive integer h, where γ is Euler's constant. See their paper for many unsolved problems: for example, if $\sigma^k(n)$ is the kth iterate of the sum of divisors function, they are unable to prove or disprove any of the following statements.

¿ for every $n > 1$, $\sigma^{k+1}(n)/\sigma^k(n) \to 1$ as $k \to \infty$?

¿ for every $n > 1$, $\sigma^{k+1}(n)/\sigma^k(n) \to \infty$ as $k \to \infty$?

¿ for every $n > 1$, $\left(\sigma^k(n)\right)^{1/k} \to \infty$ as $k \to \infty$?

¿ for every $n > 1$, there is some k with $n|\sigma^k(n)$?

¿ for every n, $m > 1$, there is some k with $m|\sigma^k(n)$?

¿ for every n, $m > 1$, there are some k, l with $\sigma^k(m) = \sigma^l(n)$?

Miriam Hausman has characterized those integers n which are solutions of the equation $n = m\phi^k(n)$; they are mainly of the form 2^a3^b.

Finucane iterated the function $\phi(n) + 1$ and asked: in how many steps does one reach a prime? Also, given a prime p, what is the distribution of the values of n whose sequences end in p? Are 5, 8, 10, 12 the only numbers which lead to 5? And 7, 9, 14, 15, 16, 18, 20, 24, 30 the only ones leading to 7?

Erdős similarly asked about the iteration of $\sigma(n) - 1$. Does it always end on a prime, or can it grow indefinitely? In none of the cases of iteration of $\sigma(n) - 1$, of $(\psi(n) + \phi(n))/2$, or of $(\phi(n) + \sigma(n))/2$ is he able to show that the growth is slower than exponential. For several results and conjectures, consult the quadruple paper cited below.

Atanassov defines some additive analogs of ϕ and σ, poses 17 questions and answers only three of them.

Krassimir T. Atanassov, New integer functions, related to ϕ and σ functions, *Bull. Number Theory Related Topics*, **11**(1987) 3–26; *MR* **90j**:11007.

P. A. Catlin, Concerning the iterated ϕ-function, *Amer. Math. Monthly*, **77**(1970) 60–61.

P. Erdős, A. Granville, C. Pomerance & C. Spiro, On the normal behavior of the iterates of some arithmetic functions, in Berndt, Diamond, Halberstam & Hildebrand (editors), Analytic Number Theory, *Proc. Conf. in honor P.T. Bateman, Allerton Park, 1989*, Birkhäuser, Boston, 1990, 165–204; *MR* **92a**:11113.

P. Erdős, Some remarks on the iterates of the ϕ and σ functions, *Colloq. Math.*, **17**(1967) 195–202; *MR* **36** #2573.

Paul Erdős & R. R. Hall, Euler's ϕ-function and its iterates, *Mathematika*, **24**(1977) 173–177; *MR* **57** #12356.

Miriam Hausman, The solution of a special arithmetic equation, *Canad. Math. Bull.*, **25**(1982) 114–117.

W. H. Mills, Iteration of the ϕ-function, *Amer. Math. Monthly*, **50**(1943) 547–549; *MR* **5**, 90.

C. A. Nicol, Some diophantine equations involving arithmetic functions, *J. Math. Anal. Appl.*, **15**(1966) 154–161.

Ivan Niven, The iteration of certain arithmetic functions, *Canad. J. Math.*, **2**(1950) 406–408; *MR* **12**, 318.

S. S. Pillai, On a function connected with $\phi(n)$, *Bull. Amer. Math. Soc.*, **35**(1929) 837–841.

Carl Pomerance, On the composition of the arithmetic functions σ and ϕ, *Colloq. Math.*, **58**(1989) 11–15; *MR* **91c**:11003.
Harold N. Shapiro, An arithmetic function arising from the ϕ-function, *Amer. Math. Monthly*, **50**(1943) 18–30; *MR* **4**, 188.
Charles R. Wall, Unbounded sequences of Euler-Dedekind means, *Amer. Math. Monthly*, **92**(1985) 587.

B42 Behavior of $\phi(\sigma(n))$ and $\sigma(\phi(n))$.

Erdős asks us to prove that $\phi(n) > \phi(n - \phi(n))$ for almost all n, but that $\phi(n) < \phi(n - \phi(n))$ for infinitely many n.

Mąkowski & Schinzel prove that $\limsup \phi(\sigma(n))/n = \infty$,

$$\limsup \phi^2(n)/n = \frac{1}{2}, \quad \text{and} \quad \liminf \sigma(\phi(n))/n \le \frac{1}{2} + \frac{1}{2^{34}-4}$$

and they ask if $\sigma(\phi(n))/n \ge \frac{1}{2}$ for all n. They point out that even $\inf \sigma(\phi(n))/n > 0$ is not proved, but Pomerance has since established this, using Brun's method.

John Selfridge, Fred Hoffman & Rich Schroeppel found 24 solutions of $\phi(\sigma(n)) = n$, namely

2^k for $k = 0$, 1, 3, 7, 15 & 31; $2^2 \cdot 3$; $2^8 \cdot 3^3$; $2^{10} \cdot 3^3 \cdot 11^2$; $2^{12} \cdot 3^3 \cdot 5 \cdot 7 \cdot 13$; $2^4 \cdot 3 \cdot 5$; $2^4 \cdot 3^2 \cdot 5$; $2^9 \cdot 3 \cdot 5^2 \cdot 31$; $2^9 \cdot 3^2 \cdot 5^2 \cdot 31$; $2^5 \cdot 3^4 \cdot 5 \cdot 11$; $2^5 \cdot 3^4 \cdot 5^2 \cdot 11$; $2^8 \cdot 3^4 \cdot 5 \cdot 11$; $2^8 \cdot 3^4 \cdot 5^2 \cdot 11$; $2^5 \cdot 3^6 \cdot 7^2 \cdot 13$; $2^6 \cdot 3^6 \cdot 7^2 \cdot 13$; $2^{13} \cdot 3^7 \cdot 5 \cdot 7^2$; $2^{13} \cdot 3^7 \cdot 5^2 \cdot 7^2$; $2^{21} \cdot 3^3 \cdot 5 \cdot 11^3 \cdot 31$; $2^{21} \cdot 3^3 \cdot 5^2 \cdot 11^3 \cdot 31$; and there are, of course, 24 corresponding solutions of $\sigma(\phi(m)) = m$. Are there others? An infinite number?

Golomb observes that if $q > 3$ and $p = 2q - 1$ are primes and $m \in \{2, 3, 8, 9, 15\}$, then $n = pm$ is a solution of $\phi(\sigma(n)) = \phi(n)$. Undoubtedly there are infinitely many such and undoubtedly no one will prove this in the foreseeable future. There are other solutions, 1, 3, 15, 45, ... ; an infinite number? He gives the solutions 1, 87, 362, 1257, 1798, 5002, 9374 to $\sigma(\phi(n)) = \sigma(n)$. He also notes that if p and $(3^p - 1)/2$ are primes (e.g., $p = 3$, 7, 13, 71, 103), then $n = 3^{p-1}$ is a solution of $\sigma(\phi(n)) = \phi(\sigma(n))$; and shows that $\sigma(\phi(n)) - \phi(\sigma(n))$ is both positive and negative infinitely often and asks what is the proportion of each?

P. Erdős, Problem P. 294, *Canad. Math. Bull.*, **23**(1980) 505.
Solomon W. Golomb, Equality among number-theoretic functions, preprint, Oct 1992; Abstract 882-11-16, *Abstracts Amer. Math. Soc.*, **14**(1993) 415–416.
A. Mąkowski & A. Schinzel, On the functions $\phi(n)$ and $\sigma(n)$, *Colloq. Math.*, **13**(1964–65) 95–99; *MR* **30** #3870.
Carl Pomerance, On the composition of the arithmetic functions σ and ϕ, *Colloq. Math.*, **58**(1989) 11–15; *MR* **91c**:11003.
József Sándor, On the composition of some arithmetic functions, *Studia Univ. Babeş-Bolyai Math.*, **34**(1989) 7–14; *MR* **91i**:11008.

B43 Alternating sums of factorials.

The numbers

$$3! - 2! + 1! = 5,$$
$$4! - 3! + 2! - 1! = 19,$$
$$5! - 4! + 3! - 2! + 1! = 101,$$
$$6! - 5! + 4! - 3! + 2! - 1! = 619,$$
$$7! - 6! + 5! - 4! + 3! - 2! + 1! = 4421,$$

and

$$8! - 7! + 6! - 5! + 4! - 3! + 2! - 1! = 35899$$

are each prime. Are there infinitely many such? Here are the factors of $A_n = n! - (n-1)! + (n-2)! - + \ldots - (-1)^n 1!$ for the next few values of n:

n	A_n	n	A_n
9	$79 \cdot 4139$	19	15578717622022981 (prime)
10	3301819 (prime)	20	$8969 \cdot 210101 \cdot 1229743351$
11	$13 \cdot 2816537$	21	$113 \cdot 167 \cdot 4511191 \cdot 572926421$
12	$29 \cdot 15254711$	22	$79 \cdot 239 \cdot 56947572104043899$
13	$47 \cdot 1427 \cdot 86249$	23	$85439 \cdot 289993909455734779$
14	$211 \cdot 1679 \cdot 229751$	24	$12203 \cdot 24281 \cdot 2010359484638233$
15	1226280710981 (prime)	25	$59 \cdot 555307 \cdot 455254005662640637$
16	$53 \cdot 6581 \cdot 56470483$	26	$1657 \cdot 234384986539153832538067$
17	$47 \cdot 7148742955723$	27	$127^2 \cdot 271 \cdot 1163 \cdot 2065633479970130593$
18	$2683 \cdot 2261044646593$	28	$61 \cdot 221171 \cdot 21820357757749410439949$

The example $n = 27$ shows that these numbers are not necessarily squarefree. Wilfrid Keller has continued the calculations for $n \leq 335$; A_n is prime for $n = 41$, 59, 61, 105 and 160.

If there is a value of n such that $n+1$ divides A_n, then $n+1$ will divide A_m for all $m > n$, and there would be only a finite number of prime values. Wagstaff established that if there is such an n, it is larger than 46340.

B44 Sums of factorials.

Đ. Kurepa defines $!n = 0! + 1! + 2! + \ldots + (n-1)!$ and asks if $!n \not\equiv 0 \bmod n$ for all $n > 2$. Slavić used a computer to establish this for $3 \leq n \leq 1000$. The conjecture is that $(!n, n!) = 2$. Wagstaff has extended the calculations and verified the conjecture for $n < 50000$, and Mijajlović for $n \leq 10^6$. He notes that for $K_n = !(n+1) - 1 = 1! + 2! + \ldots + n!$ we have $3|K_n$ for $n \geq 2$, $9|K_n$ for $n \geq 5$ and $99|K_n$ for $n \geq 10$. Wilfrid Keller has since extended this and found no new divisibilities for K_n with $n < 10^6$. In a 91-03-21 letter, Reg. Bond offers an as yet unpublished proof of the conjecture.

It is also conjectured that, except that 2^2 divides $!3$, $!n$ is squarefree. Mijajlović has confirmed that $m^2 \nmid !n$ for $m \leq 1223$.

L. Carlitz, A note on the left factorial function, *Math. Balkanika*, **5**(1975) 37–42.

Đuro Kurepa, On some new left factorial propositions, *Math. Balkanika*, **4**(1974) 383–386; *MR* **58** #10716.

Ž. Mijajlović, On some formulas involving $!n$ and the verification of the $!n$-hypothesis by use of computers, *Publ. Inst. Math.* (*Beograd*) (*N.S.*) **47(61)** (1990) 24–32; *MR* **92d**:11134.

B45 Euler numbers.

The coefficients in the expansion of $\sec x = \sum E_n(ix)^n/n!$ are the **Euler numbers**, and arise in several combinatorial contexts. $E_0 = 1$, $E_2 = -1$, $E_4 = 5$, $E_6 = -61$, $E_8 = 1385$, $E_{10} = -50521$, $E_{12} = 2702765$, $E_{14} = -199360981$, $E_{16} = 19391512145$, $E_{18} = -2404879675441$, $\ldots$. Is it true that for any prime $p \equiv 1 \bmod 8$, $E_{(p-1)/2} \not\equiv 0 \bmod p$? Is it true for $p \equiv 5 \bmod 8$?

E. Lehmer, On congruences involving Bernoulli numbers and the quotients of Fermat and Wilson, *Annals of Math.* **39**(1938) 350–360; *Zbl.* **19**, 5.

Barry J. Powell, Advanced problem 6325, *Amer. Math. Monthly*, **87**(1980) 826.

B46 The largest prime factor of n.

Erdős denotes by $P(n)$ the largest prime factor of n and asks if there are infinitely many primes p such that $(p-1)/P(p-1) = 2^k$? Or $= 2^k \cdot 3^l$?

If $n > 2$, then $P(n)$, $P(n+1)$, $P(n+2)$ are all distinct. Show that each of the six permutations of {low, medium, high} occurs infinitely often, and that they occur with equal frequency. 2^k-2, 2^k-1, 2^k show that medium, high, low occurs for infinitely many k because $P(2^k-1) \to \infty$ as $k \to \infty$ by a theorem of Bang (or Mahler). To see that low, medium high occurs infinitely often, ask if $p-1$, p, $p+1$ works for p prime. No! Try p^2-1, p^2, p^2+1. Maybe. If $P(p^2+1) < p$, try p^4-1, p^4, p^4+1. Eventually, for each prime p, there will be a value of k such that $P(p^{2^k}+1) > p$.

Selfridge settled the low, high, medium case with 2^k, 2^k+1, 2^k+2 and Tijdeman gave the following argument for medium, low, high: consider the possibilities 2^k-1, 2^k, 2^k+1; $2^{2k}-1$, 2^{2k}, $2^{2k}+1$; $2^{4k}-1$, 2^{4k}, $2^{4k}+1$; $\ldots$.

P. Erdős & Carl Pomerance, On the largest prime factors of n and $n+1$, *Aequationes Math.*, **17**(1978) 311–321; *MR* **58** #476.

B47 When does $2^a - 2^b$ divide $n^a - n^b$?

Selfridge notices that $2^2 - 2$ divides $n^2 - n$ for all n, that $2^{2^2} - 2^2$ divides $n^{2^2} - n^2$ and $2^{2^{2^2}} - 2^{2^2}$ divides $n^{2^{2^2}} - n^{2^2}$ and asks for what a and b does $2^a - 2^b$ divide $n^a - n^b$ for all n. The case $n = 3$ was proposed as E2468*, *Amer. Math. Monthly*, **81**(1974) 405 by Harry Ruderman. In his solution (**83**(1976) 288–289) Bill Vélez omits $(b, a - b) = (0, 1)$ as trivial and gives 13 other solutions, (1,1), (1,2), (2,2), (3,2), (1,4), (2,4), (3,4), (4,4), (2,6), (3,6), (2,12), (3,12), (4,12). Remarks by Pomerance (**84**(1977) 59–60) show that results of Schinzel complete Vélez's solution. The problem was also solved by Sun Qi & Zhang Ming Zhi.

A. Schinzel, On primitive prime factors of $a^n - b^n$, *Proc. Cambridge Philos. Soc.*, **58**(1962) 555–562.

Sun Qi & Zhang Ming-Zhi, Pairs where $2^a - 2^b$ divides $n^a - n^b$ for all n, *Proc. Amer. Math. Soc.*, **93**(1985) 218–220; *MR* **86c**:11004.

B48 Products taken over primes.

David Silverman noticed that if p_n is the n-th prime, then

$$\prod_{n=1}^{m} \frac{p_n + 1}{p_n - 1}$$

is an integer for $m = 1$, 2, 3, 4 and 8 and asked is it ever again an integer? Equivalently, as Mąkowski observes (reference at **B16**), for what $n = \prod_{r=1}^{m} p_r$ does $\phi(n)$ divide $\sigma(n)$? For example, if $\sigma(n) = 4\phi(n)$ then $2n$ is either perfect or abundant, $\sigma(2n) \geq 4n$.

Wagstaff asked for an elementary proof (e.g., without using properties of the Riemann ζ-function) that

$$\prod \frac{p^2 + 1}{p^2 - 1} = \frac{5}{2}$$

where the product is taken over all primes. It seems very unlikely that there is a proof which doesn't involve analytical methods. At first glance it might appear that the fractions might cancel, but none of the numerators are divisible by 3. Euler's proof is

$$\prod \frac{p^2 + 1}{p^2 - 1} = \prod \frac{p^4 - 1}{(p^2 - 1)^2} = \prod \frac{1 - p^{-4}}{(1 - p^{-2})^2} = \frac{\zeta^2(2)}{\zeta(4)} = \frac{(\pi^2/6)^2}{\pi^4/90} = \frac{5}{2}.$$

This uses $\sum n^{-k} = \prod (1 - p^{-k})^{-1}$ and $\sum n^{-2} = \pi^2/6$ and $\sum n^{-4} = \pi^4/90$. Wagstaff regards the first as elementary, but not the latter two. He would like to see a direct proof of $2(\sum n^{-2})^2 = 5\sum n^{-4}$ or of

$$4\sum_{n=1}^{\infty}\frac{1}{n^2}\sum_{m=n+1}^{\infty}\frac{1}{m^2} = 3\sum\frac{1}{n^4}$$

B49 Smith numbers.

Albert Wilansky named **Smith numbers** from his brother-in-law's telephone number

$$4937775 = 3 \cdot 5 \cdot 5 \cdot 65837,$$

the sum of whose digits is equal to the sum of the digits of its prime factors, and they soon caught the public fancy. Trivially, any prime is a Smith number: so are 4, 22, 27, 58, 85, 94, 121, ... Oltikar & Wayland gave the examples $3304(10^{317}-1)/9$ and $2 \cdot 10^{45}(10^{317}-1)/9$ and the race to find larger and larger Smith numbers was on. Yates has given

$$10^{3913210}(10^{1031}-1)(10^{4594}+3 \cdot 10^{2297}+1)^{1476}$$

with 10694985 decimal digits, but has since beaten his own record with a 13614513-digit Smith number.

Stephen K. Doig, Math Whiz makes digital discovery, *The Miami Herald*, 1986-08-22; *Coll. Math. J.*, **18**(1987) 80.

Editorial, Smith numbers ring a bell? *Fort Lauderdale Sun Sentinel*, 86-09-16, p. 8A.

Editorial, Start with 4,937,775, *New York Times*, 86-09-02.

Wayne L. McDaniel, The existence of infinitely many k-Smith numbers, *Fibonacci Quart.*, **25**(1987) 76–80.

Wayne L. McDaniel, Powerful k-Smith numbers, *Fibonacci Quart.*, **25**(1987) 225–228.

Wayne L. McDaniel, Palindromic Smith numbers, *J. Recreational Math.*, **19**(1987) 34–37.

Wayne L. McDaniel, Difference of the digital sums of an integer base b and its prime factors, *J. Number Theory*, **31**(1989) 91–98; *MR* **90e**:11021.

Wayne L. McDaniel & Samuel Yates, The sum of digits function and its application to a generalization of the Smith number problem, *Nieuw Arch. Wisk.*(4), **7**(1989) 39–51.

Sham Oltikar & Keith Wayland, Construction of Smith numbers, *Math. Mag.*, **56**(1983) 36–37.

Ivars Peterson, In search of special Smiths, *Science News*, 86-08-16, p. 105.

A. Wilansky, Smith numbers, *Two-Year Coll. Math. J.*, **13**(1982) 21.

Samuel Yates, Special sets of Smith numbers, *Math. Mag.*, **59**(1986) 293–296.

Samuel Yates, Smith numbers congruent to 4 (mod 9), *J. Recreational Math.*, **19**(1987) 139–141.

Samuel Yates, How odd the Smith are, *J. Recreational Math.*, **19**(1987) 168–174.

Samuel Yates, Digital sum sets, in R. A. Mollin (ed.), Number Theory, *Proc. 1st Canad. Number Theory Assoc. Conf., Banff, 1988*, de Gruyter, New York, 1990, pp. 627–634; *MR* **92c**:11008.

Samuel Yates, Tracking titanics, in R. K. Guy & R. E. Woodrow (eds.), The Lighter Side of Mathematics, *Proc. Strens Mem. Conf., Calgary, 1986*, Spectrum Series, Math. Assoc. of America, Washington DC, 1994.

C. Additive Number Theory

C1 Goldbach's conjecture.

One of the most infamous problems is Goldbach's conjecture that every even number greater than 4 is expressible as the sum of two odd primes. Javier Echevarria has verified it up to 2^{32}, and Matti Sinisalo to 4×10^{11}. Vinogradov proved that every *odd* number greater than $3^{3^{15}}$ is the sum of *three* primes and Chen Jing-Run has shown that all large enough even numbers are the sum of a prime and the product of at most two primes. Chen & Wang have reduced the number $3^{3^{15}}$ to $e^{e^{11.503}}$.

"Conjecture A" of Hardy & Littlewood (cf. **A1, A8**) is that the number, $N_2(n)$, of representations of an even number n as the sum of two primes, is given asymptotically by

$$N_2(n) \sim \frac{2cn}{(\ln n)^2} \prod \left(\frac{p-1}{p-2}\right),$$

where, as in **A8**, $2c \approx 1.3203$ and the product is taken over all odd prime divisors of n.

Stein & Stein have calculated $N_2(n)$ for $n < 10^5$ and have found values of n for which $N_2(n) = k$ for all $k < 1911$. It is conjectured that $N_2(n)$ takes all positive integer values. They also verified the conjecture for $n < 10^8$. Granville, van de Lune & te Riele have extended this to $2 \cdot 10^{10}$.

Let $\phi(n)$ be Euler's totient function (**B36**) so that if p is prime, $\phi(p) = p - 1$. If the Goldbach conjecture is true, then there are, for each number m, prime numbers p, q, such that

$$\phi(p) + \phi(q) = 2m.$$

If we relax the condition that p and q be prime, then it should be easier to show that there are always numbers p and q satisfying this equation. Erdős & Leo Moser ask if this can be done.

Antonio Filz defined a **prime circle** of order $2m$ to be a circular permutation of the numbers from 1 to $2m$ with each adjacent pair summing to a prime. There is essentially only one prime circle for $m = 1$, 2 and 3;

two for $m = 4$ and 48 for $m = 5$. Are there prime circles for all m? Give an asymptotic estimate of their number.

Similarly, Margaret Kenney proposed the **prime pyramid**

$$
\begin{array}{ccccccccccccc}
 & & & & & & * & & & & & & \\
 & & & & & 1 & & 2 & & & & & \\
 & & & & 1 & & 2 & & 3 & & & & \\
 & & & 1 & & 2 & & 3 & & 4 & & & \\
 & & 1 & & 4 & & 3 & & 2 & & 5 & & \\
 & 1 & & 4 & & 3 & & 2 & & 5 & & 6 & \\
1 & & - & & - & & - & & - & & - & & 7
\end{array}
$$

in which row n contains the numbers $1, 2, \ldots, n$, begins with 1, ends with n, and the sum of two consecutive entries is prime. How many ways are there of arranging the numbers in row n? This problem was also proposed by Morris Wald; the solutions given are almost certain always to work, but a proof of this may be almost as difficult as proving the Goldbach conjecture itself. The slightly less restricted problem in which the end numbers are not prescribed was earlier asked by E. T. H. Wang.

Erdős asks if there are infinitely many primes p such that every even number $\leq p-3$ can be expressed as the difference between two primes each $\leq p$. For example, $p = 13$: $10 = 13 - 3$, $8 = 11 - 3$, $6 = 11 - 5$, $4 = 7 - 3$, $2 = 5 - 3$.

J. Bohman & C.-E. Froberg, Numerical results on the Goldbach conjecture, *BIT*, **15**(1975) 239–243.

Chen Jing-Run, On the representation of a large even number as the sum of a prime and the product of at most two primes, *Sci. Sinica*, **16**(1973) 157–176; *MR* **55** #7959; II, **21**(1978) 421–430; *MR* **80e**:10037.

Chen Jing-Run & Wang Tian-Ze, On the Goldbach problem (Chinese), *Acta Math. Sinica*, **32**(1989) 702–718; *MR* **91e**:11108.

J. G. van der Corput, Sur l'hypothèse de Goldbach pour presque tous les nombres pairs, *Acta Arith.*, **2**(1937) 266–290.

N. G. Čudakov, On the density of the set of even numbers which are not representable as the sum of two odd primes, *Izv. Akad. Nauk SSSR Ser. Mat.*, **2**(1938) 25–40.

N. G. Čudakov, On Goldbach-Vinogradov's theorem, *Ann. Math.*(2), **48** (1947) 515–545; *MR***9**, 11.

Jean-Marc Deshouillers, Andrew Granville, Władysław Narkiewicz & Carl Pomerance, An upper bound in Goldbach's problem, *Math. Comput.*, **61**(1993) 209–213.

T. Estermann, On Goldbach's problem: proof that almost all even positive integers are sums of two primes, *Proc. London Math. Soc.*(2), **44**(1938) 307–314.

Antonio Filz, Problem 1046, *J. Recreational Math.*, **14**(1982) 64; **15**(1983) 71.

D. A. Goldston, Linnik's theorem on Goldbach numbers in short intervals, *Glasgow Math. J.*, **32**(1990) 285–297; *MR* **91i**:11134.

A. Granville, J. van de Lune & H. J. J. te Riele, Checking the Goldbach conjecture on a vector computer, in R. A. Mollin (ed.), Number Theory and Applications, NATO ASI Series, Kluwer, Boston, 1989, pp. 423–433; *MR* **93c**:11085.

Richard K. Guy, Prime Pyramids, *Crux Mathematicorum*, **19**(1993) 97–99.

Margaret J. Kenney, Student Math Notes, *NCTM News Bulletin*, Nov. 1986.

H. L. Montgomery & R. C. Vaughan, The exceptional set in Goldbach's problem, *Acta Arith.*, **27**(1975) 353–370.

Pan Cheng-Dong, Ding Xia-Xi & Wang Yuan, On the representation of every large even integer as the sum of a prime and an almost prime, *Sci. Sinica*, **18**(1975) 599–610; *MR* **57** #5897.

A. Perelli & J. Pintz, On the exceptional set for Goldbach's problem in short intervals, *J. London Math. Soc.*, **47**(1993) 41–49; *MR* **93m**:11104.

P. M. Ross, On Chen's theorem that every large even number has the form p_1+p_2 or $p_1+p_2p_3$, *J. London Math. Soc.*(2), **10**(1975) 500–506.

Matti K. Sinisalo, Checking the Goldbach conjecture up to $4 \cdot 10^{11}$, *Math. Comput.*, **61**(1993) 931–934.

M. L. Stein & P. R. Stein, New experimental results on the Goldbach conjecture, *Math. Mag.*, **38**(1965) 72–80; *MR* **32** #4109.

M. L. Stein & P. R. Stein, Experimental results on additive 2-bases, *Math. Comput.*, **19**(1965) 427–434.

Robert C. Vaughan, On Goldbach's problem, *Acta Arith.*, **22**(1972) 21–48.

Robert C. Vaughan, A new estimate for the exceptional set in Goldbach's problem, *Proc. Symp. Pure Math.*, (Analytic Number Theory, St. Louis), Amer. Math. Soc., **24**(1972) 315–319.

I. M. Vinogradov, Representation of an odd number as the sum of three primes, *Dokl. Akad. Nauk SSSR*, **15**(1937) 169–172.

I. M. Vinogradov, Some theorems concerning the theory of primes, *Mat. Sb. N.S.*, **2**(**44**) (1937) 179–195.

Morris Wald, Problem 1664, *J. Recreational Math.*, **20**(1987–88) 227–228; Solution **21**(1988–89) 236–237.

Edward T. H. Wang, Advanced Problem 6189, *Amer. Math. Monthly*, **85** (1978) 54 [no solution has appeared].

Wang Yuan (editor), *Goldbach Conjecture*, World Scientific, Singapore, 1984. [A collection of original papers, translated into English when other languages were used, with a 15-page Introduction.]

Dan Zwillinger, A Goldbach conjecture using twin primes, *Math. Comput.*, **33** (1979) 1071; *MR* **80b**:10071.

C2 Sums of consecutive primes.

Let $f(n)$ be the number of ways of representing n as the sum of (one or more) *consecutive* primes. For example

$$5 = 2+3 \quad \text{and} \quad 41 = 11+13+17 = 2+3+5+7+11+13$$

so that $F(5) = 2$ and $F(41) = 3$. Leo Moser has shown that

$$\lim_{x\to\infty} \frac{1}{x} \sum_{n=1}^{x} f(n) = \ln 2$$

and he asks: is $f(n) = 1$ infinitely often? Is $f(n) = k$ solvable for every k? Do the numbers for which $f(n) = k$ have a density for every k? Is $\limsup f(n) = \infty$?

Erdős asks if there is an infinite sequence of integers $1 < a_1 < a_2 < \cdots$ such that $f(n)$, the number of solutions of $a_i + a_{i+1} + \cdots + a_k = n$, tends to infinity with n. He notes that if we insist that $k > i$, then it is not even known if $f(n) > 0$ for all but finitely many n. If $a_i = i$, $f(n)$ is the number of odd divisors of n.

L. Moser, Notes on number theory III. On the sum of consecutive primes, *Canad. Math. Bull.*, **6**(1963) 159–161; *MR* **28** #75.

C3 Lucky numbers.

Gardiner and others define **lucky numbers** by modifying the sieve of Eratosthenes in the following way. From the natural numbers strike out all even ones, leaving the odd numbers. Apart from 1, the first remaining number is 3. Strike out every third member (those of shape $6k - 1$) in the new sequence, leaving

1, 3, 7, 9, 13, 15, 19, 21, 25, 27, 31, 33, ...

The next number remaining is 7. Strike out every seventh term (numbers $42k - 23$, $42k - 3$) in this sequence. Next 9 remains: strike out every ninth term from what's left, and so on, until we are left with the lucky numbers

1, 3, 7, 9, 13, 15, 21, 25, 31, 33, 37, 43, 49, 51, 63, 67, 69, 73, 75, 79, 87, 93, 99, 105, 111, 115, 127, 129, 133, 135, 141, 151, 159, 163, 169, 171, 189, 193, 195, 201, 205, 211, 219, 223, 231, ...,

Many questions arise concerning lucky numbers, parallel to the classical ones asked about primes. For example, if $L_2(n)$ is the number of solutions of $l + m = n$, where n is even and l and m are lucky, then Stein & Stein find values of n such that $L_2(n) = k$ for all $k \le 1769$, and there is a corresponding conjecture to that made in **C1**.

W. E. Briggs, Prime-like sequences generated by a sieve process, *Duke Math. J.*, **30**(1963) 297–312; *MR* **26** #6145.

R. G. Buschman & M. C. Wunderlich, Sieve-generated sequences with translated intervals, *Canad. J. Math.*, **19**(1967) 559–570; *MR* **35** #2855.

R. G. Buschman & M. C. Wunderlich, Sieves with generalized intervals, *Boll. Un. Mat. Ital.*(3), **21**(1966) 362–367.

Paul Erdős & Eri Jabotinsky, On sequences of integers generated by a sieving process, I, II, *Nerderl. Akad. Wetensch. Proc. Ser. A*, **61** = *Indag. Math.*, **20**(1958) 115–128; *MR* **21** #2628.

Verna Gardiner, R. Lazarus, N. Metropolis & S. Ulam, On certain sequences of integers generated by sieves, *Math. Mag.*, **31**(1956) 117–122; **17**, 711.

David Hawkins & W. E. Briggs, The lucky number theorem, *Math. Mag.*, **31**(1957–58) 81–84, 277–280; *MR* **21** #2629, 2630.

M. C. Wunderlich, Sieve-generated sequences, *Canad. J. Math.*, **18**(1966) 291–299; *MR* **32** #5625.

M. C. Wunderlich, A general class of sieve generated sequences, *Acta Arith.*, **16**(1969–70) 41–56; *MR* **39** #6852.

M. C. Wunderlich & W. E. Briggs, Second and third term approximations of sieve-generated sequences, *Illinois J. Math.*, **10**(1966) 694–700; *MR* **34** #153.

C4 Ulam numbers.

Ulam constructed increasing sequences of positive integers by starting from arbitrary u_1 and u_2 and continuing with those numbers which can be expressed in just one way as the sum of two distinct earlier members of the sequence. Recamán asked some of the questions which arise in connexion with the **U-numbers** ($u_1 = 1$, $u_2 = 2$).

1, 2, 3, 4, 6, 8, 11, 13, 16, 18, 26, 28, 36, 38, 47, 48, 53, 57, 62, 69, 72, 77, 82, 87, 97, 99, 102, 106, 114, 126, 131, 138, 145, 148, 155, 175, 177, 180, 182, 189, 197, 206, 209, 219, ...

(1) Can the sum of two consecutive U-numbers, apart from 1+2=3, be a U-number?

(2) Are there infinitely many numbers

23, 25, 33, 35, 43, 45, 67, 92, 94, 96, ...

which are *not* the sum of two U-numbers?

(3) (Ulam) Do the U-numbers have positive density?

(4) Are there infinitely many pairs

(1,2), (2,3), (3,4), (47,48), ...

of consecutive U-numbers?

(5) Are there arbitrarily large gaps in the sequence of U-numbers?

In answer to Question 1, Frank Owens noticed that $u_{19}+u_{20} = 62+69 = 131 = u_{31}$. In answer to Question 4, Muller calculated 20000 terms and found no further examples. On the other hand, more than 60% of these terms differed from another by exactly 2.

David Zeitlin asked if the sequence of U-numbers is **complete** in the sense that every positive number is expressible as the sum of distinct members of the sequence. Stefan Burr notes that this is so since, after 2, each term is less than twice the preceding one.

The reader is warned that the name "U-numbers" was also used by Mahler in the theory of algebraic numbers in connexion with Alan Baker's characterization of "S-numbers" and "T-numbers".

More generally, one can define s-**additive sequences** which are constructed in the same way, except that each term is the sum of two earlier terms in exactly s ways, the U-numbers corresponding to $s = 1$. If $s = 0$ the sequence is constructed from numbers which are *not* the sum of two distinct earlier members. Compare problems **C9**, **E28** and **E32** below. More generally still there are (s, t)-**additive sequences** where each term has exactly s representations as the sum of t distinct earlier members. In this notation, the U-numbers are the (1,2)-sequence initiated by $u_1 = 1$, $u_2 = 2$. Steven Finch has experimented in this area and has a number of conjectures. For example, that the sequences initiated by (u_1, u_2) with $u_1 < u_2$ and $u_1 \perp u_2$ contain only finitely many even terms in the cases (a) $(u_1, u_2) = (2, u_2)$ for $u_2 \geq 5$, (b) $(4, u_2)$, (c) (5,6), (d) $u_1 \geq 6$ and even, and (e) $u_1 \geq 7$ odd with u_2 even; but infinitely many even terms otherwise.

Steven R. Finch, Conjectures about s-additive sequences, *Fibonacci Quart.*, **29**(1991) 209–214; *MR* **92j**:11009.

Steven R. Finch, Are 0-additive sequences always regular? *Amer. Math. Monthly*, **99** (1992) 671–673.

Steven R. Finch, On the regularity of certain 1-additive sequences, *J. Combin. Theory Ser. A*, **60**(1992) 123–130; *MR* **93c**:11009.

Steven R. Finch, Patterns in 1-additive sequences, *Experiment. Math.*, **1** (1992) 57–63; *MR* **93h**11014.

P. Muller, M.Sc. thesis, University of Buffalo, 1966.

Raymond Queneau, Sur les suites s-additives, *J. Combin. Theory*, **12**(1972) 31–71; *MR* **46** #1741.

Bernardo Recamán, Questions on a sequence of Ulam, *Amer. Math. Monthly*, **80**(1973) 919–920.

S. M. Ulam, *Problems in Modern Mathematics*, Interscience, New York, 1964, p. ix.

Marvin C. Wunderlich, The improbable behaviour of Ulam's summation sequence, in *Computers and Number Theory*, Academic Press, 1971, 249–257.

C5 Sums determining members of a set.

Leo Moser asked, and Selfridge, Straus and others largely settled, to what extent the sums of all the pairs of numbers in a set determine the set. They show that if the cardinality is not a power of two, then the members are determined. Suppose that $y_1, y_2, \ldots, y_s$ are the sums $x_i + x_j$ $(i \neq j)$ of the numbers x_i, x_2, ..., x_{2^k}, so that $s = 2^{k-1}(2^k - 1)$. Are there more than two sets $\{x_i\}$ which give rise to the same set $\{y_j\}$? If $k = 3$ there may be three such sets, for example.

$$\{\pm1, \pm9, \pm15, \pm19\}, \quad \{\pm2, \pm6, \pm12, \pm22\}, \quad \{\pm3, \pm7, \pm13, \pm21\}$$

but there can't be more than three. For $k > 3$ the problem is open. The corresponding problem where sums of *triples* of elements of a set are given is also settled, except in two cases: do the sums of three distinct elements of $\{x_1, x_2, \ldots, x_n\}$ determine the set if $n = 27$ or $n = 486$? The corresponding problem for sums of *four* distinct elements was settled by Ewell.

John A. Ewell, On the determination of sets by sets of sums of fixed order, *Canad. J. Math.*, **20**(1968) 596–611.

B. Gordon, A. S. Fraenkel & E. G. Straus, On the determination of sets by the sets of sums of a certain order, *Pacific J. Math.*, **12**(1962) 187–196; *MR* **27** #3576.

J. L. Selfridge & E. G. Straus, On the determination of numbers by their sums of a fixed order, *Pacific J. Math.*, **8**(1958) 847–856; *MR* **22** #4657.

C6 Addition chains. Brauer chains. Hansen chains.

An **addition chain** for n is a sequence $1 = a_0 < a_1 < \ldots < a_r = n$ with each member after the zeroth the sum of two earlier, not necessarily distinct, members. For example

1, 1+1, 2+2, 4+2, 6+2, 8+6 and 1, 1+1, 2+2, 4+2, 4+4, 8+6

are addition chains for 14 of **length** $r = 5$. The minimal length of an addition chain for n is denoted by $l(n)$.

The main unsolved problem is the Scholz conjecture

$$¿ \qquad l(2^n - 1) \leq n - 1 + l(n) \qquad ?$$

It has been proved for $n = 2^a$, $2^a + 2^b$, $2^a + 2^b + 2^c$, $2^a + 2^b + 2^c + 2^d$ by Utz, Gioia et al, and Knuth, and demonstrated for $1 \leq n \leq 18$ by Knuth and Thurber. Brauer proved the conjecture for those n for which a shortest chain exists which is a **Brauer chain**, that is one in which each member uses the previous member as a summand. The second of the examples is not a Brauer chain, because the term 4+4 does not use the summand 6. Such an n is called a **Brauer number**. Hansen proved that there are infinitely many non-Brauer numbers, but also that the Scholz conjecture still holds if n has a shortest chain which is a **Hansen chain**, that is one for which there is a subset H of the members such that each member of the chain uses the largest element of H which is less than the member. The second example is a Hansen chain, with $H = \{1, 2, 4, 8\}$. Knuth gives the example

$$1, 2, 4, 8, 16, 17, 32, 64, 128, 256, 512, 1024, 1041, 2082, 4164, 8328, 8345, 12509$$

of a Hansen chain (H={1, 2, 4, 8, 16, 32, 64, 128, 256, 512, 1024, 1041, 2082, 4164, 8328, 8345}) for $n = 12509$ which is not a Brauer chain (32 does not use 17) and no such short Brauer chain exists for $n = 12509$.

Are there non-Hansen numbers?

It is clear that $l(2n) \leq l(n) + 1$. That strict inequality is possible was shown by Knuth with $l(382) = l(191) = 11$. The smallest even n with $l(2n) = l(n)$ is 13818, given by Thurber, who also noticed the odd adjacent pair 22453, 22455. Andrew Granville asks if there are n for which $l(4n) = l(2n) = l(n)$.

D. J. Newman considers a computer which costs 1 cent to perform each addition but nothing to perform multiplication. Then the addition chain for n costs maximally $(\log n)^{\frac{1}{2}+o(1)}$ instead of $\log n$, where log is to base 2.

For a good survey and list of problems, see Subbarao's article.

Walter Aiello & M. V. Subbarao, A conjecture in addition chains related to Scholz's conjecture, *Math. Comput.*, **61**(1993) 17–23; *MR* **93k**:11015.

A. T. Brauer, On addition chains, *Bull. Amer. Math. Soc.*, **45**(1939) 736–739.

A. Cottrell, A lower bound for the Scholz-Brauer problem, Abstract 73T-A200, *Notices Amer. Math. Soc.*, **20**(1973) A-476.

Paul Erdős, Remarks on number theory III. On addition chains, *Acta Arith.*, **6**(1960) 77–81.

R. P. Giese, PhD thesis, University of Houston, 1972.

R. P. Giese, A sequence of counterexamples of Knuth's modification of Utz's conjecture, Abstract 72T-A257, *Notices Amer. Math. Soc.*, **19**(1972) A-688.

A. A. Gioia & M. V. Subbarao, The Scholz-Brauer problem in addition chains II, *Congr. Numer. XXII*, Proc. 8th Manitoba Conf. Numer. Math. Comput. 1978, 251–274; *MR* **80i**: 10078; *Zbl.* 408.10037.

A. A. Gioia, M. V. Subbarao & M. Sugunamma, The Scholz-Brauer problem in addition chains, *Duke Math. J.*, **29**(1962) 481–487; *MR* **25** #3898.

W. Hansen, Zum Scholz-Brauerschen Problem, *J. reine angew. Math.*, **202** (1959) 129–136; *MR***25** #2027.

Kevin Hebb, Some problems on addition chains, thesis, Univ. of Alberta, 1974.

A. M. Il'in, On additive number chains (Russian), *Problemy Kibernet.*, **13** (1965) 245–248.

H. Kato, On addition chains, PhD dissertation, Univ. Southern California, June 1970.

Donald Knuth, *The Art of Computer Programming*, Vol. 2, Addison-Wesley, Reading MA, 1969, 398–422.

D. J. Newman, Computing when multiplications cost nothing, *Math. Comput.*, **46**(1986) 255–257.

Arnold Scholz, Aufgabe 253, *Jber. Deutsch. Math.-Verein. II*, **47**(1937) 41–42 (supplement).

Rolf Sonntag, Theorie der Addition Sketten, PhD Technische Universität Hannover, 1975.

K. B. Stolarsky, A lower bound for the Scholz-Brauer problem, *Canad. J. Math.*, **21**(1969) 675–683; *MR* **40** #114.

M. V. Subbarao, Addition chains – some results and problems, in R. A. Mollin (ed.) Number Theory and Applications, NATO ASI Series, Kluwer, Boston, 1989, pp. 555–574; *MR* **93a**:11105.

E. G. Straus, Addition chains of vectors, *Amer. Math. Monthly*, **71**(1964) 806–808.

E. G. Thurber, The Scholz-Brauer problem on addition chains, *Pacific J. Math.*, **49**(1973) 229–242; *MR* **49** #7233.

E. G. Thurber, On addition chains $l(mn) \le l(n) - b$ and lower bounds for $c(r)$, *Duke Math. J.*, **40**(1973) 907–913.

E. G. Thurber, Addition chains and solutions of $l(2n) = l(n)$ and $l(2^n - 1) = n + l(n) - 1$, *Discrete Math.*, **16**(1976) 279–289; *MR* **55** #5570; *Zbl.* **346**.10032.

W. R. Utz, A note on the Scholz-Brauer problem in addition chains, *Proc. Amer. Math. Soc.*, **4**(1953) 462–463; *MR* **14**, 949.

C. T. Wyburn, A note on addition chains, *Proc. Amer. Math. Soc.*, **16**(1965) 1134.

C7 The money-changing problem.

Given $n \ge 2$ integers $0 < a_1 < a_2 < \cdots < a_n$ with $(a_1, a_2, \ldots, a_n) = 1$, then $N = \sum_{i=1}^n a_i x_i$ has a solution in nonnegative integers x_i if N is large enough. The well known **coin problem** of Frobenius is to determine the greatest $N = g(a_1, a_2, \ldots, a_n)$ for which there is no solution. Sylvester showed that $g(a_1, a_2) = (a_1 - 1)(a_2 - 1) - 1$ and that the number of non-representable numbers is $(a_1 - 1)(a_2 - 1)/2$. The case $n = 3$ was first solved explicitly by Selmer & Beyer, using a continued fraction algorithm. Their result was simplified by Rödseth and later by Greenberg. No general formulas are known for $n \ge 4$. Roberts found the value of g if the a_i are in arithmetic progression.

Upper bounds for g are also sought. In 1942 Brauer showed that $g(a_1, a_2, \ldots, a_n) \le \sum_{i=1}^n a_i(d_{i-1}/d_i - 1)$ where $d_i = (a_1, a_2, \ldots, a_i)$. Erdős & Graham showed that

$$g(a_1, a_2, \ldots, a_n) \le 2a_{n-1}\lfloor a_n/n \rfloor - a_n$$

(which is best possible if $n = 2$ and a_2 is odd). They define

$$\gamma(n, t) = \max_{\{a_i\}} g(a_1, a_2, \ldots, a_n)$$

where the maximum is taken over all $0 < a_1 < a_2 < \cdots < a_n \le t$ with $(a_1, a_2, \ldots, a_n) = 1$. Their theorem shows that $\gamma(n, t) < 2t^2/n$ and they proved that $\gamma(n, t) \ge t^2/(n-1) - 5t$. Lewin showed that $\gamma(3, t) = \lfloor (t-2)^2/2 \rfloor - 1$ and generally that $g(a_1, a_2, \ldots, a_n) \le \lfloor (a_{n-1} - 1)(a_n - 2)/2 \rfloor - 1$ for $n \ge 3$.

Ernst Selmer was of considerable help in rewriting sections **C7** and **C12**; his paper quoted below contains all relevant references up to 1976.

E. R. Berlekamp, J. H. Conway & R. K. Guy, *Winning Ways for your Mathematical Plays*, Academic Press, London, 1982, Chap. 18.

Alfred Brauer, On a problem of partitions, *Amer. J. Math.*, **64**(1942) 299–312.

F. Curtis, On formulas for the Frobenius number of a numerical semigroup, *Math. Scand.*, **67**(1990) 190–192.

J. Dixmier, Proof of a conjecture of Erdős and Graham concerning the problem of Frobenius, *J. Number Theory*, **34**(1990) 198–209; *MR* **91g**:11007.

P. Erdős & R. L. Graham, On a linear diophantine problem of Frobenius, *Acta Arith.*, **21**(1972) 399–408.

Harold Greenberg, Solution to a linear diophantine equation for nonnegative integers, *J. Algorithms*, **9**(1988) 343–353; *MR* **89j**:11122.

Mordechai Lewin, On a linear diophantine problem, *Bull. London Math. Soc.*, **5**(1973) 75–78; *MR* **47** #3311.

Niu Xue-Feng, A formula for the largest integer which cannot be represented as $\sum_{i=1}^{s} a_i x_i$, *Heilongjiang Daxue Ziran Kexue Xuebao*, **9**(1992) 28–32; *MR* **93k**:11019.

J. B. Roberts, Note on linear forms, *Proc. Amer. Math. Soc.*, **7**(1956) 465–469.

Ø. J. Rødseth, On a linear Diophantine problem of Frobenius, *J. reine angew. Math.*, **301**(1978) 171–178; *MR* **58** #27741.

Ernst S. Selmer, On the linear diophantine problem of Frobenius, *J. reine angew. Math.*, **293/294**(1977) 1–17; *MR* **56** #246.

E. S. Selmer & Ö. Beyer, On the linear diophantine problem of Frobenius in three variables, *J. reine angew. Math.*, **301**(1978) 161–170.

J. J. Sylvester, *Math. Quest. Educ. Times*, **41**(1884) 21.

Herbert S. Wilf, A circle of lights algorithm for the "money changing problem", *Amer. Math. Monthly*, **85**(1978) 562–565.

C8 Sets with distinct sums of subsets.

The set of integers $\{2^i : 0 \le i \le k\}$, of cardinality $k+1$, has the sums of all its 2^{k+1} subsets distinct. Erdős has asked for the maximum number, m, of positive integers $a_1 < a_2 < \cdots < a_m \le 2^k$, with all sums of subsets distinct. With Leo Moser he showed that $k+1 \le m < k + \frac{1}{2}\log k + 2$ where the logarithm is to base 2. Noam Elkies improved the constant 2 on the right to $\frac{1}{2}\log\pi < 0.826$.

Conway & Guy have given a sequence, $u_0 = 0$, $u_1 = 1$, $u_{n+1} = 2u_n - u_{n-r}$ $(n \ge 1)$ where r is the nearest integer to $\sqrt{2n}$, from which may be derived the set of $k+2$ integers

$$A = \{a_i = u_{k+2} - u_{k+2-i} : 1 \le i \le k+2\}.$$

They conjecture that this set has subsets with distinct sums (established by Mike Guy for $k \le 40$ and by Fred Lunnon for $n \le 79$). For $k \ge 21$, $u_{k+2} < 2^k$, so that $m \ge k+2$ for $k \ge 21$, since once a set with the desired cardinality is found, its cardinality may be increased by doubling the size of each member and adjoining the member 1 (or any odd number). Conway

& Guy conjecture that A gives, in essence, the best possible solution, $m = k+2$, to the problem, though Lunnon defines a class of generalized Conway-Guy sequences, some of which give a smaller limit (e.g., 0.220963) than that of $u_n/2^n$ (≈ 0.23512531). Erdős offers \$500.00 for a proof or disproof of $m = k + O(1)$.

J. H. Conway & R. K. Guy, Sets of natural numbers with distinct sums, *Notices Amer. Math. Soc.*, **15**(1968) 345.

J. H. Conway & R. K. Guy, Solution of a problem of P. Erdős, *Colloq. Math.*, **20**(1969) 307.

P. Erdős, Problems and results in additive number theory, *Colloq. Théorie des Nombres, Bruxelles*, 1955, Liège & Paris, 1956, 127–137, esp. p. 137.

Noam Elkies, An improved lower bound on the greatest element of a sum-distinct set of fixed order, *J. Combin. Theory Ser. A*, **41**(1986) 89–94; *MR* **87b**:05012.

Martin Gardner, Number 5, *Science Fiction Puzzle Tales*, Penguin, 1981.

Hansraj Gupta, Some sequences with distinct sums, *Indian J. Pure Math.*, **5**(1974) 1093–1109; *MR* **57** #12440.

Richard K. Guy, Sets of integers whose subsets have distinct sums, *Ann. Discrete Math.* **12**(1982) 141–154.

B. Lindström, On a combinatorial problem in number theory, *Canad. Math. Bull.*, **8**(1965) 477–490.

B. Lindström, Om et problem av Erdős for talfoljder, *Nordisk. Mat. Tidskrift*, **16**1-2(1968) 29–30, 80.

W. Fred Lunnon, Integer sets with distinct subset-sums, *Math. Comput.*, **50**(1988) 297–320.

Paul Smith, Problem E 2536*, *Amer. Math. Monthly*, **82**(1975) 300. Solutions and comments, **83**(1976) 484.

C9 Packing sums of pairs.

Suppose that m is the maximum number of integers $1 \le a_1 < a_2 < \ldots < a_m \le n$ in a **Sidon sequence**, i.e., one in which the sums of pairs, $a_i + a_j$, are all different. It is known that

$$n^{1/2}(1-\epsilon) < m \le n^{1/2} + n^{1/4} + 1.$$

The upper bound is due to Lindstrom, improving a result of Erdős & Turán. The lower bound is due to Singer. Erdős & Turán ask, is $m = n^{1/2} + O(1)$? Erdős offers \$500 for settling this question.

Cameron & Erdős ask for an estimate of $F(n)$, the number of Sidon sequences whose members are at most n. With m as above, it is not even known if $F(n)/2^m \to \infty$, only that the upper limit is infinite. They believe that $F(n) < n^{\epsilon\sqrt{n}}$. They would also like an estimate of the number of maximal Sidon sequences (those to which no further $a \le n$ can be adjoined).

If $\{a_i\}$ continues as an infinite sequence, Erdős & Turán proved that $\limsup a_k/k^2 = \infty$ and gave a sequence with $\liminf a_k/k^2 < \infty$. Ajtai, Komlós & Szemerédi have shown that there is such a sequence with $a_k < ck^3/\ln n$.

Erdős & Rényi proved that there is a sequence satisfying $a_k < k^{2+\epsilon}$ for which the number of solutions of $a_i + a_j = t$ is $\leq c$.

Erdős notes that $\sum_{i=1}^{x} a_i^{-1/2} < c(\ln x)^{1/2}$ and asks if this is best possible. He asks if it is true that

$$\frac{1}{\ln x} \sum_{a_i+a_j \leq x} \frac{1}{a_i + a_j} \to 0$$

as $x \to \infty$ and suggests that perhaps

$$\sum_{a_i+a_j<x} \frac{1}{a_i + a_j} < c_1 \ln \ln x.$$

It is known that it can be $> c_2 \ln \ln x$.

Erdős also asks if a Sidon sequence $a_1 < a_2 < \ldots < a_k$ can be prolonged to a perfect difference set (see **C10**), i.e.,

$$a_1 < a_2 < \ldots < a_k < a_{k+1} < \ldots < a_{p+1} = p^2 + p + 1$$

with the differences $a_u - a_v$, $1 \leq u, v \leq p + 1$, $u \neq v$, representing every nonzero residue mod $p^2 + p + 1$ exactly once?

He could not even decide if it can be prolonged to

$$a_1 < a_2 < \ldots < a_k < a_{k+1} < \ldots < a_n, \quad a_n < (1 + o(1))n^2,$$

i.e., if it can be made as dense as possible asymptotically.

Let $a_1 < a_2 < \ldots < a_n$ be any sequence of integers. Is it true that it contains a Sidon subsequence $a_{i_1}, \ldots, a_{i_m}$ with $m = (1+o(1))n^{\frac{1}{2}}$? Komlós, Sulyok & Szemerédi (see **E11**) proved this with $m > cn^{\frac{1}{2}}$.

If $f(n)$ is the number of solutions of $n = a_i + a_j$, is there a sequence with

$$\lim f(n)/\ln n = c?$$

Erdős & Turán conjecture that if $f(n) > 0$ for all sufficiently large n, or if $a_k < ck^2$ for all k, then $\limsup f(n) = \infty$; Erdős also offers \$500 for settling this question.

Graham & Sloane rephrase the question in two more obviously packing forms:

Let $v_\alpha(k)$ [respectively $v_\beta(k)$] be the smallest v such that there is a k-element set $A = \{0 = a_1 < a_2 < \ldots < a_k\}$ of integers with the property that the sums $a_i + a_j$ for $i < j$ [respectively $i \leq j$] belong to $[0, v]$ and represent each element of $[0, v]$ at most once. The set A associated with v_β is often called a B_2**-sequence** (compare **E28**).

They give the values of v_α and v_β displayed in Table 3 and note that the bounds

$$2k^2 - O(k^{3/2}) < v_\alpha, v_\beta < 2k^2 + O(k^{36/23})$$

follow from a modification of the Erdős-Turán argument.

Table 3. Values of v_α, v_β and Exemplary Sets.

k	$v_\alpha(k)$	Example of A	$v_\beta(k)$	Example of A
2	1	{0,1}	2	{0,1}
3	3	{0,1,2}	6	{0,1,3}
4	6	{0,1,2,4}	12	{0,1,4,6}
5	11	{0,1,2,4,7}	22	{0,1,4,9,11}
6	19	{0,1,2,4,7,12}	34	{0,1,4,10,12,17}
7	31	{0,1,2,4,8,13,18}	50	{0,1,4,10,18,23,25}
8	43	{0,1,2,4,8,14,19,24}	68	{0,1,4,9,15,22,32,34}
9	63	{0,1,2,4,8,15,24,29,34}	88	{0,1,5,12,25,27,35,41,44}
10	80	{0,1,2,4,8,15,24,29,34,46}	110	{0,1,6,10,23,26,34,41,53,55}

Cilleruelo has shown that there is a sequence $\{a_k\}$, $a_k \ll k^2$ such that the sums $a_i^2 + a_j^2$ are all different.

If $g(m)$ is the largest integer n such that every set of integers of size m contains a subset of size n whose pairwise sums are distinct, then Abbott has shown that $g(m) > cm^{1/2}$ for any constant $c < \frac{2}{25}$ and all sufficiently large m.

In 1956 Erdős proved the existence of a sequence S such that all sufficiently large integers n are represented between $c_1 \ln n$ and $c_2 \ln n$ times as the sum of two members of S, and more recently Erdős & Tetali have obtained the corresponding result for the sum of k members of S.

Harvey L. Abbott, Sidon sets, *Canad. Math. Bull.*, **33**(1990) 335–341; *MR* **91k**: 11022.

Miklós Ajtai, János Komlós & Endre Szemerédi, A dense infinite Sidon sequence, *European J. Combin.*, **2**(1981) 1–11; *MR* **83f**:10056.

R. C. Bose & S. Chowla, Theorems in the additive theory of numbers, *Comment. Math. Helv.*, **37**(1962-63) 141–147.

Javier Cilleruelo, B_2-sequences whose terms are squares, *Acta Arith.*, **55** (1990) 261–265; *MR* **91i**:11023.

Javier Cilleruelo & Antonio Córdoba, $B_2[\infty]$-sequences of square numbers, *Acta Arith.*, **61**(1992) 265–270.

P. Erdős, Some of my forgotten problems in number theory, *Hardy-Ramanujan J.*, **15** (1992) 34–50.

P. Erdős & R. Freud, On sums of a Sidon-sequence, *J. Number Theory*, **38**(1991) 196–205.

P. Erdős & R. Freud, On Sidon-sequences and related problems (Hungarian), *Mat. Lapok*, **2**(1991) 1–44.

P. Erdős & W. H. J. Fuchs, On a problem of additive number theory, *J. London Math. Soc.*, **31**(1956) 67–73.

P. Erdős & E. Szemerédi, The number of solutions of $m = \sum_{i=1}^{k} x_i^k$, *Proc. Symp. Pure Math. Amer. Math. Soc.*, **24**(1973) 83–90.

Paul Erdős, Melvyn B. Nathanson & Prasad Tetali, Independence of solution sets and minimal asymptotic bases (preprint, 1993).

Paul Erdős & Prasad Tetali, Representations of integers as the sum of k terms, *Random Structures Algorithms*, **1**(1990) 245–261; *MR* **92c**:11012.

P. Erdős & P. Turán, On a problem of Sidon in additive number theory, and on some related problems, *J. London Math. Soc.*, **16**(1941) 212–215; *MR* **3**, 270. Addendum, **19**(1944) 208; *MR* **7**, 242.

R. L. Graham & N. J. A. Sloane, On additive bases and harmonious graphs, *SIAM J. Alg. Discrete Math.*, **1**(1980) 382–404.

H. Halberstam & K. F. Roth, *Sequences*, 2nd Edition, Springer, New York, 1982, Chapter 2.

Jia Xing-De, Some problems and results on subsets of asymptotic bases, *Qufu Shifan Daxue Xuebao Ziran Kexue Ban*, **13**(1987) 45–49; *MR* **88k**:11015.

Jia Xing-De, On the distribution of a B_2-sequence, *Qufu Shifan Daxue Xuebao Ziran Kexue Ban*, **14**(1988) 12–18; *MR* **89j**:11023.

Jia Xing-De, On finite Sidon sequences, *J. Number Thoery*, **44**(1993) 84–92.

F. Krückeberg, B_2-Folgen und verwandte Zahlenfolgen, *J. reine angew. Math.*, **206**(1961) 53–60.

B. Lindström, An inequality for B_2-sequences, *J. Combin. Theory*, **6**(1969) 211–212; *MR* **38** #4436.

Imre Z. Ruzsa, A just basis, *Monatsh. Math.*, **109**(1990) 145–151; *MR* **91e**: 11016.

J. Singer, A theorem in finite projective geometry and some applications to number theory, *Trans. Amer. Math. Soc.*, **43**(1938) 377–385; *Zbl* **19**, 5.

Vera T. Sós, An additive problem in different structures, *Graph Theory, Combinatorics, Algorithms, and Applications*, (SIAM Conf., San Francisco, 1989), 1991, 486–510; *MR* **92k**:11026.

C10 Modular difference sets and error correcting codes.

Singer's result, mentioned in **C9**, is based on **perfect difference sets**, i.e., a set of residues $a_1, a_2, \ldots, a_{k+1}$ (mod n) such that every nonzero residue (mod n) can be expressed uniquely in the form $a_i - a_j$. For example, $\{1,2,4\}$ mod 7 and $\{1,2,5,7\}$ mod 13. Perfect difference sets can exist only if $n = k^2 + k + 1$, and Singer proved that such a set exists whenever k is a prime power. Marshall Hall has shown that numerous non-prime-powers *cannot* serve as values of k and Evans & Mann that there is no such $k < 1600$ that is not a prime power. It is conjectured that no perfect difference set exists unless k is a prime power.

Can a given finite sequence, which contains no repeated differences, always be extended to form a perfect difference set?

Perfect difference sets may be used to make **Golomb rulers**. Subtract one from the elements of the difference set, e.g., {0,1,4,6} and take these as marks on a ruler of length 6, which can be used to measure all the lengths 1, 2, 3, 4, 5, 6. More generally we may look for less than perfect rulers of length n with $k+1$ marks $\{0, a_1, \ldots, a_{k-1}, n\}$ subject to various conditions. E.g., (a) all $\binom{k+1}{2}$ distances distinct, (b) maximum number of distinct distances for given n and k, (c) all integer distances from 1 up to some maximum e to be measurable. We cannot satisfy all of these conditions of $k \geq 4$, but Leech has found examples of perfect 'jointed' rulers. The trees

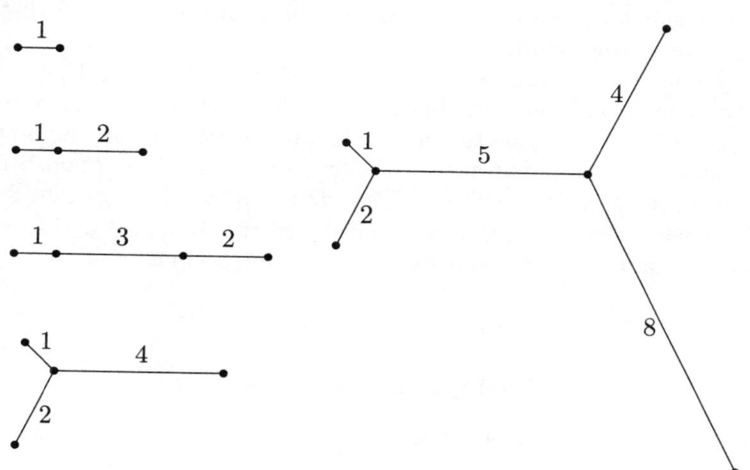

have edges with the lengths shown, and may be used to measure all lengths from 1 up to 1, 3, 6, 6, 15.

Gibbs & Slater, Herbert Taylor and Yang Yuan-Sheng have improved Leech's results for paths and for more general trees to

n	2	3	4	5	6	7	8	9	10	11	12
paths	1	3	6	9	13	18	24	29	37	45	(51)
trees	1	3	6	9	15	20	26	34	41	(48)	(55)

where the entries in parentheses are not necessarily best possible. There are connexions with the graceful labelling and harmonious labelling of graphs; see **C13** and a possibly forthcoming combinatorics volume in this series.

Dean Hickerson asks for the maximum number m such that the integers $1 \leq a_1 < a_2 < \cdots < a_m \leq n$ have differences $a_j - a_i$, $j > i$, among which the integer s occurs at most $2s$ times.

Graham & Sloane exhibit the problem of difference sets as the modular version of the packing problems of **C9**. They define $v_\gamma(k)$ [respectively $v_\delta(k)$] as the smallest number v such that there exists a subset $A = \{0 = a_1 < a_2 < \cdots < a_k\}$ of the integers (mod v) with the property that each

r can be written in at most one way as $r \equiv a_i + a_j \bmod v$ with $i < j$ [respectively $i \le j$].

Their interest in v_γ is in its application to **error-correcting codes**. If $A(k, 2d, w)$ is the maximum number of binary vectors with w ones and $k - w$ zeros (**words** of **length** k and **weight** w) such that any two vectors differ in at least $2d$ places, then (for $d = 3$)

$$A(k,6,w) \ge \binom{k}{w} \Big/ v_\gamma(k)$$

(and the result for general d uses sets for which all sums of $d-1$ distinct elements are distinct modulo v).

They note that $A(k, 2d, w)$ has been studied by Erdős & Hanani, by Schönheim, and by Stanton, Kalbfleisch & Mullin in the context of extremal set theory. Let $D(t, k, v)$ be the maximum number of k-element subsets of a v-element set S such that every t-element subset of S is contained in at most one of the k-element subsets. Then $D(t, k, v) = A(v, 2k - 2t + 2, k)$.

The values of v_δ in Table 4 are from Baumert's Table 6.1 and those of v_γ from Graham & Sloane who give the following bounds

$$k^2 - O(k) < v_\gamma(k) < k^2 + O(k^{36/23}),$$

$$k^2 - k + 1 \le v_\delta(k) < k^2 + O(k^{36/23}).$$

Equality holds on the left of the latter whenever $k - 1$ is a prime power.

Table 4. Values of v_γ, v_δ and Exemplary Sets.

k	$v_\gamma(k)$	Example of A	$v_\delta(k)$	Example of A
2	2	{0,1}	3	{0,1}
3	3	{0,1,2}	7	{0,1,3}
4	6	{0,1,2,4}	13	{0,1,3,9}
5	11	{0,1,2,4,7}	21	{0,1,4,14,16}
6	19	{0,1,2,4,7,12}	31	{0,1,3,8,12,18}
7	28	{0,1,2,4,8,15,20}	48	{0,1,3,15,20,38,42}
8	40	{0,1,5,7,9,20,23,35}	57	{0,1,3,13,32,36,43,52}
9	56	{0,1,2,4,7,13,24,32,42}	73	{0,1,3,7,15,31,36,54,63}
10	72	{0,1,2,4,7,13,23,31,39,59}	91	{0,1,3,9,27,49,56,61,77,81}

L. D. Baumert, *Cyclic Difference Sets*, Springer Lect. Notes Math. **182**, New York, 1971.

M. R. Best, A. E. Brouwer, F. J. MacWilliams, A. M. Odlyzko & N. J. A. Sloane, Bounds for binary codes of length less than 25, *IEEE Trans. Inform. Theory*, **IT-24**(1978) 81–93.

F. T. Boesch & Li Xiao-Ming, On the length of Golomb's rulers, *Math. Appl.*, **2**(1989) 57–61; *MR* **91h**:11015.

J. H. Conway & N. J. A. Sloane, *Sphere Packings, Lattices and Groups*, Springer-Verlag, New York, 1988.

P. Erdős & H. Hanani, On a limit theorem in combinatorical analysis, *Publ. Math. Debrecen*, **10**(1963) 10–13.

T. A. Evans & H. Mann, On simple difference sets, *Sankhyā*, **11**(1951) 357–364; *MR* **13**, 899.

Richard A. Gibbs & Peter J. Slater, Distinct distance sets in a graph, *Discrete Math.*, **93**(1991) 155–165.

M. J. E. Golay, Note on the representation of 1, 2, ..., n by differences, *J. London Math. Soc.*(2), **4**(1972) 729–734; *MR* **45** #6784.

R. L. Graham & N. J. A. Sloane, Lower bounds for constant weight codes, *IEEE Trans. Inform. Theory*, **IT-26**(1980) 37–43; *MR* **81d**:94026.

R. L. Graham & N. J. A. Sloane, On additive bases and harmonious graphs, *SIAM J. Algebraic Discrete Methods*, **1**(1982) 382–404; *MR* **82f**:10067.

M. Hall, Cyclic projective planes, *Duke Math. J.*, **14**(1947) 1079–1090; *MR* **9**, 370.

John Leech, On the representation of 1, 2, ..., n by differences, *J. London Math. Soc.*, **31**(1956) 160–169; *MR* **19**, 942f.

John Leech, Another tree labelling problem, *Amer. Math. Monthly*, **82**(1975) 923–925.

F. J. MacWilliams & N. J. A. Sloane, *The Theory of Error-Correcting Codes*, North-Holland, 1977.

J. C. P. Miller, Difference bases: three problems in the additive theory of numbers, A. O. L. Atkin & B. J. Birch (editors) *Computers in Number Theory*, Academic Press, London, 1971, pp. 299–322; *MR* **47** #4817.

J. Schönheim, On maximal systems of k-tuples, *Stud. Sci. Math. Hungar.*, **1**(1966) 363–368.

R. G. Stanton, J. G. Kalbfleisch & R. C. Mullin, Covering and packing designs, *Proc. 2nd Conf. Combin. Math. Appl.*, Chapel Hill, 1970, 428–450.

Herbert Taylor, A distinct distance set of 9 nodes in a tree of diameter 36, *Discrete Math.*, **93**(1991) 167–168.

B. Wichmann, A note on restricted difference bases, *J. London Math. Soc.*, **38**(1963) 465–466; *MR* **28** #2080.

C11 Three-subsets with distinct sums.

One can generalize the ideas of **C9** and **E28** and define a B_h-**sequence** to be one in which the sums of h terms are distinct. Bose & Chowla showed that if $A_h(n)$ is the largest cardinality of a B_h-sequence in $[1, n]$, then

$$A_h(n) \geq n^{1/h}(1 + o(1)).$$

In the opposite direction, Jia has shown that if $h = 2k$, then

$$A_h(n) \leq k^{1/2k}(k!)^{1/k} n^{1/h}(1 + o(1)).$$

For $h = 2k - 1$, Chen and Graham independently proved the bound

$$A_h(n) \le (k!)^{2/(2k-1)} n^{1/h}(1 + o(1)).$$

For $h = 3$ Graham obtained the further small improvement

$$A_3(n) \le \left(4 - \tfrac{1}{228}\right)^{\frac{1}{3}} n^{\frac{1}{3}}(1 + o(1)).$$

In the infinite case, Erdős offers \$500.00 for a proof or disproof of

$$¿ \quad \liminf \frac{A_h(n)}{n^{1/h}} = 0 \quad ?$$

For $h = 2$ this was proved by Erdős himself, and for $h = 4$ by Nash. The case $h = 6$ was treated by Jia, and more generally Chen showed that if $h = 2k$ was even, then

$$\liminf A_h(n) \left(\frac{\ln n}{n}\right)^{\frac{1}{h}} < \infty$$

It remains an open problem to prove this when h odd.

The paper of Chen & Kløve gives references to the electrical engineering literature on B_h-sequences.

W. C. Babcock, Intermodulation interference in radio systems, *Bell Systems Tech. J.*, **32**(1953) 63–73.

R. C. Bose & S. Chowla, *Report Inst. Theory of Numbers*, Univ. of Colorado, Boulder, 1959, p. 335.

R. C. Bose & S. Chowla, Theorems in the additive theory of numbers, *Comment. Math. Helvet.*, **37**(1962-63) 141–147.

Sheng Chen, On size of finite Sidon sequences, *Proc. Amer. Math. Soc.*, (to appear).

Sheng Chen, A note on B_{2k}-sequences, *J. Number Theory*, (1994)

Sheng Chen, On Sidon sequences of even orders, *Acta Arith.*, **64**(1993) 325–330.

Chen Wen-De & Torliev Kløve, Lower bounds on multiple difference sets, *Discrete Math.*, **98**(1991) 9–21; *MR* 93a:05028.

S. W. Graham, Upper bounds for B_3-sequences, Abstract 882-11-25, *Abstracts Amer. Math. Soc.*, **14**(1993) 416.

S. W. Graham, Upper bounds for Sidon sequences, (preprint 1993).

D. Hajela, Some remarks on $B_h[g]$-sequences, *J. Number Theory*, **29**(1988) 311–323; *MR* **90d**:11022.

M. Helm, On B_{2k}-sequences, *Acta Arith.*, **63**(1993) 367–371.

M. Helm, A remark on B_{2k}-sequences, *J. Number Theory*, (to appear).

Jia Xing-De, On B_6-sequences, *Qufu Shifan Daxue Xuebao Ziran Kexue Ban*, **15**(1989) 7–11; *MR* **90j**:11022.

Jia Xing-De, on B_{2k}-sequences, *J. Number Theory*, (to appear).

T. Kløve, Constructions of $B_h[g]$-sequences, *Acta Arith.*, **58**(1991) 65–78; *MR* **92f**:11033.

Li An-Ping, On B_3-sequences, *Acta Math. Sinica*, **34**(1991) 67–71; *MR* **92f**: 11037.

B. Lindström, A remark on B_4-sequences, *J. Combin. Theory*, **7**(1969) 276–277.

John C. M. Nash, On B_4-sequences, *Canad. Math. Bull.*, **32**(1989) 446–449; *MR* **91e**: 11025.

C12 The postage stamp problem.

The covering problem which is dual to the packing problem **C9** goes back at least to Rohrbach. A popular form of it concerns the design of a set of integer denominations of postage stamp, $A_k = \{a_1, a_2, \ldots, a_k\}$ with $1 = a_1 < a_2 < \cdots < a_k$ to be used on envelopes with room for at most h stamps, so that all integer amounts of postage up to a given bound can be affixed. What is the smallest integer $N(h, A_k)$ which is *not* representable by a linear combination $\sum_{i=1}^k x_i a_i$ with $x_i \geq 0$ and$\sum_{i=1}^k x_i \leq h$? The number of consecutive possible amounts of postage, $n(h, A_k) = N(h, A_k) - 1$ is called the h-**range** (German: h-Reichweite) of A_k. In this context A_k is called an **additive basis of order** h or h-**basis**. At first the main interest was in the 'global' problem: Given h and k, find an **extremal basis** A_k^* with largest possible h-range, $n(h,k) = n(h, A_k^*) = \max_{A_k} n(h, A_k)$. More recently the 'local' aspect has come more into focus: Find $n(h, A_k)$ when h and a particular basis A_k are given.

The local problem is completely solved only for $k = 2$ and $k = 3$. Trivially $n(h, A_2) = (h + 3 - a_2)a_2 - 2$ for $h \geq a_2 - 2$. Rødseth developed a general method, based on a continued fraction algorithm, for determining $n(h, A_3)$. From this, Selmer derived explicit formulas covering (asymptotically) about 99% of all A_3.

From the formula for $n(h, A_2)$ Stöhr concluded that

$$n(h,2) = \lfloor (h^2 + 6h + 1)/4 \rfloor.$$

The global problem for $k = 3$ was solved by Hofmeister, who showed in particular that, for $h \geq 20$,

$$n(h,3) = \tfrac{4}{3}(\tfrac{h}{3})^3 + 6(\tfrac{h}{3})^2 + Ah + B$$

where A and B depend on the residue of h modulo 9. Mossige showed that

$$n(h,4) \geq 2.008 \left(\tfrac{h}{4}\right)^4 + O(h^3)$$

and, together with Kirfel (so far unpublished) that this bound is sharp. Kirfel has also shown that the limit

$$c_k = \lim_{h \to \infty} \frac{n(h,k)}{(h/k)^k}$$

exists for all $k \geq 1$. Kolsdorf showed that

$$n(h,5) \geq 3.06\left(\tfrac{h}{5}\right)^5 + O(h^4).$$

Mrose showed that

$$n(h_1+h_2,\, k_1+k_2) \geq (n(h_1,\, k_1)+1)\cdot(n(h_2,\, k_2)+1) \qquad (*)$$

for all positive integers $h-1$, h_2, k_1, k_2 and deduced that if k_i $(i=1,\,2)$ are fixed and

$$n(h,\, k_i) \geq \alpha_i \left(\frac{h}{k_i}\right)^{k_i} + O(h^{k_i-1})$$

then

$$n(h,\, k_1+k_2) \geq \alpha_1\alpha_2\left(\frac{h}{k_1+k_2}\right)^{k_1+k_2} + O(h^{k_1+k_2-1}).$$

Thus if x_i are fixed nonnegative integers with $k=\sum_{i=1}^5 ix_i$ then

$$n(h,\, k) \geq (3.06)^{x_5}(2.008)^{x_4}\left(\frac{4}{3}\right)^{x_3}\left(\frac{h}{k}\right)^k + O(h^{k-1})$$

The best general upper bound for fixed k is due to Rødseth:

$$n(h,k) \leq \frac{(k-1)^{k-2}}{(k-2)!}\left(\frac{h}{k}\right)^k + O(h^{k-1}).$$

For fixed h, emphasis has been on the case $h=2$. In 1937 Rohrbach showed that

$$c_1\left(\frac{k}{2}\right)^2 + O(k) \leq n(2,k) \leq c_2\left(\frac{k}{2}\right)^2 + O(k)$$

with $c_1=1$ and $c_2=1.9968$. After several improvements the best known results are $c_1=\frac{8}{7}$ (Mrose) and $c_2=1.9208$ (Klotz). Windecker showed that

$$n(3,\, k) \geq \frac{4}{3}\left(\frac{k}{3}\right)^3 + \frac{16}{3}\left(\frac{k}{3}\right)^2 + O(k).$$

Again, with $(*)$, if y_i are fixed nonnegative integers with $h=\sum_{i=1}^3 iy_i$ then

$$n(h,\, k) \geq \left(\frac{4}{3}\right)^{y_3}\left(\frac{8}{7}\right)^{y_2}\left(\frac{k}{h}\right)^h + O(k^{h-1}).$$

Graham & Sloane (compare **C9, C10**) define $n_{\alpha(k)}$ [respectively $n_{\beta(k)}$] as the largest number n such that there is a k-element set $A=\{0=a_1<a_2<\cdots<a_k\}$ of the integers with the property that each r in $[1,n]$ can be written in at least one way as $r=a_i+a_j$ with $i<j$ [respectively $i\leq j$], so that their $n_{\beta(k)}$ is here written $n(2,k-1)$, and their $n_{\alpha(k)}$

corresponds to the problem of two stamps of different denominations, with a zero denomination included.

They give the values for $n_{\alpha(k)}$ and $n_{\beta(k)}$ in Table 5.

Table 5. Values for $n_{\alpha(k)}$ and $n_{\beta(k)}$ and Exemplary Sets.

k	$n_\alpha(k)$	Example of A	$n_\beta(k)$	Example of A
2	1	{0,1}	2	{0,1}
3	3	{0,1,2}	4	{0,1,2}
4	6	{0,1,2,4}	8	{0,1,3,4}
5	9	{0,1,2,3,6}	12	{0,1,3,5,6}
6	13	{0,1,2,3,6,10}	16	{0,1,3,5,7,8}
7	17	{0,1,2,3,4,8,13}	20	{0,1,2,5,8,9,10}
8	22	{0,1,2,3,4,8,13,18}	26	{0,1,2,5,8,11,12,13}
9	27	{0,1,2,3,4,5,10,16,22}	32	{0,1,2,5,8,11,14,15,16}
10	33	{0,1,2,3,4,5,10,16,22,28}	40	{0,1,3,4,9,11,16,17,19,20}
11	40	{0,1,2,4,5,6,10,13,20,27,34}	46	{0,1,2,3,7,11,15,19,21,22,24}
12	47	{0,1,2,3,6,10,14,18,21,22,23,24}	54	{0,1,2,3,7,11,15,19,23,25,26,28}
13	56	{0,1,2,4,6,7,12,14,17,21,30,39,48}	64	{0,1,3,4,9,11,16,21,23,28,29,31,32}
14	65	{0,1,2,4,6,7,12,14,17,21,30,39,48,57}	72	{0,1,3,4,9,11,16,20,25,27,32,33,35,36}

Tables for more general $n(h,k)$ were computed by Lunnon and extended by Mossige and recently by Challis.

Serious students of these problems will consult the encyclopædic three volumes of Selmer, which contain 121 references. A useful summary, with 47 references, has been prepared by Djawadi & Hofmeister.

M. F. Challis, Two new techniques for computing extremal h-bases A_k, *Comput. J.*, **36**(1993) 117–126. [An updating of the appendix is available from the author.]

Mehdi Djawadi & Gerd Hofmeister, The postage stamp problem *Mainzer Seminarberichte, Additive Zahlentheorie*, **3**(1993) 187–195.

Paul Erdős & Melvyn B. Nathanson, Additive bases with many representations, *Acta Arith.*, **52**(1989) 399–406; *MR* **91e**:11015.

N. Hämmerer & G. Hofmeister, Zu einer Vermutung von Rohrbach, *J. reine angew. Math.*, **286/287**(1976) 239–247; *MR* **54** #10181.

G. Hofmeister, Asymptotische Abschätzungen für dreielementige extremalbasen in natürlichen Zahlen, *J. reine angew. Math.*, **232**(1968) 77–101; *MR* **38** #1068.

G. Hofmeister, Die dreielementigen Extremalbasen, *J. reine angew. Math.*, **339** (1983) 207–214.

G. Hofmeister, C. Kirfel & H. Kolsdorf, Extremale Reichweitenbasen, No. **60**, Dept. Pure Math., Univ. Bergen, 1991.

Jia Xing-De, On a combinatorial problem of Erdős and Nathanson, *Chinese Ann. Math. Ser. A*, **9**(1988) 555–560; *MR* **90i**:11018.

Jia Xing-De, On the order of subsets of asymptotic bases, *J. Number Theory* **37**(1991) 37–46; *MR* **92d**:11006.

Jia Xing-De & Melvyn B. Nathanson, A simple construction of minimal asymptotic bases, *Acta Arith.*, **52**(1989) 95–101; *MR* **90g**:11020.

Christoph Kirfel, Extremale asymptotische Reichweitenbasen, *Acta Arith.*, **60** (1992) 279–288; *MR* **92m**:11012.

W. Klotz, Extremalbasen mit fester Elementeanzahl, *J. reine angew. Math.*, **237**(1969) 194–220.

W. Klotz, Eine obere Schranke für die Reichweite einer Extremalbasis zweiter Ordnung, *J. reine angew. Math.*, **238**(1969) 161–168 (and see 194–220); *MR* **40** #117, 116.

W. F. Lunnon, A postage stamp problem, *Comput. J.*, **12**(1969) 377–380; *MR* **40** #6745.

L. Moser, On the representation of 1, 2, ..., n by sums, *Acta Arith.*, **6**(1960) 11–13; *MR* **23** #A133.

L. Moser, J. R. Pounder & J. Riddell, On the cardinality of h-bases for n, *J. London Math. Soc.*, **44**(1969) 397–407; *MR* **39** #162.

S. Mossige, Algorithms for computing the h-range of the postage stamp problem, *Math. Comput.*, **36**(1981) 575–582; *MR* **82e**:10095.

S. Mossige, On extremal h-bases A_4, *Math. Scand.*, **61**(1987) 5–16; *MR* **89e**:11008.

A. Mrose, Untere Schranken für die Reichweiten von Extremalbasen fester Ordnung, *Abh. Math. Sem. Univ. Hamburg*, **48**(1979) 118–124; *MR* **80g**:10058.

Melvyn B. Nathanson, Extremal properties for bases in additive number theory, Number Theory, Vol. I (Budapest, 1987), *Colloq. Math. Soc. János Bolyai* **51**(1990) 437–446; *MR* **91h**:11009.

J. Riddell & C. Chan, Some extremal 2-bases, *Math. Comput.*, **32**(1978) 630–634; *MR* **57** #16244.

Ø. Rødseth, On h-bases for n, *Math. Scand.*, **48**(1981) 165–183; *MR* **82m**: 10034.

Øystein J. Rødseth, An upper bound for the h-range of the postage stamp problem, *Acta Arith.* **54**(1990) 301–306; *MR* **91h**:11013.

H. Rohrbach, Ein Beitrag zur additiven Zahlentheorie, *Math. Z.*, **42**(1937) 1–30; *Zbl.* **15**, 200.

H. Rohrbach, Anwendung eines Satzes der additiven Zahlentheorie auf eine gruppentheoretische Frage, *Math. Z.*, **42**(1937) 538–542; *Zbl.* **16**, 156.

Ernst S. Selmer, On the postage stamp problem with three stamp denominations, *Math. Scand.*, **47**(1980) 29–71; *MR* **82d**:10046.

Ernst S. Selmer, The Local Postage Stamp Problem, Part I General Theory, Part II The Bases A_3 and A_4, Part III Supplementary Volume, No. 42, 44, 57, Dept. Pure Math., Univ. Bergen, (86-04-15, 86-09-15, 90-06-12) ISSN 0332-5047. (Available on request.)

Ernst S. Selmer, On Stöhr's recurrent h-bases for N, *Kongel. Norske Vidensk. Selsk.*, Skr. 3, 1986, 15 pp.

Ernst S. Selmer, Associate bases in the postage stamp problem, *J. Number Theory*, **42**(1992) 320–336.

Ernst S. Selmer & Arne Rödne, On the postage stamp problem with three stamp denominations, II, *Math. Scand.*, **53**(1983) 145–156; *MR* **85j**:11075.

Ernst S. Selmer & Björg Kristin Selvik, On Rödseth's h-bases $A_k = \{1, a_2, 2a_2, \dots, (k-2)a_2, a_k\}$, *Math. Scand.*, **68**(1991) 180–186; *MR* **92k**: 11010.

Alfred Stöhr, Gelöste und ungelöste Fragen über Basen der natürlichen Zahlenreihe I, II, *J. reine angew. Math.*, **194**(1955) 40–65, 111–140; *MR* **17**, 713.

R. Windecker, Eine Abschnittsbasis dritter Ordnung, *Kongel. Norske Vidensk. Selsk. Skrifter*, **9**(1976) 1–3.

C13 The corresponding modular covering problem. Harmonious labelling of graphs.

Just as **C10** was the modular version of the packing problem **C9**, so we can propose the modular version of the corresponding covering problem **C12**.

Graham & Sloane complete their octad of definitions with $n_\gamma(k)$ [respectively $n_\delta(k)$] as the largest number n such that there is a subset $A = \{0 = a_1 < a_2 < \cdots < a_k\}$ of the residue classes modulo n with the property that each r can be written in at least one way as $r \equiv a_i + a_j \bmod n$ with $i < j$ [respectively $i \le j$].

They call a connected graph with v vertices and $e \ge v$ edges **harmonious** if there is a labelling of the vertices x with distinct labels $l(x)$ so that when an edge xy is labelled with $l(x) + l(y)$, the edge labels form a complete system of residues (mod e). Trees (for which $e = v - 1$) are also called harmonious if just one vertex label is duplicated and the edge labels form a complete system (mod $v - 1$). The connexion with the present problem is that $n_\gamma(v)$ is the greatest number of edges in any harmonious graph on v vertices.

For example, from Table 6 we note that $n_\gamma(5) = 9$ is attained by the set {0,1,2,4,7} so that a maximum of 9 edges can occur in a harmonious graph on 5 vertices (Figure 6).

Table 6. Values of n_γ, n_δ and Exemplary Sets.

k	$n_\gamma(k)$	Example of A	$n_\delta(k)$	Example of A
2	1	—	3	{0,1}
3	3	{0,1,2}	5	{0,1,2}
4	6	{0,1,2,4}	9	{0,1,3,4}
5	9	{0,1,2,4,7}	13	{0,1,2,6,9}
6	13	{0,1,2,3,6,10}	19	{0,1,3,12,14,15}
7	17	{0,1,2,3,4,8,13}	21	{0,1,2,3,4,10,15}
8	24	{0,1,2,4,8,13,18,22}	30	{0,1,3,9,11,12,16,26}
9	30	{0,1,2,4,10,15,17,22,28}	35	{0,1,2,7,8,11,26,29,30}
10	36	{0,1,2,3,6,12,19,20,27,33}		

Graham & Sloane compare and contrast harmonious graphs with graceful graphs, which will be discussed in the graph theory chapter of a later volume in this series. A graph is **graceful** if, when the vertex labels are chosen from $[0, e]$ and the edge labels are calculated by $|l(x) - l(y)|$, the latter are all distinct (i.e., take the values $[1, e]$).

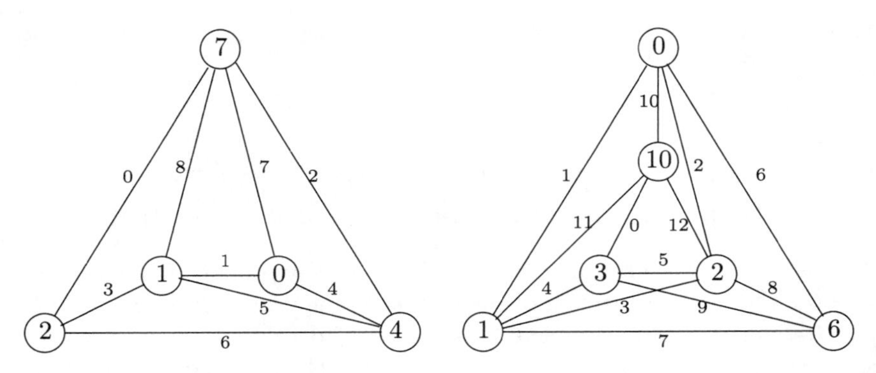

Figure 6. Maximal Harmonious Graphs.

Trees are conjectured to be both harmonious and graceful, but these are open questions. A cycle C_n is harmonious just if n is odd, and graceful just if $n \equiv 0$ or 3 mod4. The friendship graph or windmill is harmonious just if $n \not\equiv 2 \bmod 4$ and graceful just if $n \equiv 0$ or 1 mod4. Fans and wheels are both harmonious and graceful, as is the Petersen graph. The graphs of the five Platonic solids are naturally graceful and one would expect them to be harmonious, but this is not so for the cube or octahedron. Joseph Gallian maintains a bibliography of graph labelling.

Joseph A. Gallian, A survey – recent results, conjectures and open problems in labeling graphs, *J. Graph Theory*, **13**(1989) 491–504; *MR* **90k**:05132.

C14 Maximal sum-free sets.

Denote by $l(n)$ the largest l so that if $a_1, a_2, \ldots, a_n$ are any distinct natural numbers, one can always find l of them so that $a_{i_j} + a_{i_k} \neq a_m$ for $1 \leq j < k \leq l$, $1 \leq m \leq n$. Note that $j \neq k$, else the set $\{a_i = 2^i \mid 1 \leq i \leq n\}$ would imply that $l(n) = 0$. A remark of Klarner shows that $l(n) > c \ln n$. On the other hand, the set $\{2^i + 0, \pm 1 \mid 1 < i \leq s+1\}$ implies that $l(3s) < s+3$, so $l(n) < \frac{1}{3}n + 3$. Selfridge extends this by using the set $\{(3m+t)2^{m-i} \mid -i < t < i,\ 1 \leq i \leq m\}$ to show that $l(m^2) < 2m$. Choi, using sieve methods, has further improved this to $l(n) \ll n^{0.4+\epsilon}$.

The problem can be generalized to ask if, for every l, there is an $n_0 = n_0(l)$ so that if $n > n_0$ and $a_1, a_2, \ldots, a_n$ are any n elements of a group with no product $a_{i_1}a_{i_2} = e$, the unit [here i_1, i_2 may be equal, so there is no a_i of order 1 or 2, and no a_i whose inverse is also an a_i] then there are l of the a_i such that $a_{i_j}a_{i_k} \neq a_m$, $1 \leq j < k \leq l$, $1 \leq m \leq n$. This has not even been proved for $l = 3$.

For the generalization to sets containing no solution of $a_1x_1 + \ldots + a_kx_k = x_{k+1}$ see the papers of Funar and Moree.

Harvey L. Abbott, On a conjecture of Funar concerning generalized sum-free sets, *Niew Arch. Wisk.*(4), **4**(1991) 249–252.
S. L. G. Choi, On sequences not containing a large sum-free subsequence, *Proc. Amer. Math. Soc.*, **41**(1973) 415–418; *MR* **48** #3910.
S. L. G. Choi, On a combinatorial problem in number theory, *Proc. London Math. Soc.*,(3) **23**(1971) 629–642; *MR* **45** #1867.
P. H. Diananda & Yap H.-P., Maximal sum-free sets of elements of finite groups, *Proc. Japan Acad.*, **45**(1969) 1–5; *MR* **39** #6968.
Louis Funar, Generalized sum-free sets of integers, *Nieuw Arch. Wisk.*(4) **8**(1990) 49–54; *MR* **91e**:11012.
Pieter Moree, On a conjecture of Funar, *Nieuw Arch. Wisk.*(4) **8**(1990) 55–60; *MR* **91e**:11013.
Leo Moser, Advanced problem 4317, *Amer. Math. Monthly*, **55**(1948) 586; solution Robert Steinberg, **57**(1950) 345.
Anne Penfold Street, A maximal sum-free set in A_5, *Utilitas Math.*, **5**(1974) 85–91; *MR* **49** #7156.
Anne Penfold Street, Maximal sum-free sets in abelian groups of order divisible by three, *Bull. Austral. Math. Soc.*, **6**(1972) 317–318; *MR* **47** #5147.
P. Varnavides, On certain sets of positive density, *J. London Math. Soc.*, **34**(1959) 358–360; *MR* **21** #5595.
Edward T. H. Wang, On double-free set of integers, *Ars Combin.*, **28**(1989) 97–100; *MR* **90d**:11011.
Yap Hian-Poh, Maximal sum-free sets in finite abelian groups, *Bull. Austral. Math. Soc.*, **4**(1971) 217–223; *MR* **43** #2081 [and see *ibid*, **5**(1971) 43–54; *MR* **45** #3574; *Nanta Math.*, **2**(1968) 68–71; *MR* **38** #3345; *Canad. J. Math.*, **22**(1970) 1185–1195; *MR* **42** #1897; *J. Number Theory*, **5**(1973) 293–300; *MR* **48** #11356].

C15 Maximal zero-sum-free sets.

Erdős & Heilbronn asked for the largest number $k = k(m)$ of distinct residue classes, modulo m, so that no subset has sum zero. For example, the set

$$1,\ -2,\ 3,\ 4,\ 5,\ 6$$

shows that $k(20) \geq 6$, and in fact equality holds. The pattern of this example shows that

$$k \geq \lfloor(-1+\sqrt{8m+9})/2\rfloor \qquad (m \geq 5)$$

Equality holds for $5 \leq m \leq 24$. However, Selfridge observes that if m is of the form $2(l^2+l+1)$, the set

$$1,\ 2,\ \ldots,\ l-1,\ l,\ \tfrac{1}{2}m,\ \tfrac{1}{2}m+1,\ \ldots,\ \tfrac{1}{2}m+l$$

implies that

$$k \geq 2l+1 = \sqrt{2m-3}$$

In fact he conjectures that, for any even m, this set or the set with l deleted always gives the best result. For example, $k(42) \geq 9$.

On the other hand, if p is a prime in the interval

$$\tfrac{1}{2}k(k+1) < p < \tfrac{1}{2}(k+1)(k+2)$$

he conjectures that $k(p) = k$, where the set can be simply

$$1,\ 2,\ \ldots,\ k$$

The case $k(43) = 8$ was confirmed by Clement Lam, so k is not a monotonic function of m.

The only case where a better inequality is known than $k \geq \lfloor\sqrt{2m-3}\rfloor$ is $k(25) \geq \sqrt{50-1} = 7$, as is shown by the set 1, 6, 11, 16, 21, 5, 10. If m is of the form $25l(l+1)/2$ and *odd*, then it is possible to improve on the set 1, -2, 3, 4, ..., but if m is of that form and *even*, then the construction already given for m even is always better.

Is $k = \lfloor(-1+\sqrt{8m+9})/2\rfloor$ for an infinity of values of m?

For which values of m are there realizing sets none of whose members are prime to m? For example, $m = 12$: {3,4,6,10} or {4,6,9,10}. Is there a value of m for which *all* realizing sets are of this type?

Erdős & Heilbronn proved that if $a_1, a_2, \ldots, a_k$, $k \geq 3(6p)^{1/2}$, are distinct residues mod$p$, where p is prime, then every residue mod$p$ can be written in the form $\sum_{i=1}^{k} \epsilon_i a_i$, $\epsilon_i = 0$ or 1. They conjectured that the same holds for $k > 2\sqrt{p}$ and that this is best possible and Olsen proved this. They further conjectured that the number, s, of distinct residues of the form $a_i + a_j$, $1 \leq i < j \leq k$, is at least $\min\{p, 2k-3\}$. Partial results have been obtained by Mansfield, by Rødseth and by Freiman, Low & Pitman. Dias da Silva & Hamidoune gave a complete proof of the Erdős-Heilbronn conjecture, and in fact proved that if A^h denotes the set of all sums of h distinct elements of A, $A \subseteq \mathbb{Z}/p\,\mathbb{Z}$, $|A| = k$, then $|A^h| \geq \min\{p, hk - h^2 + 1\}$. Nathanson simplifies their proof and, with Ruzsa, shows that if A, $B \subseteq \mathbb{Z}/p\,\mathbb{Z}$, $|A| = k > l = |B|$, then $|\{a+b : a \in A, b \in B, a \neq b\}| \geq \min\{p, k+l-2\}$.

W. Brakemeier, *Ein Beitrag zur additiven Zahlentheorie*, Dissertation, Tech. Univ. Braunschweig 1973.

W. Brakemeier, Eine Anzahlformel von Zahlen modulo n, *Monatsh. Math.*, **85** (1978) 277–282.

A. L. Cauchy, Recherches sur les nombres, *J. École Polytech.*, **9**(1813) 99–116.

H. Davenport, On the addition of residue classes, *J. London Math. Soc.*, **10**(1935) 30–32.

H. Davenport, A historical note, *J. London Math. Soc.*, **22**(1947) 100–101.

J. A. Dias da Silva & Y. O. Hamidoune, Cyclic spaces for Grassmann derivatives and additive theory, *Bull. London Math. Soc.*, (to appear).

P. Erdős, Some problems in number theory, in *Computers in Number Theory*, Academic Press, London & New York, 1971, 405–413.

P. Erdős & H. Heilbronn, On the addition of residue clesses mod p, *Acta Arith.*, **9**(1969) 149–159.

G. A. Freiman, *Foundations of a Structural Theory of Set Addition*, Transl. Math. Monographs, **37**(1973), Amer. Math. Soc., Providence RI.

G. A. Freiman, L. Low & J. Pitman, 1993 preprint.

Henry B. Mann & John E. Olsen, Sums of sets in the elementary abelian group of type (p, p), *J. Combin. Theory*, **2**(1967) 275–284.

R. Mansfield, How many slopes in a polygon? *Israel J. Math.*, **39**(1981) 265–272; *MR* **84j**:03074.

M. B. Nathanson, *Additive Number Theory: Inverse Problems and the Geometry of Sumsets*, Springer-Verlag, 1994.

M. B. Nathanson & I. Z. Ruzsa, Sums of different residues modulo a prime, 1993 preprint.

John E. Olsen, An addition theorem modulo p, *J. Combin. Theory*, **5**(1968) 45–52.

John E. Olsen, An addition theorem for the elementary abelian group, *J. Combin. Theory*, **5**(1968) 53–58.

J. M. Pollard, A generalisation of the theorem of Cauchy and Davenport, *J. London Math. Soc.*, **8**(1974) 460–462.

J. M. Pollard, Addition properties of residue classes, *J. London Math. Soc.*, **11**(1975) 147–152.

U.-W. Rickert, *Über eine Vermutung in der additiven Zahlentheorie*, Dissertation, Tech. Univ. Braunschweig 1976.

Øystein J. Rødseth, Sums of distinct residues mod p, *Acta Arith.*, **65**(1993) 181–184.

C. Ryavec, The addition of residue classes modulo n, *Pacific J. Math.*, **26** (1968) 367–373.

E. Szemerédi, On a conjecture of Erdős and Heilbronn, *Acta Arith.*, **17**(1970-71) 227–229.

C16 Nonaveraging sets. Nondividing sets.

A **nonaveraging set** A of integers $0 \le a_1 < a_2 < \cdots < a_n \le x$ was defined by Erdős & Straus by the property that no a_i shall be the arithmetic mean of any subset of A with more than one element. Denote by $f(x)$ the maximum number of elements in such a set, and by $g(x)$ the maximum number of elements in a subset B of the integers $[0, x]$ such that no two distinct subsets of B have the same arithmetic mean, and by $h(x)$ the corresponding maximum where the subsets of B have different cardinality. Erdős & Straus and Abbott showed (by using Szemerédi's **E10** result) that

$$\frac{1}{10}\log x + O(1) < \log f(x) < \frac{2}{3}\log x + O(1)$$

$$\frac{1}{2}\log x - 1 < g(x) < \log x + O(\ln\ln x)$$

$$\sqrt{\ln x} - 1 + O(1/\sqrt{\ln x}) < \log h(x) < 2\log\ln x + O(1)$$

and conjecture that $f(x) = \exp(c\sqrt{\ln x}) = o(x^\epsilon)$ and that $h(x) = (1 + o(1))\log x$. [$\log x = (\ln x)/(\ln 2)$ is the logarithm to base 2.] Since then Abbott has improved the constant $\frac{1}{10}$ to $\frac{1}{5}$ and Bosznay to $\frac{1}{4}$ and Erdős & Sárközy have lowered the upper bound on $\log f(x)$ to $\frac{1}{2}(\log x + \log\ln x) + O(1)$.

Erdős originally asked for the maximum number, $k(x)$, of integers in $[0, x]$ so that no one divides the sum of any others. Such **nondividing sets** are obviously nonaveraging, so $k(x) \le f(x)$. Straus showed that $k(x) \ge \max\{f(x/f(x)), f(\sqrt{x})\}$.

Abbott has shown that if $l(n)$ is the largest m such that *every* set of n integers contains a nonaveraging subset of size m, then $l(n) > n^{1/13-\epsilon}$.

Compare **C14–16** with **E10–14**.

H. L. Abbott, On a conjecture of Erdős and Straus on non-averaging sets of integers, *Congr. Numer. XV, Proc. 5th Brit. Combin. Conf., Aberdeen*, 1975, 1–4.

H. L. Abbott, Extremal problems on non-averaging and non-dividing sets, *Pacific J. Math.*, **91**(1980) 1–12.

H. L. Abbott, On the Erdős–Straus non-averaging set problem, *Acta Math. Hungar.* **47**(1986) 117–119.

Á. P. Bosznay, On the lower estimation of non-averaging sets, *Acta Math. Hungar.*, **53**(1989) 155–157; *MR* **90d**:11016.

P. Erdős & A. Sárközy, On a problem of Straus, *Disorder in physical systems*, Oxford Univ. Press, New York, 1990, pp. 55–66; *MR* **91i**:11012.

P. Erdős & E. G. Straus, Non-averaging sets II, in *Combinatorial Theory and its Applications* II, *Colloq. Math. Soc. János Bolyai* **4**, North-Holland, 1970, 405–411; *MR* **47** #4804.

E. G. Straus, Non-averaging sets, *Proc. Symp. Pure Math.*, **19**, Amer. Math. Soc., Providence 1971, 215–222.

C17 The minimum overlap problem.

Let $\{a_i\}$ be an arbitrary set of n distinct integers, $1 \le a_i \le 2n$, and $\{b_j\}$ be the complementary set $1 \le b_j \le 2n$, with $b_j \ne a_i$. M_k is the number of solutions of $a_i - b_j = k$ $(-2n < k < 2n)$ and $M = \min\max_k M_k$, where the minimum is taken over all sequences $\{a_i\}$. Erdős proved that $M > n/4$; Scherk improved this to $M > (1 - 2^{-1/2})n$ and Swierczkowski to $M > (4 - \sqrt{6})n/5$. Leo Moser obtained the further improvements $M > \sqrt{2}(n-1)/4$ and $M > \sqrt{4 - \sqrt{15}}(n-1)$. In the other direction, Motzkin, Ralston & Selfridge obtained examples to show that $M < 2n/5$, contrary to Erdős's conjecture that $M = \frac{1}{2}n$. Is there a number c such that $M \sim cn$?

Leo Moser asks the corresponding question where the cardinality of $\{a_i\}$ is not n, but k, where $k = \lfloor \alpha n \rfloor$ for some real α, $0 < \alpha < 1$.

A closely related problem is attributed to J. Czipszer: Let $A_k = \{a_1+k, a_2+k, \ldots, a_n+k\}$ where $a_1 < a_2 < \ldots < a_n$ are arbitrary integers and $k \geq 0$. Let M_k be the number of elements of A_k not in $A-0$ and $M = \min_{A_0} \max_{0<k\leq n} M_k$. Czipszer proved that $n/2 \leq M \leq 2n/3$ and conjectured that $M = 2n/3$. Katz & Schnitzer showed that $M > 0.6n$ for $n \geq 26$. Moser & Murdeshwar considered the continuous analog.

P. Erdős, Some remarks on number theory (Hebrew, English summary), *Riveon Lematematika*, **9**(1955) 45–48; *MR* **17**, 460.

M. Katz & F. Schnitzer, On a problem of J. Czipszer, *Rend. Sem. Mat. Univ. Padova* **44**(1970) 85–90; *MR* **45** #8540.

L. Moser, On the minimum overlap problem of Erdős, *Acta Arith.*, **5**(1959) 117–119; *MR* **21** #5594.

L. Moser & M. G. Murdeshwar, On the overlap of a function with its translates, *Nieuw Arch. Wisk.*(3), **14**(1966) 15–18; *MR* **33** #4218.

T. S. Motzkin, K. E. Ralston & J. L. Selfridge, Minimum overlappings under translation, *Bull. Amer. Math. Soc.*, **62**(1956) 558.

S. Swierczkowski, On the intersection of a linear set with the translation of its complement, *Colloq. Math.*, **5**(1958) 185-197; *MR* **21** #1955.

C18 The n queens problem.

What is the minimum number of Queens which can be placed on an $n \times n$ chessboard so that every square is either occupied or attacked by a Queen? Gerge noted that, in graph theory language, this is the same as finding the minimum externally stable set for a graph on 64 vertices with two vertices joined just if they are on the same rank, file or diagonal. In his notation $\beta = 5$ for Queens (Figure 7a), 8 for Bishops (Figure 7b) and 12 for Knights (Figure 7c).

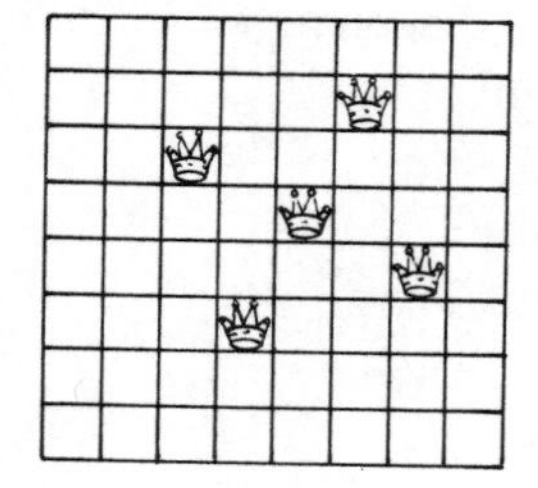
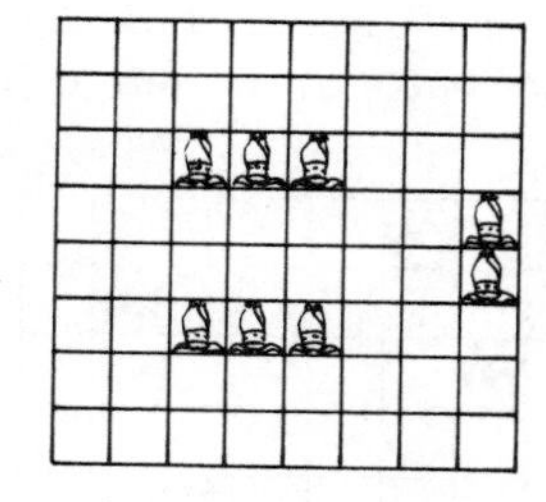
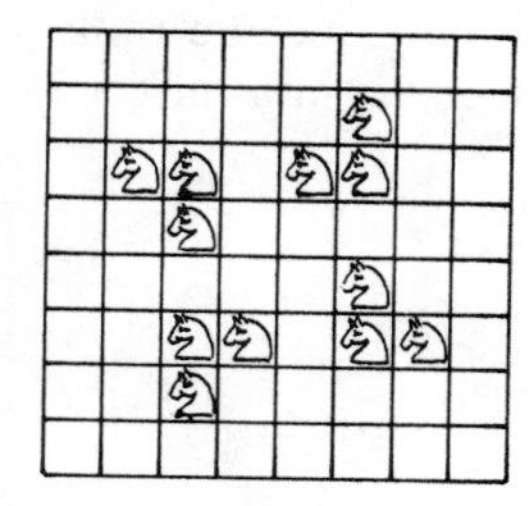

Figure 7. Minimum Covers of the Chessboard by Queens, Bishops and Knights.

Although there is no condition that a piece may not guard another piece, this condition is satisfied by the Queens and Bishops, but not by the Knights. Since in Chess a piece does not attack the square that it stands on, there are in fact two sets of problems. For example, Victor Meally notes that, if queens are allowed to guard one another, 3 queens (at a6, c2, e4) suffice on a 6×6 board, and 4 on a 7×7 board.

For the Queens, Kraitchik gave the following table for an $n \times n$ board.

n	5	6	7	8	9	10	11	12	13	14	15	16	17
number of Queens	3	4	5	5	5	5	5	6	7	8	9	9	9

Corresponding configurations for $n = 5$, 6, 11 are shown in Figure 8.

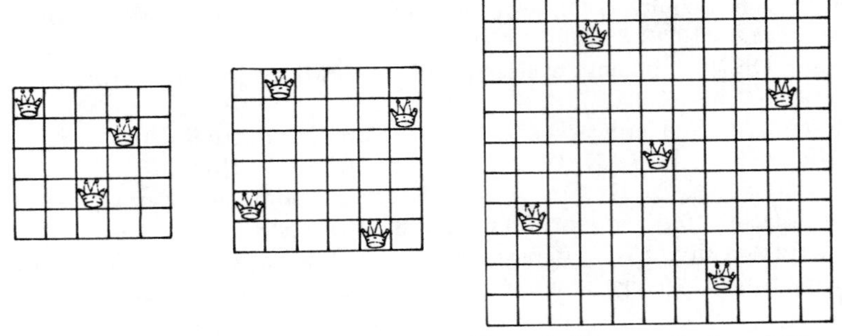

Figure 8. Queens Covering $n \times n$ Boards for $n = 5$, 6 and 11.

If we try to partition the numbers from 1 to $2n$ into n pairs a_i, b_i so that the $2n$ numbers $a_i \pm b_i$ fall one in each residue class modulo $2n$, then it is found to be impossible. Less restrictively, Shen & Shen asked that the $2n$ numbers $a_i \pm b_i$ be distinct. They gave examples for $n = 3$: 1, 5; 2, 3; 4, 6; for $n = 6$: 1, 10; 2, 6; 3, 9; 4, 11; 5, 8; 7, 12; and $n = 8$: 1, 10; 2, 14; 3, 16; 4, 11; 5, 9; 6, 12; 7, 15; 8, 13; and Selfridge showed that there was always a solution for $n \geq 3$. How many solutions are there for each n?

If the condition $b_i = i$ $(1 \leq i \leq n)$ is added, we have the reflecting Queens problem: place n queens on an $n \times n$ chessboard so that no two are on the same rank, file or diagonal, where, on a diagonal, we include reflexions in a mirror in the centre of the zero column (Fig. 9).

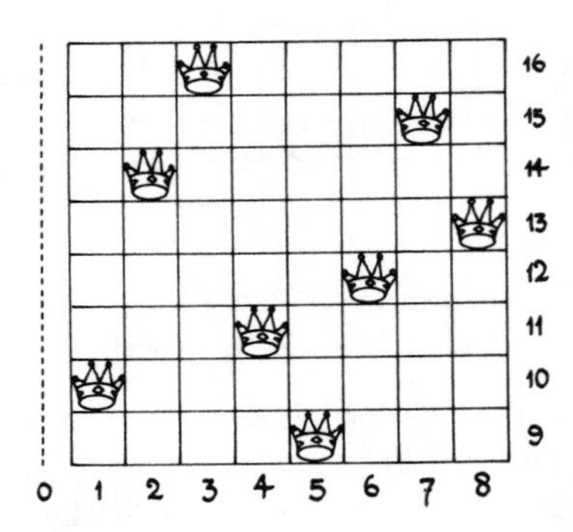

Figure 9. A Solution of the Reflecting Queens Problem.

We can again ask for the number of solutions for each n, both in the case where we distinguish between solutions obtained by rotation and reflexion and where we do not.

Claude Berge, *Theory of Graphs*, Methuen, 1962, p. 41.

Paul Berman, Problem 122, *Pi Mu Epsilon J.*, **3**(1959-64) 118, 412.

A. Bruen & R. Dixon, The n-queens problem, *Discrete Math.*, **12**(1975) 393–395.

E. J. Cockayne & S. T. Hedetniemi, A note on the diagonal queens domination problem, DM-301-IR, Univ. of Victoria, BC, March 1984.

B. Gamble, B. Shepherd & E. T. Cockayne, Domination of chessboards by queens on a column, DM-318-IR, Univ. of Victoria, BC, July 1984.

Solomon W. Golomb & Herbert Taylor, Constructions and properties of Costas arrays, Dept. Elec. Eng., Univ. S. California, July 1981–Oct. 1983.

B. Hansche & W. Vucenic, On the n-queens problem, *Notices Amer. Math. Soc.*, **20**(1973) A-568.

G. B. Huff, On pairings of the first $2n$ natural numbers, *ActaArith.*, **23**(1973) 117–126.

D. A. Klarner, The problem of the reflecting queens, *Amer. Math. Monthly*, **74**(1967) 953–955; *MR* **40** #7123.

Maurice Kraitchik, *Mathematical Recreations*, Norton, New York, 1942, 247–256.

J. D. Sebastian, Some computer solutions to the reflecting queens problem, *Amer. Math. Monthly*, **76**(1969) 399–400; *MR* **39** #4018.

J. L. Selfridge, Pairings of the first $2n$ integers so that sums and differences are all distinct, *Notices Amer. Math. Soc.*, **10**(1963) 195.

Shen Mok-Kong & Shen Tsen-Pao, Research Problem 39, *Bull. Amer. Math. Soc.*, **68** (1962) 557.

B. Shepherd, B. Gamble & E. T. Cockayne, Domination parameters for the bishops graph, DM-327-IR, Univ. of Victoria, BC, Oct. 1884.

M. Slater, Problem 1, *Bull. Amer. Math. Coc.*, **69**(1963) 333.

P. H. Spencer & E. J. Cockayne, An upper bound for the domination number of the queens graph, DM-376-IR, Univ. of Victoria, BC, June 1985.

Ilan Vardi, Computational Recreations in *Mathematica*©, Addison-Wesley, Redwood City CA, 1991, Chap. 6.

C19 Is a weakly independent sequence the finite union of strongly independent ones?

Selfridge calls a set of positive integers $a_1 < a_2 < \cdots < a_k$ **independent** if $\sum c_i a_i = 0$ (where the c_i are integers, not all 0) implies that at least one of the c_i is < -1. By using the pigeonhole principle it is easy to show that if k positive integers are independent, then a_1 is at least 2^{k-1}. He offers \$10.00 for an answer to the question: is the set of k independent integers $a_i = 2^k - 2^{k-i}$ $(1 \le i \le k)$ the only set with largest member less than 2^k? It is the only such set with $a_1 = 2^{k-1}$.

Call a(n infinite) sequence $\{a_i\}$ of positive integers **weakly independent** if any relation $\sum \epsilon_i a_i$ with $\epsilon_i = 0$ or ± 1 and $\epsilon_i = 0$ except finitely often, implies $\epsilon_i = 0$ for all i, and call it **strongly independent** if the same is true with $\epsilon_i = 0$, ± 1, or ± 2. Richard Hall asks if every weakly independent sequence is a finite union of strongly independent sequences.

J. L. Selfridge, Problem 123, *Pi Mu Epsilon J.*, **3**(1959-64) 118, 413–414.

C20 Sums of squares.

Paul Turán asks for a characterization of those positive integers n which can be expressed as the sum of four pairwise coprime squares, i.e., $n = x_1^2 + x_2^2 + x_3^2 + x_4^2$ with $x_i \perp x_j$ $(1 \leq i < j \leq 4)$. Leech notes that at most one of the x_i is even, so that $n \equiv 3$, 4 or 7 mod 8. Similarly, numbers $n \equiv 2 \bmod 3$ are not so representable.

Turán also conjectured that all positive integers can be represented as the sum of at most five pairwise coprime squares, but Mąkowski (see reference at **B5**) notes that numbers $4^k(24l+15)$ with $k \geq 2$ can't be represented in this way. There are arbitrarily large integers that are not representable as the sum of *exactly* five coprime squares, since $3n = x_1^2 + \ldots + x_5^2$ implies that 3 divides two of the x_i. In fact, the product of two distinct primes of shape $24k + 7$ cannot be the sum of fewer than ten pairwise coprime squares, and Leech asks if this is the record.

Apart from 256 examples, the largest of which is 1167, every number can be expressed as the sum of at most five *composite* numbers.

Chowla conjectures that every positive integer is the sum of at most four elements of the set $\{(p^2 - 1)/24\}$ where p is prime and $p \geq 5$. The smallest number which requires four such summands is 33.

Compare these problems with some earlier results of Wright, who showed, for example, that if $\lambda_1, \ldots, \lambda_4$ were given real numbers with sum 1, then every n with a sufficiently large odd factor is expressible as $n = m_1^2 + \ldots + m_4^2$ with $|m_i^2 - \lambda_i n| = o(n)$. He has similar results for five or more squares, and for three squares (provided, of course, that n is not of the form $4_k(8l + 7)$ in this last case).

Bohman, Fröberg & Riesel showed that there are just 31 numbers which *can't* be expressed as the sum of *distinct* squares, and that all numbers greater than 188 *can* be expressed as the sum of at most five distinct squares. Only two numbers, 124 and 188, require six distinct squares. The results are implicit in Gordon Pall's Theorem 3, and in Sprague's papers, which give the more general result for any powers. If an exact number of squares is required, then Halter-Koch showed that every integer > 412 and not divisible by 8 is a sum of four distinct nonzero squares, and that every odd integer > 157 is so representable. He also showed that every integer > 245 is the sum of five distinct nonzero squares, every integer > 333 is the sum of six such, every integer > 390 the sum of seven such, every integer > 462 the sum of eight such, and so on, up to all > 1036 being the sum of twelve such squares. Bateman, Hildebrand & Purdy have produced a sequel to Halter-Koch's paper.

Denote by s_n the largest integer that is *not* the sum of distinct nth powers of positive integers. Sprague showed that $s_2 = 128$; Graham reported that $s_3 = 12758$ and Lin used his method to obtain $s_4 = 5134240$. Cam Patterson used his sieve, and a result of Richert, to obtain $s_5 = 67898771$.

Štefan Porubský used a result of Cassels to give an affirmative answer to a question of R. E. Dressler: for each positive integer k, is every sufficiently large positive integer the sum of distinct k-th powers of primes?

One can also ask for every number to be expressible as the sum of as few as possible polygonal numbers of various kinds. For example there is Gauß's famous 1796-07-10 diary entry:

$$\text{ΕΥΡΗΚΑ!} \qquad \text{num} = \triangle + \triangle + \triangle,$$

i.e., every number is expressible as the sum of three triangular numbers. For hexagonal numbers $r(2r-1)$, the answer is the same if you allow hexagonal numbers of negative rank, $r(2r+1)$, but if these are excluded, then 11 and 26 require six hexagonal numbers of positive rank to represent them. Is it possible that every sufficiently large number is expressible as the sum of three such numbers? Equivalently, is every sufficiently large number $8n+3$ expressible as the sum of three squares of numbers of shape $4r-1$, with r positive? The corresponding question for pentagonal numbers $\frac{1}{2}r(3r-1)$ is to ask if every sufficiently large number of shape $24n+3$ is expressible as the sum of three squares of numbers of shape $6r-1$?

G. E. Andrews, ΕΥΡΗΚΑ! num $= \triangle + \triangle + \triangle$, *J. Number Theory*, **23**(1986) 285–293.

Paul T. Bateman, Adolf Hildebrand & George B. Purdy, Expressing a positive integer as a sum of a given number of distinct squares of positive integers, *Acta Arith.*, (see Abstract 882-11-136, *Abstracts Amer. Math. Soc.*, **14**(1993) 420).

Jan Bohman, Carl-Erik Fröberg & Hans Riesel, Partitions in squares, *BIT*, **19**(1979) 297–301; *MR* **80k**:10043.

J. W. S. Cassels, On the representation of integers as the sums of distinct summands taken from a fixed set, *Acta Sci. Math. Szeged*, **21**(1960) 111–124.

John A. Ewell, On sums of triangular numbers and sums of squares, *Amer. Math. Monthly*, **99**(1992) 752–757; *MR* **93j**:11021.

R. L. Graham, Complete sequences of polynomial values, *Duke Math. J.*, **31**(1964) 275–285.

Andrew Granville & Zhu Yi-Liang, Representing binomial coefficients as sums of squares, *Amer. Math. Monthly*, **97**(1990) 486–493.

Emil Grosswald & L. E. Mattics, Solutions to problem E 3262, *Amer. Math. Monthly*, **97**(1990) 240–242.

Richard K. Guy, Every number is expressible as the sum of how many polygonal numbers? *Amer. Math. Monthly*, **101**(1994) 169–172.

F. Halter-Koch, Darstellung natürliche Zahlen als Summe von Quadraten, *Acta Arith.*, **42**(1982) 11–20; *MR* **84b**:10025.

Lin Shen, Computer experiments on sequences which form integral bases, in J. Leech (editor) *Computational Problems in Abstract Algebra*, 365–370, Pergamon, 1970.

G. Pall, On sums of squares, *Amer. Math. Monthly*, **40**(1933) 10–18.

Cameron Douglas Patterson, *The Derivation of a High Speed Sieve Device*, PhD thesis, The Univ. of Calgary, March 1992.

V. A. Plaksin, Representation of numbers as a sum of four squares of integers, two of which are prime, *Soviet Math. Dokl.*, **23**(1981) 421–424; transl. of *Dokl. Akad. Nauk SSSR*, **257**(1981) 1064–1066; *MR* **82h**:10027.

Štefan Porubský, Sums of prime powers, *Monatsh. Math.*, **86**(1979) 301–303.

H. E. Richert, Über zerlegungen in paarweise verschiedene zahlen, *Norsk Mat. Tidskrift*, **31**(1949) 120–122.

R. P. Sprague, Über zerlegungen in ungleiche Quadratzahlen, *Math. Z.*, **51** (1949) 289–290; *MR* **10** 283.

R. P. Sprague, Über zerlegungen in n-te Potenzen mit lauter verschiedenen Grundzahlen, *Math. Z.*, **51**(1948) 466–468; *MR* **10** 514.

E. M. Wright, The representation of a number as a sum of five or more squares, *Quart. J. Math. Oxford*, **4**(1933) 37–51, 228–232.

E. M. Wright, The representation of a number as a sum of four 'almost proportional' squares, *Quart. J. Math. Oxford*, **7**(1936) 230–240.

E. M. Wright, Representation of a number as a sum of three or four squares, *Proc. London Math. Soc.*(2), **42**(1937) 481–500.

D. Diophantine Equations

"A subject which can be described briefly by saying that a great part of it is concerned with the discussion of the rational or integer solutions of a polynomial equation $f(x_1, x_2, \ldots, x_n) = 0$, with integer coefficients. It is well known that for many centuries, no other topic has engaged the attention of so many mathematicians, both professional and amateur, or has resulted in so many published papers."

This quotation from the preface of Mordell's book, *Diophantine Equations*, Academic Press, London, 1969, indicates that in this section we shall have to be even more eclectic than elsewhere. If you're interested in the subject, consult Mordell's book, which is a thoroughgoing but readable account of what is known, together with a great number of unsolved problems. There are well-developed theories of rational points on algebraic curves, so we mainly confine ourselves to higher dimensions, for which standard methods have not yet been developed.

D1 Sums of like powers. Euler's conjecture.

"It has seemed to many Geometers that this theorem [Fermat's Last Theorem] may be generalized. Just as there do not exist two cubes whose sum or difference is a cube, it is certain that it is impossible to exhibit three biquadrates whose sum is a biquadrate, but that at least four biquadrates are needed if their sum is to be a biquadrate, although no one has been able up to the present to assign four such biquadrates. In the same manner it would seem to be impossible to exhibit four fifth powers whose sum is a fifth power, and similarly for higher powers."

No advance was made on Euler's statement until 1911 when R. Norrie assigned four such biquadrates:

$$30^4 + 120^4 + 272^4 + 315^4 = 353^4$$

Fifty-five years later Lander & Parkin gave a counter-example to Euler's more general conjecture:

$$27^5 + 84^5 + 110^5 + 133^5 = 144^5$$

It's now well known, since his discovery hit the national newspapers, that Noam Elkies has disproved Euler's conjecture for fourth powers. The infinite family of solutions, of which the first member is

$$2682440^4 + 15365639^4 + 18796760^4 = 20615673^4,$$

comes from an elliptic curve given by $u = -5/8$ in the parametric solution of $x^4 - y^4 = z^4 + t^2$ given by Dem'janenko. The smallest solution,

$$95800^4 + 217519^4 + 414560^4 = 422481^4,$$

corresponding to $u = -9/20$, was subsequently found by Roger Frye. There are infinitely many values of u which give curves of positive rank, and further families of solutions.

The solution of Jan Kubiček, making the sum of three cubes a cube, coincides with that of F. Vieta (see Dickson's *History* [**A17**], Vol. 2, pp. 550–551).

Simcha Brudno asks the following questions. Is there a parametric solution to $a^5 + b^5 + c^5 + d^5 = e^5$? [The above solution is the only one with $e \leq 765$.] Is there a parametric solution to $a^4 + b^4 + c^4 + d^4 = e^4$? Are there counterexamples to Euler's conjecture with powers higher than the fifth? Is there a solution of $a^6 + b^6 + c^6 + d^6 + e^6 = f^6$? Although there are solutions of $a_1^s + \ldots + a_{s-1}^s = b^s$ for $s = 4$ and 5, there is no known solution, even of $a_1^s + \ldots + a_s^s = b^s$, for $n \geq 6$.

Parametric solutions are known for equal sums of equal numbers of like powers,

$$\sum_{i=1}^{m} a_i^s = \sum_{i=1}^{m} b_i^s$$

with $a_i > 0$, $b_i > 0$, for $2 \leq s \leq 4$ and $m = 2$ and for $s = 5$, 6 and $m = 3$. Can a solution be found for $s = 7$ and $m = 4$? For $s = 5$, $m = 2$, it is not known if there is any nontrivial solution of $a^5 + b^5 = c^5 + d^5$. Dick Lehmer once thought that there might be a solution with a sum of about 25 decimal digits, but a search by Blair Kelly III yielded no nontrivial solution with sum $\leq 1.02 \times 10^{26}$.

The "Hardy-Ramanujan number" $1729 = 1^3 + 12^3 = 9^3 + 10^3$ was first found by Bernard Frénicle de Bessy in 1657; Leech, in 1957, found

$$87539319 = 167^3 + 436^3 = 228^3 + 423^3 = 255^3 + 414^3$$

and Rosenstiel, Dardis & Rosenstiel recently found $6963472309248 =$

$$2421^3 + 19083^3 = 5436^3 + 18948^3 = 10200^3 + 18072^3 = 13322^3 + 15530^3.$$

Theorem 412 in Hardy & Wright shows that the number of such sums can be made arbitrarily large, but the least example is not known for five or more equal sums. If negative integers are allowed, then Randall Rathbun supplies the examples:

$$\begin{aligned} 6017193 &= 166^3 + 113^3 = 180^3 + 57^3 = 185^3 - (68)^3 \\ &= 209^3 - (146)^3 = 246^3 - (207)^3 \\ 1412774811 &= 963^3 + 804^3 = 1134^3 - (357)^3 = 1155^3 - (504)^3 \\ &= 1246^3 - (805)^3 = 2115^3 - (2004)^3 = 4746^3 - (4725)^3 \\ 11302198488 &= 1926^3 + 1608^3 = 1939^3 + 1589^3 = 2268^3 - (714)^3 = 2310^3 - (1008)^3 \\ &= 2492^3 - (1610)^3 = 4230^3 - (4008)^3 = 9492^3 - (9450)^3 \end{aligned}$$

Mordell and Mahler proved that the number of solutions of $n = x^3 + y^3$ can be $> c(\ln n)^\alpha$ and Silverman has improved their value of α from $\frac{1}{4}$ to $\frac{1}{3}$ and has shown that, if it is required that the pairs of cubes are to be mutually prime, then there is a constant c such that the number of such solutions is $< c^{r(n)}$, where $r(n)$ is the rank of the elliptic curve $x^3 + y^3 = n$. It is much harder to find these solutions. The largest number of representations so far found is three, by P. Vojta in 1983:

$$15170835645 = 517^3 + 2468^3 = 709^3 + 2456^3 = 1733^3 + 2152^3$$

but present-day insight and computing power may soon beat this. If negative integers are allowed, Randall Rathbun gives the cubefree example

$$16776487 = 220^3 + 183^3 = 255^3 + 58^3 = 256^3 + (-9)^3 = 292^3 + (-201)^3$$

The number 1729 is also the third Carmichael number (see **A13**) and Pomerance has observed that the second Carmichael number, 1105, is expressible as the sum of two squares in more ways than any smaller number. Granville invites the reader to make a corresponding statement about the first Carmichael number, 561.

Euler knew that $635318657 = 133^4 + 134^4 = 59^4 + 158^4$, and Leech showed this to be the smallest example. No one knows of three such equal sums.

A method is known for generating parametric solutions of $a^4 + b^4 = c^4 + d^4$ which will generate all published solutions from the trivial one $(\lambda, 1, \lambda, 1)$; it will only produce solutions of degree $6n+1$. Here, in answer to a question of Brudno, $6n+1$ need not be prime. Although degree 25 does not appear, 49 does. More recently, Ajai Choudhry has found a parametric solution of degree 25.

Swinnerton-Dyer has a second method for generating new solutions from old and can show that the two methods, together with the symmetries, generate all *nonsingular* parametric solutions, i.e., all solutions which correspond to points on curves with no singular points. [A point on a curve with homogeneous equation $F(x, y, z) = 0$ is **singular** just if $\frac{\partial F}{\partial x} = \frac{\partial F}{\partial y} = \frac{\partial F}{\partial z} = 0$ there.] Moreover, the process is constructive in the sense that he can give a finite procedure for finding all nonsingular solutions of given degree. All nonsingular solutions have odd degree, and all sufficiently large odd degrees do occur. Unfortunately, singular solutions do exist. Swinnerton-Dyer has a process for generating them, but has no reason to believe that it gives them all. The problem of describing them all needs completely new ideas. Some of the singular solutions have even degree and he conjectures (and could probably prove) that all sufficiently large even degrees occur in this way.

In the same sense, Andrew Bremner can find "all" parametric solutions of $a^6+b^6+c^6=d^6+e^6+f^6$ which also satisfy the equations

$$\begin{aligned} a^2+ad-d^2 &= f^2+fc-c^2 \\ b^2+be-e^2 &= d^2+da-a^2 \\ c^2+cf-f^2 &= e^2+eb-b^2 \end{aligned}$$

(this is not such a restriction as might at first appear). Many solutions of $a^6+b^6+c^6=d^6+e^6+f^6$ also satisfy $a^2+b^2+c^2=d^2+e^2+f^2$ and all known simultaneous solutions (with appropriately chosen signs) of these two equations also satisfy the previous three equations, e.g., the smallest solution, found by Subba-Rao, $(a,b,c,d,e,f)=(3,19,22,-23,10,-15)$. Is there a counter-example? Peter Montgomery has listed 18 equal sums of three sixth powers where the corresponding sums of squares are not equal. The least is

$$25^6+62^6+138^6=82^6+92^6+135^6.$$

Bremner can also find "all" parametric solutions of $a^5+b^5+c^5=d^5+e^5+f^5$ which also satisfy $a+b+c=d+e+f$ and $a-b=d-e$.

Bob Scher calls a sum $\sum_{i=1}^m a_i^p=0$ (where p is prime) "perfect" if, for any a_i, there is also a unique a_i' such that $a_i+a_i'\equiv 0 \bmod p$. If $a_i\equiv 0 \bmod p$, then $a_i'=a_i$. He shows that if $p=3$ and $m<9$ or if $p=5$ and $m<7$, then every such sum is perfect.

There has been some interest in the simultaneous equations

$$\begin{aligned} a^n+b^n+c^n &= d^n+e^n+f^n \\ a^nb^nc^n &= d^ne^nf^n \end{aligned}$$

For $n=2$ the problem goes back at least to Bini, with partial solutions by Dubouis and Mathieu; a neat general solution has recently been given by John B. Kelly. Stephane Vandemergel has found 62 solutions for $n=3$ and three solutions for $n=4$: (29,66,124;22,93,116), (54,61,196;28,122,189) and (19,217,657;9,511,589). He notes that if $r^n+s^n=u^n+v^n$, then $(ru,su,v^2;rv,sv,u^2)$ is a solution, which shows that there are infinitely many solutions for $n\le 4$. Most of his solutions are not of this form.

There is a considerable history (see Dickson [**A17**], II, Ch. 24) of the

Tarry–Escott problem, and a book by Gloden on (systems of) **multigrade equations**:

$$\sum_{i=1}^{l} n_i^j = \sum_{i=1}^{l} m_i^j \quad (j = 1, \ldots, k)$$

A spectacular example, with $k = 9$, $l = 10$, is due to Letac, with $(n_i; m_i) =$

$$\pm 12, \pm 11881, \pm 20231, \pm 20885, \pm 23738; \ \pm 436, \pm 11857, \pm 20449, \pm 20667, \pm 23750.$$

Smyth has shown that this is a member of an infinite family of independent solutions.

U. Bini, Problem 3424, *L'Intermédiaire des Math.*, **15**(1908) 193.

B. J. Birch & H. P. F. Swinnerton-Dyer, Notes on elliptic curves, II, *J. reine angew. Math.*, **218**(1965) 79–108.

Andrew Bremner, Pythagorean triangles and a quartic surface, *J. reine angew. Math.*, **318**(1980) 120–125.

Andrew Bremner, A geometric approach to equal sums of sixth powers, *Proc. London Math. Soc.*(3), **43**(1981) 544–581; *MR* **83g**:14018.

Andrew Bremner, A geometric approach to equal sums of fifth powers, *J. Number Theory*, **13**(1981) 337–354; *MR* **83g**:14017.

Andrew Bremner & Richard K. Guy, A dozen difficult Diophantine dilemmas, *Amer. Math. Monthly*, **95**(1988) 31–36.

S. Brudno, Some new results on equal sums of like powers, *Math. Comput.*, **23**(1969) 877–880.

S. Brudno, On generating infinitely many solutions of the diophantine equation $A^6 + B^6 + C^6 = D^6 + E^6 + F^6$, *Math. Comput.*, **24**(1970) 453–454.

S. Brudno, Problem 4, *Proc. Number Theory Conf. Univ. of Colorado*, Boulder, 1972, 256–257.

Simcha Brudno, Triples of sixth powers with equal sums, *Math. Comput.*, **30**(1976) 646–648.

S. Brudno & I. Kaplansky, Equal sums of sixth powers, *J. Number Theory*, **6**(1974) 401–403.

Ajai Choudhry, The Diophantine equation $A^4 + B^4 = C^4 + D^4$, *Indian J. Pure Appl. Math.*, **22**(1991) 9–11; *MR* **92c**:11024.

Ajai Choudhry, Symmetric Diophantine systems, *Acta. Arith.*, **59**(1991) 291–307; *MR* **92g**:11030.

Jean-Joël Delorme, On the Diophantine equation $x_1^6 + x_2^6 + x_3^6 = y_1^6 + y_2^6 + y_3^6$, *Math. Comput.*, **59**(1992) 703–715; *MR* **93a**:11023.

V. A. Dem'janenko, L. Euler's conjecture (Russian), *Acta Arith.*, **25**(1973/74) 127–135; *MR* **50** #12912.

Dubouis & Mathieu, Réponse 3424, *L'Intermédiaire des Math.*, **16**(1909) 41–42, 112.

Noam Elkies, On $A^4 + B^4 + C^4 = D^4$, *Math. Comput.*, **51**(1988) 825–835; and see Ivars Peterson, *Science News*, **133**, #5(88-01-30) 70; Barry Cipra, *Science*, **239**(1988) 464; and James Gleick, *New York Times*, 88-04-17.

A. Gérardin, *L'Intermédiare des Math.*, **15**(1908) 182; *Sphinx-Œdipe*, 1906-07 80, 128.

A. Gloden, *Mehrgradige Gleichungen*, Noordhoff, Groningen, 1944.

John B. Kelly, Two equal sums of three squares with equal products, *Amer. Math. Monthly*, **98**(1991) 527–529; *MR* **92j**:11025.

Jan Kubiček, A simple new solution to the diophantine equation $A^3+B^3+C^3=D^3$ (Czech, German summary), *Časopis Pěst. Mat.*, **99**(1974) 177–178.

L. J. Lander, Geometric aspects of diophantine equations involving equal sums of like powers, *Amer. Math. Monthly*, **75**(1968) 1061–1073.

L. J. Lander & T. R. Parkin, Counterexample to Euler's conjecture on sums of like powers, *Bull. Amer. Math. Soc.*, **72**(1966) 1079; *MR* **33** #5554.

L. J. Lander, T. R. Parkin & J. L. Selfridge, A survey of equal sums of like powers, *Math. Comput.*, **21**(1967) 446–459; *MR* **36** #5060.

Leon J. Lander, Equal sums of unlike powers, *Fibonacci Quart.*, **28**(1990) 141–150; *MR* **91e**:11033.

John Leech, Some solutions of Diophantine equations, *Proc. Cambridge Philos. Soc.* **53**(1957) 778–780; *MR* **19** 837f.

R. Norrie, in University of Saint Andrews 500th Anniversary Memorial Volume of Scientific Papers, published by the University of St. Andrews, 1911, 89.

E. Rosenstiel, J. A. Dardis & C. R. Rosenstiel, The four least solutions in distinct positive integers of the Diophantine equation $s=x^3+y^3=z^3+w^3=u^3+v^3=m^3+n^3$, *Bull. Inst. Math. Appl.*, **27**(1991) 155–157; *MR* **92i**:11134.

Joseph H. Silverman, Integer points on curves of genus 1, *J. London Math. Soc.*(2), **28**(1983) 1–7; *MR* **84g**:10033.

Joseph H. Silverman, Taxicabs and sums of two cubes, *Amer. Math. Monthly*, **100**(1993) 331–340; *MR* **93m**:11025.

C. J. Smyth, Ideal 9th-order multigrades and Letac's elliptic curve, *Math. Comput.*, **57**(1991) 817–823.

K. Subba-Rao, On sums of sixth powers, *J. London Math. Soc.*, **9**(1934) 172–173.

Ju. D. Trusov, New series of solutions of the thirty second problem of the fifth book of Diophantus (Russian). *Ivanov. Gos. Ped. Inst. Učen. Zap.*, **44**(1969) 119–121 (and see 122–123); *MR* **47** #4925(-6).

Morgan Ward, Euler's three biquadrate problem, *Proc. Nat. Acad. Sci. U.S.A.*, **31**(1945) 125–127; *MR* **6**, 259.

Morgan Ward, Euler's problem on sums of three fourth powers, *Duke Math. J.*, **15**(1948) 827–837; *MR* **10**, 283.

A. S. Werebrusow, *L'Intermédiare des Math.*, **12**(1905) 268; **25**(1918) 139.

Aurel J. Zajta, Solutions of the diophantine equation $A^4+B^4=C^4+D^4$, *Math. Comput.*, **41**(1983) 635–659.

D2 The Fermat problem.

It seems that Fermat's Last Theorem, that the equation

$$x^p+y^p=z^p$$

is impossible in positive integers for an odd prime p, is at last indeed a theorem. But the proof is long and, at the time of writing, leans on the work of others. Indeed, as I write, a hole is reported to have been found, to repair which may be possible, but "may take one or two years." Ribet had shown that the theorem follows from the Taniyama-Weil conjectures on elliptic curves. For as much of the proof as our margin will hold, see **B19**. It raises philosophical problems, such as: are we running out of 'comprehensible' proofs and must we teach computers to check our results for us?

Kummer proved the theorem for all regular primes, where a prime p is **regular** if it doesn't divide h_p, the number of equivalence classes of ideals in the cyclotomic field $\mathbb{Q}(\zeta_p)$. He also showed that a prime was regular just if it did not divide the numerators of the **Bernoulli numbers** B_2, B_4, ..., B_{p-3}, where

$$B_{2k} = (-1)^{k-1}\frac{2(2k)!}{(2\pi)^{2k}}\zeta(2k)$$

($\zeta(2k) = \sum_{n=1}^{\infty} n^{-2k}$ and see **A17**). Of the 78497 primes less than 10^6, 47627 are regular, agreeing well with their conjectured density, $e^{-1/2}$. However, it has not even been proved that there are infinitely many. On the other hand, Jensen has proved that there are infinitely many irregular primes. The prime $p = 16843$ divides B_{p-3} and Richard McIntosh, as well as Buhler, Crandall, Ernvall & Metsänkylä, observes that this is also true for $p = 2124679$.

Some of J. M. Gandhi's subsidiary problems, mentioned in the first edition, are covered by Wiles's method (if it holds up under scrutiny), which shows the unsolvability of

$$x^p + y^p + \gamma z^p = 0$$

if $p \geq 11$ and γ is a power of one of the primes 3, 5, 7, 11, 13, 17, 19, 23, 29, 53, 59, ...

For what integers c are there integer solutions of $x^4 + y^4 = cz^4$ with $x \perp y$ and$z > 1$? Leech gives a method for finding non-trivial solutions for any $z = a^4 + b^4$: the smallest he has found is $25^4 + 149^4 = 5906 \cdot 17^4$. Bremner & Morton show that 5906 is the least integer that is the sum of two rational fourth powers, but not the sum of two integer fourth powers.

Prove that $x^n + y^n = n!z^n$ has no integer solutions with $n > 2$. Erdős & Obláth showed that $x^p \pm y^p = n!$ has none with $p > 2$, and Erdős states that $x^4 + y^4 = n!$ has no solutions with $x \perp y$. Indeed, even without this last condition, Leech notes that for $n > 3$ there's a prime $\equiv 3 \bmod 4$ in the interval $[n+1, 2n]$ and hence a simple (i.e. not repeated) such prime divisor of $n!$ for $n > 6$, so $n!$ is not even the sum of two squares for $n > 6$ (apart from $n = 0$, 1, 2, the only solution is $6! = 24^2 + 12^2$). [Compare **D25**.]

Granville's paper, in which the reviewer felt "a new powerful wave", had already related the Fermat problem to numerous other conjectures, including the *ABC* conjecture (**B19**) and Erdős's powerful numbers conjecture (**B16**).

Andrew Bremner & Patrick Morton, A new characterization of the integer 5906, *Manuscripta Math.*, **44**(1983) 187–229; *MR* **84i**:10016.

J. P. Buhler, R. E. Crandall & R. W. Sompolski, Irregular primes to one million, *Math. Comput.*, **59**(1992) 717–722; *MR* **93a**:11106.

J. P. Buhler, R. E. Crandall, R. Ernvall & T. Metsänkylä, Irregular primes and cyclotomic invariants to four million, *Math. Comput.*, **61**(1993) 151–153; *MR* **93k**:11014.

M. Chellali, Accélération de calcul de nombres de Bernoulli, *J. Number Theory*, **28**(1988) 347–362.

Don Coppersmith, Fermat's last theorem (case 1) and the Wieferich criterion, *Math. Comput.*, **54**(1990) 895–902; *MR* **90h**:11024.

Harold M. Edwards, *Fermat's Last Theorem, a Genetic Introduction to Algebraic Number Theory*, Springer-Verlag, New York, 1977.

P. Erdős & R. Obláth, Über diophantische Gleichungen der form $n! = x^p \pm y^p$ and $n! \pm m! = x^p$, *Acta Litt. Sci. Szeged*, **8**(1937) 241–255; *Zbl.* **17**.004.

Andrew Granville, Some conjectures related to Fermat's last theorem, *Number Theory (Banff)*, 1988, de Gruyter, 1990, 177–192; *MR* **92k**:11036.

Andrew Granville, Some conjectures in analytic number theory and their connection with Fermat's last theorem, in *Analytic Number Theory (Proc. Conf. Honor Bateman*, 1989), Birkhäuser, 1990, 311–326; *MR* **92a**:11031.

K. Inkeri & A. J. van der Poorten, Some remarks on Fermat's conjecture, *Acta Arith.*, **36**(1980) 107-111.

Wells Johnson, Irregular primes and cyclotomic invariants, *Math. Comput.*, **29** (1975) 113–120; *MR* **51** #12781.

D. H. Lehmer, E. Lehmer & H. S. Vandiver, An application of high-speed computing to Fermat's Last Theorem, *Proc. Nat. Acad. Sci. U.S.A.*, **40**(1954), 25–33.

Paulo Ribenboim, *13 Lectures on Fermat's Last Theorem*, Springer-Verlag, New York, Heidelberg, Berlin, 1979; see *Bull. Amer. Math. Soc.*, **4**(1981) 218–222; *MR* **81f**:10023.

K. Ribet, On modular representations of Gal($^-\mathbf{Q}/\mathbf{Q}$) arising from modular forms, *Invent. Math.*, **100**(1990) 431–476.

J. L. Selfridge, C. A. Nicol & H. S. Vandiver, Proof of Fermat's last theorem for all prime exponents less than 4002, *Proc. Nat. Acad. Sci., U.S.A.* **41**(1955) 970–973; *MR* **17**, 348.

Daniel Shanks & H.C. Williams, Gunderson's function in Fermat's last theorem, *Math. Comput.*, **36**(1981) 291–295.

R. W. Sompolski, The second case of Fermat's last theorem for fixed irregular prime exponents, Ph.D. thesis, Univ. Illinois Chicago, 1991.

Jonathan W. Tanner & Samuel S. Wagstaff, New bound for the first case of Fermat's last theorem, *Math. Comput.*, **53**(1989) 743–750; *MR* **90h**:11028.

Samuel S. Wagstaff, The irregular primes to 125000, *Math. Comput.*, **32**(1978) 583–591; *MR* **58** #10711.

D3 Figurate numbers.

Mordell, on p. 259 of his book, asks if the only integer solutions of

$$6y^2 = (x+1)(x^2-x+6)$$

are given by $x = -1$, 0, 2, 7, 15 and 74? By Theorem 1 of Mordell's Chapter 27 there are only finitely many. The equation arises from

$$y^2 = \binom{x}{0} + \binom{x}{1} + \binom{x}{2} + \binom{x}{3}.$$

Andrew Bremner gave the additional solution $(x,y) = (767, 8672)$ and showed that that is all, but the result was already given by Ljunggren in 1971.

Similarly, Martin Gardner took the figurate numbers: triangle, square, tetrahedron and square pyramid; and equated them in pairs. Of the six resulting problems, he noted that they were all solved except "triangle = square pyramid", which leads to the equation

$$3(2y+1)^2 = 8x^3 + 12x^2 + 4x + 3.$$

The number of solutions is again finite. Are they all given by $x = -1$, 0, 1, 5, 6 and 85? Schinzel sends Avanesov's affirmative answer, rediscovered by Uchiyama.

The "triangle = tetrahedron" problem is a special case of a more general question about equality of binomial coefficients (see **B31**) — the only nontrivial examples of $\binom{n}{2} = \binom{m}{3}$ are $(m,n) =$ (10,16), (22,56) and (36,120). Are there nontrivial examples of $\binom{n}{2} = \binom{m}{4}$ other than (10,21)?

The case "square pyramid = square" is Lucas's problem. Is $x = 24$, $y = 70$ the only nontrivial solution of the diophantine equation

$$y^2 = x(x+1)(2x+1)/6?$$

This was solved affirmatively by Watson, using elliptic functions, and by Ljunggren, using a Pell equation in a quadratic field. Mordell asked if there was an elementary proof, and affirmative answers have been given by Ma, by Xu & Cao, by Anglin and by Pintér.

The same equation in disguise is to ask if (48, 140) is the unique nontrivial solution to the case "square = tetrahedron", since the previous equation may be written

$$(2y)^2 = 2x(2x+1)(2x+2)/6,$$

though, as Peter Montgomery notes, this doesn't eliminate the possibility of an odd square. A more modern treatment is to put $12x = X - 6$, $72y = Y$ and note that $Y^2 = X^3 - 36X$ is curve 576H2 in John Cremona's tables. The point (12,36) (which gives an odd square) serves as a generator. There's an infinity of rational solutions, but the only nontrivial integer solution to the original problem is given by the point $(294, 5040)$.

More general than asking for the sum of the first n squares to be square, we can ask for the sum of any n consecutive squares to be square. If S is the set of n for which this is possible, then it is known that S is infinite, but has density zero, and that if n is a nonsquare member of S, then there are infinitely many solutions for such an n. If $N(x)$ is the number of members of S less than x, then the best that seems to be known is that

$$c\sqrt{x} < N(x) = O\left(\frac{x}{\ln x}\right).$$

The elements of S, $1 < n < 73$, and the corresponding least values of a for which the sum of n squares starting with a is square, are

n	2	11	23	24	26	33	47	49	50	59
a	3	18	7	1	25	7	539	25	7	22

More generally still, one can ask that the sums of the squares of the members of an arbitrary arithmetic progression should be square. K. R. S. Sastry notes that this can occur if the number of terms in the progression is square.

In answer to the question: which triangular numbers are the product of three consecutive integers, Tzanakis & de Weger gave the (only) answers 6, 120, 210, 990, 185136 and 258474216. Unfortunately, Mohanty's elementary proof of the same result is erroneous.

Other examples of elliptic curves were treated by Bremner & Tzanakis who showed that there are just 26 integer points on $y^2 = x^3 - 7x + 10$ and they also examined $y^2 = x^3 - bx + c$ for $(b, c) = (172, 505)$, (172,820) and (112,2320).

There are infinitely many solutions of $\binom{a}{k} - \binom{b}{k} = c^k$ for $k = 3$. Are there any for $k = 4$? And is $(a, b, c) = (18, 12, 6)$ an isolated example for $k = 5$?

H. L. Abbott, P. Erdős & D. Hanson, On the number of times an integer occurs as a binomial coefficient, *Amer. Math. Monthly*, **81**(1974) 256–261.

S. C. Althoen & C. Lacampagne, Tetrahedral numbers as sums of square numbers, *Math. Mag.*, **64**(1991) 104–108.

W. S. Anglin, The square pyramid puzzle, *Amer. Math. Monthly*, **97**(1990) 120–124; *MR* **91e**:11026.

È. T. Avanesov, The Diophantine equation $3y(y+1) = x(x+1)(2x+1)$ (Russian), *Volž. Mat. Sb. Vyp.*, **8**(1971) 3–6; *MR* **46** #8967.

È. T. Avanesov, Solution of a problem on figurate numbers (Russian), *Acta Arith.*, **12**(1966/67) 409–420; *MR* **35** #6619.

E. Barbette, Les sommes de p-ièmes puissances distinctes égales à une p-ième puissance, Liège, 1910, 77–104.

Laurent Beeckmans, Squares expressible as the sum of consecutive squares, *Amer. Math. Monthly*, **100**(1993)

Andrew Bremner, An equation of Mordell, *Math. Comput.*, **29**(1975) 925–928; *MR* **51** #10219.

Andrew Bremner & Nicholas Tzanakis, Integer points on $y^2 = x^3 - 7x + 10$, *Math. Comput.*, **41**(1983) 731–741.

J. E. Cremona, *Algorithms for Modular Elliptic Curves*, Cambridge Univ. Press, 1992. [Information for conductor 702, inadvertently omitted from the tables, is obtainable from the author.]

Ion Cucurezeanu, An elementary solution of Lucas' problem, *J. Number Theory*, **44**(1993) 9-12.

H. E. Dudeney, *Amusements in Mathematics*, Nelson, 1917, 26, 167.

P. Erdős, On a Diophantine equation, *J. London Math. Soc.*, **26**(1951) 176–178; *MR* **12**, 804d.

Raphael Finkelstein, On a Diophantine equation with no non-trivial integral solution, *Amer. Math. Monthly*, **73**(1966) 471–477; *MR* **33** #4004.

Martin Gardner, Mathematical games, On the patterns and the unusual properties of figurate numbers, *Sci. Amer.*, **231** No. 1 (July, 1974) 116–120.

Charles M. Grinstead, On a method of solving a class of Diophantine equations, *Math. Comput.*, **32**(1978) 936–940; *MR* **58** #10724.

Heiko Harborth, Fermat-like binomial equations, in Applications of Fibonacci Numbers, Kluwer, 1988, 1–5.

Moshe Laub, O. P. Lossers & L. E. Mattics, Problem 6552 and solution, *Amer. Math. Monthly*, **97**(1990) 622–625.

D. A. Lind, The quadratic field $\mathcal{Q}(\sqrt{5})$ and a certain Diophantine equation, *Fibonacci Quart.*, **6**(1968) 86–93.

W. Ljunggren, New solution of a problem proposed by E. Lucas, *Norsk Mat. Tidskr.*, **34**(1952) 65–72; *MR* **14**, 353h.

W. Ljunggren, A diophantine problem, *J. London Math. Soc.* (2), **3**(1971) 385–391; *MR* **45** #171.

E. Lucas, Problem 1180, *Nouv. Ann. Math.*(2), **14**(1875) 336.

Ma De-Gang, An elementary proof of the solution to the Diophantine equation $6y^2 = x(x+1)(2x+1)$, *Sichuan Daxue Xuebao*, **4**(1985) 107–116; *MR* **87e**:11039.

S. P. Mohanty, Integer points of $y^2 = x^3 - 4x + 1$, *J. Number Theory*, **30**(1988) 86-93; *MR* **90e**:11041a; but see A. Bremner, **31**(1989) 373; **90e**:11041b.

Ákos Pintér, A new solution of two old diophantine equations, Technical Report No. 92 (1993), Department of Mathematics, Kossuth University, Debrecen, Hungary.

David Singmaster, How often does an integer occur as a binomial coefficient? *Amer. Math. Monthly*, **78**(1971) 385–386.

David Singmaster, Repeated binomial coefficients and Fibonacci numbers, *Fibonacci Quart.*, **13** (1975) 295–298.

Craig A. Tovey, Multiple occurrences of binomial coefficients, *Fibonacci Quart.*, **23**(1985) 356–358.

N. Tzanakis & B. M. M. de Weger, On the practical solution of the Thue equation, *J. Number Theory*, **31**(1989) 99-132; *MR* **90c**:11018.

Saburô Uchiyama, Solution of a Diophantine problem, *Tsukuba J. Math.*, **8**(1984) 131–137; *MR* **86i**:11010.

G. N. Watson, The problem of the square pyramid, *Messenger of Math.*, **48** (1918/19) 1–22.

Z. Y. Xu & Cao Zhen-Fu, On a problem of Mordell, *Kexue Tongbao*, **30**(1985) 558–559.

D4 Sums of l kth powers.

Let $r_{k,l}(n)$ be the number of solutions of $n = \sum_{i=1}^{l} x_i^k$ in *positive* integers x_i. Hardy & Littlewood's Hypothesis K is that $\epsilon > 0$ implies that $r_{k,k}(n) = O(n^\epsilon)$. This is well-known for $k = 2$; in fact, for sufficiently large n,

$$r_{2,2}(n) < n(1+\epsilon)\ln 2/\ln\ln n$$

and this does not hold if $\ln 2$ is replaced by anything smaller. Mahler disproved the hypothesis for $k = 3$ by showing that $r_{3,3} > c_1 n^{1/12}$ for infinitely many n.

Erdős thinks it possible that for all n, $r_{3,3} < c_2 n^{1/12}$ but nothing is known. Probably Hypothesis K fails for every $k \geq 3$, but also it's probable that $\sum_{n=1}^{x}(r_{k,k}(n))^2 < x^{1+\epsilon}$ for sufficiently large x.

S. Chowla proved that for $k \geq 5$, $r_{k,k}(n) \neq O(1)$ and, with Erdős, that for every $k \geq 2$ and for infinitely many n,

$$r_{k,k} > \exp(c_k \ln n/\ln\ln n).$$

Mordell proved that $r_{3,2}(n) \neq O(1)$ and Mahler that $r_{3,2}(n) > (\ln n)^{1/4}$ for infinitely many n. No nontrivial upper bound for $r_{3,2}(n)$ is known. Jean Lagrange has shown that $\limsup r_{4,2}(n) \geq 2$ and that $\limsup r_{4,3}(n) = \infty$.

Another tough problem is to estimate $A_{k,l}(x)$, the number of $n \leq x$ which are expressible as the sum of l k-th powers. Landau showed that

$$A_{2,2}(x) = (c + o(1))x/(\ln x)^{1/2},$$

Erdős & Mahler proved that if $k > 2$, then $A_{k,2} > c_k x^{2/k}$, and Hooley that $A_{k,2} > (c_k + o(1))x^{2/k}$. It seems certain that if $l < k$, then $A_{k,l} > c_{k,l}x^{l/k}$ and that $A_{k,k} > x^{1-\epsilon}$ for every ϵ, but these have not been established.

It follows from the Chowla-Erdős result that for all k there is an n_k such that the number of solutions of $n_k = p^3 + q^3 + r^3$ is greater than k. No corresponding result is known for more than three summands.

S. Chowla, The number of representations of a large number as a sum of non-negative nth powers, *Indian Phys.-Math. J.*, **6**(1935) 65–68; *Zbl.* **12**.339.

H. Davenport, Sums of three positive cubes, *J. London Math. Soc.*, **25**(1950) 339–343; *MR* **12**, 393.

P. Erdős, On the representation of an integer as the sum of k kth powers, *J. London Math. Soc.*, **11**(1936) 133–136; *Zbl.* **13**.390.

P. Erdős, On the sum and difference of squares of primes, I, II, *J. London Math. Soc.*, **12**(1937) 133–136, 168–171; *Zbl.* **16**.201, **17**.103.

P. Erdős & K. Mahler, On the number of integers that can be represented by a binary form, *J. London Math. Soc.*, **13**(1938) 134–139; *Zbl.* **13**.390.

P. Erdős & E. Szemerédi, On the number of solutions of $m = \sum_{i=1}^{k} x_i^k$, *Proc. Symp. Pure Math.*, **24** Amer. Math. Soc., Providence, 1972, 83–90.

W. Gorzkowski, On the equation $x_0^2 + x_1^2 + \ldots + x_n^2 = x_{n+1}^k$, *Ann. Soc. Math. Polon. Ser. I Comment. Math. Prace Mat.*, **10**(1966) 75–79; *MR* **32** #7495.

G. H. Hardy & J. E. Littlewood, Partitio Numerorum VI: Further researches in Waring's problem, *Math. Z.*, **23**(1925) 1–37.

Jean Lagrange, Thèse d'État de l'Université de Reims, 1976.

K. Mahler, On the lattice points on curves of genus 1, *Proc. London Math. Soc.*, **39**(1935) 431–466.

K. Mahler, Note on Hypothesis K of Hardy and Littlewood, *J. London Math. Soc.*, **11**(1936) 136–138.

D5 Sum of four cubes.

Is every number the sum of four cubes? This has been proved for all numbers except possibly those of the form $9n \pm 4$.

More demanding is to ask if every number is the sum of four cubes with two of them equal. Specifically, is there a solution of $76 = x^3+y^3+2z^3$? The other numbers less than 1000 which may still be in doubt are 148, 183, 230, 356, 418, 428, 445, 482, 491, 580, 671, 788, 931 and 967. A 70-01-20 letter from M. Lal to A. Mąkowski reports J. C. Littlejohn's observations that $253 = 0^3+5^3+2{\cdot}4^3$, $519 = 0^3+17^3+2(-13)^3$ and $734 = (-520)^3+(-700)^3+2{\cdot}623^3$. Andrew Bremner tells me that $923 = 27512^3+(-27517)^3+2{\cdot}1784^3$.

Are all numbers which are not of the form $9n \pm 4$ the sum of *three* cubes? Computer searches have found representations for all numbers less than 1000, except for

30	33	42	52	74	75	84	110	114
156	165	195	290	318	366	390	420	435
444	452	462	478	501	530	534	564	579
588	600	606	609	618	627	633	732	735
758	767	786	789	795	830	834	861	894
903	906	912	921	933	948	964	969	975

In a 93-05-25 email message Andrew Bremner tells me that

$$75 = 435203083^3 + (-435203231)^3 + 4381159^3$$

(which gives an imprimitive solution for 600) while Conn & Vaserstein have discovered that

$$84 = 41639611^3 + (-41531726)^3 + (-8241191)^3.$$

The equation $3 = x^3 + y^3 + z^3$ has the solutions $(1, 1, 1)$ and $(4, 4, -5)$. Are there any others?

J. W. S. Cassels, A note on the Diophantine equation $x^3 + y^3 + z^3 = 3$, *Math. Comput.*, **44**(1985) 265–266; *MR* **86d**:11021.

W. Conn & L. N. Vaserstein, On sums of three integral cubes, *Contemporary Math.*, (to appear).

S. W. Dolan, On expressing numbers as the sum of two cubes, *Math. Gaz.*, **66**(1982) 31–38.

W. J. Ellison, Waring's problem, *Amer. Math. Monthly*, **78**(1971) 10–36.

V. L. Gardiner, R. B. Lazarus & P. R. Stein, Solutions of the diophantine equation $x^3 + y^3 = z^3 - d$, *Math. Comput.*, **18**(1964) 408–413; *MR* **31** #119.

D. R. Heath-Brown, W. M. Lioen & H. J. J. te Riele, On solving the Diophantine equation $x^3 + y^3 + z^3 = k$ on a vector computer, *Math. Comput.*, **61**(1993) 235–244.

Chao Ko, Decompositions into four cubes, *J. London Math. Soc.*, **11**(1936) 218–219.

Kenyi Koyama, On the solutions of the Diophantine equation $x^3 + y^3 + z^3 = n$ (preprint).

M. Lal, W. Russell & W. J. Blundon, A note on sums of four cubes, *Math. Comput.*, **23**(1969) 423–424; *MR* **39** #6819.

A. Mąkowski, Sur quelques problèmes concernant les sommes de quatre cubes, *Acta Arith.*, **5**(1959) 121–123; *MR* **21** #5609.

J. C. P. Miller & M. F. C. Woollett, Solutions of the diophantine equation $x^3 + y^3 + z^3 = k$, *J. London Math. Soc.*, **30**(1955) 101–110; *MR* **16**, 797e.

W. Scarowsky & A. Boyarsky, A note on the Diophantine equation $x^n + y^n + z^n = 3$, *Math. Comput.*, **42**(1984) 235–237; *MR* **85c**:11029.

A. Schinzel & W. Sierpiński, Sur les sommes de quatre cubes, *Acta Arith.*, **4**(1958) 20–30.

Sun Qi, On Diophantine equation $x^3 + y^3 + z^3 = n$, *Kexue Tongbao*, **33**(1988) 2007–2010; *MR* **90g**:11033 (see also **32**(1987) 1285–1287).

D6 An elementary solution of $x^2 = 2y^4 - 1$.

Ljunggren has shown that the only solutions of $y^2 = 2x^4 - 1$ in positive integers are (1,1) and (239,13) but his proof is difficult. Mordell asks if it is possible to find a simple or elementary proof. Whether Steiner & Tzanakis have simplified the solution may be a matter of taste; they use the theory of linear forms in logarithms of algebraic numbers.

Ljunggren and others have made considerable investigations into equations of similar type. For references see the first edition. Cohn has considered the equation $y^2 = Dx^4 + 1$ for all $D \leq 400$. This will have rational solutions just when there are rational points on the curves $y^2 = x(x^2 - 4D)$ and $Y^2 = X(X^2 + 16D)$, which are nonsingular provided $\pm D$ is not square.

J. H. E. Cohn, The diophantine equation $y^2 = Dx^4 + 1$, I, *J. London Math. Soc.*, **42**(1967) 475–476; *MR* **35** #4158; II, *Acta Arith.*, **28**(1975/76) 273–275; *MR* **52** #8029; III, *Math. Scand.*, **42**(1978) 180–188; *MR* **80a**:10031.
W. Ljunggren, Some remarks on the diophantine equations $x^2 - \mathcal{D}y^4 = 1$ and $x^4 - \mathcal{D}y^2 = 1$, *J. London Math. Soc.*, **41**(1966) 542–544; *MR* **33** #5555.
L. J. Mordell, The diophantine equation $y^2 = Dx^4 + 1$, *J. London Math. Soc.*, **39**(1964) 161–164; *MR* **29** #65.
Ray Steiner & Nikos Tzanakis, Simplifying the solution of Ljunggren's equation $X^2 + 1 = 2Y^4$, *J. Number Theory*, **37**(1991) 123–132; *MR* **91m**:11018.

D7 Sum of consecutive powers made a power.

Rufus Bowen conjectured that the equation

$$1^n + 2^n + \cdots + m^n = (m+1)^n$$

has no nontrivial solutions, and Leo Moser showed that there were none with $m \le 10^{1000000}$ and none with n odd. Zhou & Kang raised the bound to $m \le 10^{2000000}$. Van de Lune & te Riele showed that the equation is almost never solvable. Note that $n \sim m \ln 2$.

Tijdeman observes that general results on the equation

$$1^n + 2^n + \cdots + k^n = m^n$$

do not appear to have any implications for the special equation.

Erdős proposed the problem to prove that if m, n are integers satisfying (K), then (L′) and (M′) are true, where

$$\text{(K)} \qquad \left(1 - \tfrac{1}{m}\right)^n > \tfrac{1}{2} > \left(1 - \tfrac{1}{m-1}\right)^n,$$
$$\text{(L}'\text{)} \qquad 1^n + 2^n + \cdots + (m-2)^n < (m-1)^n,$$
$$\text{(M}'\text{)} \qquad 1^n + 2^n + \cdots + m^n > (m+1)^n,$$

and that (L) and (M) are each true infinitely often, where

$$\text{(L)} \qquad 1^n + 2^n + \cdots + (m-1)^n < m^n,$$
$$\text{(M)} \qquad 1^n + 2^n + \cdots + (m-1)^n > m^n.$$

Van de Lune proved that (K) implies (L′) and Best & te Riele proved that (K) implies (M′) and that (M) holds for at most $c \ln x$ values of $m \le x$. Van de Lune & te Riele proved that (L) is true for almost all pairs (m, n). Best & te Riele computed 33 pairs (m, n) for which (K) and (M) both hold, the smallest being

$$m = 1121626023352385, \qquad n = 777451915729368.$$

Are there infinitely many such pairs?

More recently Pieter Moree, te Riele & Urbanowicz have shown that, in the original equation, n must be divisible by the l.c.m. of all integers up to 200, and that m is not divisible by any regular prime (see **D2**), nor by any irregular prime < 1000.

M. R. Best & H. J. J. te Riele, On a conjecture of Erdős concerning sums of powers of integers, *Report NW 23/76*, Mathematisch Centrum Amsterdam, 1976.

P. Erdős, Advanced problem 4347, *Amer. Math. Monthly*, **56**(1949) 343.

K. Győry, R. Tijdeman & M. Voorhoeve, On the equation $1^k+2^k+\cdots+x^k=y^z$, *Acta Arith.*, **37**(1980) 233–240.

J. van de Lune, On a conjecture of Erdős (I), *Report ZW 54/75*, Mathematisch Centrum Amsterdam, 1975.

J. van de Lune & H. J. J. te Riele, On a conjecture of Erdős (II), *Report ZW 56/75*, Mathematisch Centrum Amsterdam, 1975.

J. van de Lune & H. J. J. te Riele, A note on the solvability of the diophantine equation $1^n+2^n+\cdots+m^n=G(m+1)^n$, *Report ZW 59/75*, Mathematisch Centrum Amsterdam, 1975.

P. Moree, H. J. J. te Riele & J. Urbanowicz, Divisibility properties of integers x and k satisfying $1^k+2^k+\cdots+(x-1)^k=x^k$, *Math. Comput.*, **62**(1994).

L. Moser, On the diophantine equation $1^n+2^n+\cdots+(m-1)^n=m^n$, *Scripta Math.*, **19**(1953) 84–88; *MR* **14**, 950.

J. J. Schäffer, The equation $1^p+2^p+\cdots+n^p=m^q$, *Acta Math.*, **95**(1956) 155–189; *MR* **17**, 1187.

Jerzy Urbanowicz, Remarks on the equation $1^k+2^k+\cdots+(x-1)^k=x^k$, *Nederl. Akad. Wetensch. Indag. Math.*, **50**(1988) 343–348; *MR* **90b**:11026.

M. Voorhoeve, K. Győry & R. Tijdeman, On the diophantine equation $1^k+2^k+\cdots+x^k+R(x)=y^z$, *Acta Math.*, **143**(1979) 1–8; *MR* **80e**:10020.

Zhou Guo-Fu & Kang Ji-Ding, On the diophantine equation $\sum_{k=1}^{m} k^n=(m+1)^n$, *J. Math. Res. Exposition*, **3**(1983) 47–48; *MR* **85m**:11020.

D8 A pyramidal diophantine equation.

Wunderlich asks for (a parametric representation of) *all* solutions of the equation $x^3+y^3+z^3=x+y+z$. Bernstein, S. Chowla, Edgar, Fraenkel, Oppenheim, Segal, and Sierpiński have given solutions, some of them parametric, so there are certainly infinitely many. Eighty-eight of them have unknowns less than 13000. Bremner has effectively determined all parametric solutions.

Leon Bernstein, Explicit solutions of pyramidal Diophantine equations, *Canad. Math. Bull.*, **15**(1972) 177–184; *MR* **46** #3442.

Andrew Bremner, Integer points on a special cubic surface, *Duke Math. J.*, **44**(1977) 757-765; *MR* **58** #27745.

Hugh Maxwell Edgar, Some remarks on the Diophantine equation $x^3+y^3+z^3=x+y+z$, *Proc. Amer. Math. Soc.*, **16**(1965) 148–153; *MR* **30** #1094.

A. S. Fraenkel, Diophantine equations involving generalized triangular and tetrahedral numbers, in *Computers in Number Theory, Proc. Atlas Symp. No. 2,* Oxford, 1969, Academic Press, 1971, 99–114; *MR* **48** #231.
A. Oppenheim, On the Diophantine equation $x^3+y^3+z^3=x+y+z$, *Proc. Amer. Math. Soc.*, **17**(1966) 493–496; *MR* **32** #5590.
A. Oppenheim, On the diophantine equation $x^3+y^3+z^3=px+py-qz$, *Univ. Beograd Publ. Elektrotehn. Fac. Ser.* #235(1968); *MR* **39** #126.
S. L. Segal, A note on pyramidal numbers, *Amer. Math. Monthly*, **69**(1962) 637–638; *Zbl.* **105**, 36.
W. Sierpiński, Sur une propriété des nombres tétraédraux, *Elem. Math.*, **17**(1962) 29–30; *MR* **24** #A3118.
W. Sierpiński, Trois nombres tétraédraux en progression arithmetique, *Elem. Math.*, **18**(1963) 54–55; *MR* **26** #4957.
M. Wunderlich, Certain properties of pyramidal and figurate numbers, *Math. Comput.*, **16**(1962) 482–486; *MR* **26** #6115.

D9 Difference of two powers.

Except that there remains a finite amount of computation, Tijdeman has settled the old conjecture of Catalan, that the only consecutive powers, higher than the first, are 2^3 and 3^2. But this finite amount of computation is far beyond computer range and will not be achieved without some additional theoretical ideas. Langevin has deduced from Tijdeman's proof that if n, $n+1$ are powers, then $n < \exp\exp\exp\exp 730$, and Aaltonen & Inkeri have shown that $x^p - y^q = 1$ and $x,\ y > 2$ imply that $x,\ y > 10^{500}$. Mignotte has shown that $p < 1.21 \times 10^{26}$, $q < 1.31 \times 10^{18}$. If p and q are prime and $q \equiv 3 \bmod 4$, these bounds can be reduced to 2.7×10^{24} and 1.23×10^{18}. There is no solution if $\min\{p,q\} = 2$ or 3. Glass & others have shown this to be true for $\min\{p,q\} \in \{5,7,11\}$. Mignotte shows that there are no solutions for $p = 19$ and that the only possibility for $p = 53$ is $q = 4889$.

Bennett has shown that $4 \le N \le k \cdot 3^k$ implies that

$$\left\| \left(\frac{N+1}{N} \right)^k \right\| > 3^{-k}$$

where $\|x\|$ is the distance from x to the nearest integer.

Leech asks if there are any solutions of $|a^m - b^n| < |a-b|$ with $m,\ n \ge 3$. With equality he notes $|5^3 - 2^7| = 5-2$ and $|13^3 - 3^7| = 13-3$. Are these all? are the shared exponents 3, 7 significant?

If $a_1 = 4$, $a_2 = 8$, $a_3 = 9$, ... is the sequence of powers higher than the first, Chudnowsky claims to have proved that $a_{n+1} - a_n$ tends to infinity with n. Erdős conjectures that $a_{n+1} - a_n > c'n^c$, but says that there is no present hope of proof.

Erdős asks if there are infinitely many numbers not of the form $x^k - y^l$, $k > 1$, $l > 1$.

Carl Rudnick denotes by $N(r)$ the number of positive solutions of $x^4 - y^4 = r$, and asks if $N(r)$ is bounded. Hansraj Gupta observes that Hardy & Wright (p. 201) give Swinnerton-Dyer's version of Euler's parametric solution of $x^4 - y^4 = u^4 - v^4$, which establishes that $N(r)$ is 0, 1 or 2 infinitely often. For example $133^4 - 59^4 = 158^4 - 134^4 = 300783360$. For an example with $N(r) = 3$, Zajta gives

$$401168^4 - 17228^4 = 415137^4 - 248289^4 = 421296^4 - 273588^4.$$

There can hardly be any doubt that $N(r)$ is bounded.

Hugh Edgar asks how many solutions (m, n) does $p^m - q^n = 2^h$ have, for primes p and q and h an integer? Examples are $3^2 - 2^3 = 2^0$; $3^3 - 5^2 = 2^1$; $5^3 - 11^2 = 2^2$; $5^2 - 3^2 = 2^4$; $3^4 - 7^2 = 2^5$; are there others? Andrzej Schinzel writes that work of Gelfond and of Rumsey & Posner implies that the equation has only finitely many solutions. Reese Scott goes a fair way towards settling the question. He observes that the finiteness of the number of solutions for given $(p, q, c(= 2^h))$ follows from a result of Pillai, and proves that this number is often at most one, with a small number of specifically listed exceptional cases, where it is two or possibly three.

M. Aaltonen & K. Inkeri, Catalan's equation $x^p - y^q = 1$ and related congruences, *Math. Comput.*, **56**(1991) 359–370; *MR* **91g**:11025.

David M. Battany, Advanced Problem 6110*, *Amer. Math. Monthly*, **83** (1976) 661.

M. Bennett, Fractional parts of powers of rational numbers, *Math. Proc. Cambridge Philos. Soc.*, **114**(1993) 191–201.

Cao Zhen-Fu, Hugh Edgar's problem on exponential Diophantine equations (Chinese), *J. Math. (Wuhan)*, **9**(1989) 173–178; *MR* **90i**:11035.

Cao Zhen-Fu & Wang Du-Zheng, On the Diophantine equation $a^x - b^y = (2p)^z$ (Chinese). *Yangzhou Shiyuan Xuebao Ziran Kexue Ban*, **1987** no. 4 25–30; *MR* **90c**:11020.

P. S. Dyer, A solution of $A^4 + B^4 = C^4 + D^4$, *J. London Math. Soc.*, **18**(1943) 2–4; *MR* **5**, 89e.

A. O. Gelfond, Sur la divisibilité de la différence des puissances de deux nombres entiers par une puissance d'un idéal premier, *Mat. Sbornik*, **7**(1940) 724.

A. M. W. Glass, D. B. Meronk, T. Okada & R. P. Steiner, A small contribution to Catalan's equation, 1992 (to appear).

Aaron Herschfeld, The equation $2^x - 3^y = d$, *Bull. Amer. Math. Soc.*, **42** (1936) 231–234; *Zbl.* **14** 8a.

K. Inkeri, On Catalan's conjecture, *J. Number Theory*, **34** (1990) 142–152; *MR* **91e**:11030.

Michel Langevin, Quelques applications de nouveaux résultats de Van der Poorten, *Sém. Delange-Pisot-Poitou*, (1975/76). *Théorie des nombres: Fasc.* 2, *Exp. No.* G12, Paris, 1977; *MR* **58** #16550.

Le Mao-Hua, A note on the diophantine equation $ax^m - by^n = k$, *Indag. Math. (N.S.)*, **3**(1992) 185–191; *MR* **93c**:11016.

Maurice Mignotte, Sur l'équation de Catalan, *C.R. Acad. Sci. Paris Sér. I Math.*, **314**(1992) 165–168; *MR* **93e**:11044.

M. Mignotte, Sur l'équation de Catalan, II, *Theor. Comput. Sci.*, **123**(1994) 145–149.

Trygve Nagell, Sur une classe d'équations exponentielles, *Ark. Mat.*, **3**(1958) 569–582; *MR* **21** #2621.

S. Sivasankaranarayana Pillai, On the inequality "$0 < a^x - b^y \le n$", *J. Indian Math. Soc.*, **19**(1931) 1–11; Zbl. **1** 268b.

S. S. Pillai, On $a^x - b^y = c$, *ibid.* (N.S.) **2**(1936) 119–122; Zbl. **14** 392e.

S. S. Pillai, A correction to the paper "On $A^x - B^y = C$", *ibid.* (N.S.) **2**(1937) 215; Zbl. **16** 348b.

Howard Rumsey & Edward C. Posner, On a class of exponential equations, *Proc. Amer. Math. Soc.*, **15**(1964) 974–978.

Reese Scott, On the equations $p^x - b^y = c$ and $a^x + b^y = c^z$, *J. Number Theory*, **44**(1993) 153–165.

R. Tijdeman, On the equation of Catalan, *Acta Arith.*, **29**(1976) 197–209; *MR* **53** #7941.

Aurel J. Zajta, Solutions of the Diophantine equation $A^4 + B^4 = C^4 + D^4$, *Math. Comput.*, **41**(1983) 635–659, esp. p. 652; *MR* **85d**:11025.

D10 Exponential diophantine equations.

Brenner & Foster pose the following general problem. Let $\{p_i\}$ be a finite set of primes and $\epsilon = \pm 1$. When can the exponential diophantine equation $\sum \epsilon_i p_i^{x_i} = 0$ be solved by elementary methods (e.g., by modular arithmetic)? More exactly, given p_i,ϵ_i, what criteria determine whether there exists a modulus M such that the given equation is equivalent to the congruence $\sum \epsilon_i p_i^{x_i} \equiv 0 \bmod M$? They solve many particular cases, mostly where the p_i are four in number and less than 108. In a few cases elementary methods avail, even if two of the primes are equal, but in general they do not. In fact, neither $3^a = 1 + 2^b + 2^c$ nor $2^a + 3^b = 2^c + 3^d$ can be reduced to a single congruence. Tijdeman notes that another approach to these purely exponential diophantine equations (which play a role in group theory) is by Baker's method (compare **F23**). This makes it possible to solve these last two equations.

Hugh Edgar asks if there is a solution, other than $1+3+3^2+3^3+3^4 = 11^2$, of the equation $1 + q + q^2 + \ldots + q^{x-1} = p^y$ with p, q odd primes and $x \ge 5$, $y \ge 2$. An important breakthrough in this area is the paper of Reese Scott (see **D9**).

Leo J. Alex, Problem E2880, *Amer. Math. Monthly*, **88**(1981) 291.

Leo J. Alex, Problem 6411, *Amer. Math. Monthly*, **89**(1982) 788.

Leo J. Alex, The diophantine equation $3^a + 5^b = 7^c + 11^d$, *Math. Student*, **48**(1980) 134–138 (1984); *MR* **86e**:11022.

Leo J. Alex & Lorraine L. Foster, Exponential Diophantine equations, in *Théorie des nombres, Québec 1987*, de Gruyter, Berlin – New York (1989) 1–6; *MR* **90d**:11030.

J. L. Brenner & Lorraine L. Foster, Exponential Diophantine equations, *Pacific J. Math.*, **101**(1982) 263–301; *MR* **83k**:10035.

Lorraine L. Foster, Solution to Problem S31 [1980, 403], *Amer. Math. Monthly*, **89**(1982) 62.

Guo Zhi-Tang & Wu Yun-Fei, On the equation $1 + q + q^2 + \cdots + q^{n-1} = p^m$ (Chinese), *J. Harbin Inst. Tech.*, **24**(1992) 6–9; *MR* **93k**:11024.

Le Mao-Hua, A note on the Diophantine equation $(x^m - 1)/(x - 1) = y^n$, *Acta Arith.*, **64**(1993) 19–28.

A. Mąkowski & A. Schinzel, Sur l'équation indéterminée de R. Goormachtigh, *Mathesis*, **68**(1959) 128–142; *MR* **22** #9472.

Mo De-Ze & R. Tijdeman, Exponential Diophantine equations with four terms, *Indag. Math.(N.S.)*, **3**(1992) 47–57; *MR* **93d**:11035.

A. Rotkiewicz & W. Złotkowski, On the Diophantine equation $1 + p^{\alpha_1} + p^{\alpha_2} + \ldots + p^{\alpha_k} = y^2$, in Number Theory, Vol. II (Budapest, 1987), North-Holland, *Colloq. Math. Soc. János Bolyai*, **51**(1990) 917–937; *MR* **91e**:11032.

T. N. Shorey, Integers with identical digits, *Acta Arith.*, **53** (1989) 187–205; *MR* **90j**:11027.

Christopher M. Skinner, On the Diophantine equation $ap^x + bq^v = c + dp^z q^w$, *J. Number Theory*, **35**(1990) 194–207; *MR* **91h**:11021.

Robert Styer, Small two-variable exponential Diophantine equations, *Math. Comput.*, **60**(1993) 811–816.

B. M. M. de Weger, Solving exponential Diophantine equations using lattice basis reduction algorithms, *J. Number Theory*, **26**(1987) 325–367; erratum **31**(1989) 88-89; *MR* **88k**:11097 **90a**:11040.

D11 Egyptian fractions.

The Rhind papyrus is amongst the oldest written mathematics that has come down to us; it concerns the representation of rational numbers as the sum of unit fractions,

$$\frac{m}{n} = \frac{1}{x_1} + \frac{1}{x_2} + \ldots + \frac{1}{x_k}$$

This has suggested numerous problems, many of which are unsolved, and continues to suggest new problems, so the interest in Egyptian fractions is as great as it has ever been. Our bibliography shows only a fraction of what has been written. Bleicher has given a careful survey of the subject and draws attention to the various algorithms that have been proposed for constructing representations of the given type: the Fibonacci-Sylvester algorithm, Erdős's algorithm, Golomb's algorithm and two of his own, the Farey series algorithm and the continued fraction algorithm. See also the extensive Section 4 of the collection of problems by Erdős & Graham mentioned at the beginning of this volume, and the bibliography obtainable from Paul Campbell.

Erdős & Straus conjectured that the equation

$$\frac{4}{n}=\frac{1}{x}+\frac{1}{y}+\frac{1}{z}$$

could be solved in positive integers for all $n>1$. There is a good account of the problem in Mordell's book, where it is shown that the conjecture is true, except possibly in cases where n is congruent to 1^2, 11^2, 13^2, 17^2, 19^2 or 23^2 mod 840. Several have worked on the problem, including Bernstein, Obláth, Rosati, Shapiro, Straus, Yamamoto, and Nicola Franceschine who has verified the conjecture for $n<10^8$. Schinzel has observed that one can express

$$\frac{4}{at+b}=\frac{1}{x(t)}+\frac{1}{y(t)}+\frac{1}{z(t)}$$

with $x(t)$, $y(t)$, $z(t)$ integer polynomials in t with positive leading coefficients and $a\perp b$, only if b is *not* a quadratic residue of a.

Sierpiński made the corresponding conjecture concerning

$$\frac{5}{n}=\frac{1}{x}+\frac{1}{y}+\frac{1}{z}.$$

Palamà confirmed it for all $n\le 922321$ and Stewart has extended this to $n\le 1057438801$ and for all n not of the form $278460k+1$.

Schinzel relaxed the condition that the integers x, y, z should be positive, replaced the numerators 4 and 5 by a general m and required the truth only for $n>n_m$. That n_m may be greater than m is exemplified by $n_{18}=23$. The conjecture has been established for successively larger values of m by Schinzel, Sierpiński, Sedláček, Palamà and Stewart & Webb, who prove it for $m<36$. Breusch and Stewart independently showed that if $m/n>0$ and n is odd, then m/n is the sum of a finite number of reciprocals of odd integers. See also Graham's papers. Vaughan has shown that if $E_m(N)$ is the number of $n\le N$ for which $m/n=1/x+1/y+1/z$ has no solution, then

$$E_m(N)\ll N\exp\{-c(\ln N)^{2/3}\}$$

where c depends only on m. Hofmeister & Stoll have shown that if $F_m(N)$ is the number of $n\le N$ for which $m/n=1/x+1/y$ has no solution, then

$$F_m(N)\ll N(\ln N)^{-1/\phi(m)}$$

where $\phi(m)$ is Euler's totient function (**B36**).

Hofmeister notes that this implies that $A_m(N)/N \to 1$ as $N \to \infty$ where $A_m(N)$ is the number of b, $1 \le b \le N$ for which there is a representation $m/b = 1/n_1 + 1/n_2$, so that almost *all* lattice points on the *line* $y = m$, $x \ge 1$ have such a representation. Paradoxically, Mittelbach lets $B(N)$ be the number of lattice points (a, b), $1 \le a \le b \le N$ for which there is a representation $a/b = 1/n_1 + 1/n_2$ and proves that $B(N)/\frac{1}{2}N(N+1) \to 0$. I.e., almost *no* lattice points in the *triangle* (1,1), (1,N), (N,N) have a representation for large N.

In contrast to the result of Breusch and Stewart, the following problem, asked by Stein, Selfridge, Graham and others, has not been solved. If m/n, a rational number (n odd), is expressed as $\sum 1/x_i$, where the x_i are successively chosen to be the least possible odd integers which leave a nonnegative remainder, is the sum always finite? For example,

$$\frac{2}{7} = \frac{1}{5} + \frac{1}{13} + \frac{1}{115} + \frac{1}{10465}$$

John Leech, in a 77-03-14 letter, asks what is known about sets of unequal odd integers whose reciprocals add to 1, such as

$$\frac{1}{3} + \frac{1}{5} + \frac{1}{7} + \frac{1}{9} + \frac{1}{15} + \frac{1}{21} + \frac{1}{27} + \frac{1}{35} + \frac{1}{63} + \frac{1}{105} + \frac{1}{135} = 1$$

He says that you need at least nine in the set, while on the other hand the largest denominator must be at least 105. Notice the connexion with Sierpiński's pseudoperfect numbers (**B2**).

$$945 = 315 + 189 + 135 + 105 + 63 + 45 + 35 + 27 + 15 + 9 + 7$$

It is known that if n is odd, then m/n is always expressible as a sum of distinct odd unit fractions.

Tenenbaum & Yokota show that m/n can be expressed as the sum of r unit fractions with denominators $\le 4n(\ln n)^2 \log_2 n$ where $r \le (1+\epsilon)$ $\ln n/\log_2 n$ but that $1+\epsilon$ cannot be replaced by $1-\epsilon$.

Victor Meally ordered the rationals a/b, $a \perp b$, between 0 and 1 by size of $a+b$ and of a: $\frac{1}{2}$, $\frac{1}{3}$, $\frac{1}{4}$, $\frac{2}{3}$, $\frac{1}{5}$, $\frac{1}{6}$, $\frac{2}{5}$, $\frac{3}{4}$, $\ldots$ and noted that $\frac{2}{3}$, $\frac{4}{5}$ and $\frac{8}{11}$ are the earliest members of the sequence that need 2, 3 and 4 unitary fractions to represent them. Which are the earliest that need 5? 6? 7? Stephane Vandemergel, in a 93-04-28 letter, states that $\frac{16}{17}$ requires 5 unitary fractions, and $\frac{77}{79}$ needs 6.

Barbeau expressed 1 as the sum of the reciprocals of 101 distinct positive integers, no one dividing another. Erdős & Graham showed that if n is squarefree, then m/n can always be written as a finite sum of reciprocals of squarefree integers each having exactly ω distinct prime factors, for $\omega \ge 3$. There are many cases in which ω can be taken as 2. For $m = n = 1$ at least 38 integers are required: Allan Johnson manages it with $\omega = 2$ and the 48 numbers

6	21	34	46	58	77	87	115	155	215	287	391
10	22	35	51	62	82	91	119	187	221	299	689
14	26	38	55	65	85	93	123	203	247	319	731
15	33	39	57	69	86	95	133	209	265	323	901

Is this the smallest possible set? Richard Stong also solved this problem, but used a larger set.

Erdős, in a 72-01-14 letter, sets $\frac{1}{2}+\frac{1}{3}+\ldots+\frac{1}{n}=\frac{a}{b}$, where $b=[2,3,\ldots,n]$, the l.c.m. of $2,3,\ldots,n$. He observes that $\frac{1}{2}+\frac{1}{3}=\frac{5}{6}$ and $\frac{1}{2}+\frac{1}{3}+\frac{1}{4}=\frac{13}{12}$ are such that $a\pm 1\equiv 0 \bmod b$ and asks if this occurs again: he conjectures not. Is $a\perp b$ infinitely often?

If $\sum_{i=1}^{t} 1/x_i = 1$ with $x_1<x_2<x_3<\ldots$ *distinct* positive integers Erdős & Graham ask what is $m(t)$, the min max x_i, where the minimum is taken over all sets $\{x_i\}$. For example, $m(3)=6$, $m(4)=12$, $m(12)=120$. Is $m(t)<ct$ for some constant c? In this notation, is it possible to have $x_{i+1}-x_i\le 2$ for all i? Erdős conjectures that it is not and offers \$10.00 for a solution.

Erdős & Graham ask if it is true that any coloring of the integers with c colors gives a monochromatic solution of

$$\sum\frac{1}{x_i}=1,\quad x_1<x_2<\ldots\quad \text{(finite sum)}.$$

This is open even for $c=2$. If the answer is affirmative, let $f(c)$ be the smallest integer for which every c-coloring of the integers $1\le t\le f(c)$ contains a monochromatic solution. Determine or estimate $f(c)$.

Erdős also asks that if

$$\frac{1}{x_1}+\frac{1}{x_2}+\ldots+\frac{1}{x_k}=1,\quad x_1<x_2<\ldots<x_k$$

and k is fixed, what is $\max x_1$? If k varies, what integers can be equal to x_k, the largest denominator? Not primes, and not several other integers; do the excluded integers have positive density? Density 1 even? Which integers can be x_k or x_{k-1} ? Which can be x_k or x_{k-1} or x_{k-2} ? Is $\liminf\frac{x_k}{x_1}>e$? It is trivial that the limit is $\ge e$. In fact perhaps it is infinite. If $m(k)$ is $\max(x_k)$ for each k, then Yokota improves a result of Erdős & Graham by proving that there is an increasing sequence of integers k for which $m(k)/k\le(\ln\ln k)^3$. Is there a sequence of k such that $m(k)/k$ is bounded?

Erdős further asks if it is true that for every solution of

$$\frac{1}{x_1}+\frac{1}{x_2}+\ldots+\frac{1}{x_k}=1,$$

$\max(x_{i+1}-x_i)\ge 3$? {2,3,6} shows that >3 is not true but perhaps this is the only counterexample. Perhaps $\max(x_{i+1}-x_i)\le c$ has only a finite number of solutions. If the x_i are the union of r blocks of consecutive integers, the number of solutions is finite and depends only on r. That the sequence $\frac{1}{2}+\frac{1}{3}+\frac{1}{7}+\frac{1}{43}+\ldots$ gives the $\max x_k$ as a function of r was proved by Curtiss.

If $N(n)$ is the set of integers that can be written as a sum of distinct reciprocals of integers $\leq n$, then Yokota shows that every natural number up to

$$\frac{\ln n}{2}\left(1-\frac{2\ln\ln n}{\ln n}\right)$$

is in $N(n)$, so that $\#N(n) \geq (\frac{1}{2}+o(1))\ln n$. Erdős asks for an estimate for the size of the smallest integer not in $N(n)$, and of the largest in $N(n)$. How many integers $< \sum_{i=1}^{n} 1/i$ cannot be so expressed? Of the ones that can, what is their distribution? Do they come in bunches?

Given a sequence x_1, x_2, ... of positive density, is there always a finite subset with $\sum 1/x_{i_k} = 1$? If $x_i < ci$ for all i, is there such a finite subset? Erdős again offers \$10.00 for a solution. If $\liminf x_i/i < \infty$, he strongly conjectures that the answer is negative and offers only \$5.00 for a solution.

Denote by $N(t)$ the number of solutions $\{x_1, x_2, \ldots, x_t\}$ of $1 = \sum 1/x_i$ and by $M(t)$ the number of distinct solutions $x_1 \leq x_2 \leq \ldots \leq x_t$. Singmaster calculated

t	1	2	3	4	5	6
$M(t)$	1	1	3	14	147	3462
$N(t)$	1	1	10	215	12231	2025462

and Erdős asked for an asymptotic formula for $M(t)$ or $N(t)$.

Graham has shown that if $n > 77$ we can partition $n = x_1+x_2+\ldots+x_t$ into t distinct positive integers so that $\sum 1/x_i = 1$. More generally, that for any positive rational numbers α, β, there is an integer $r(\alpha,\beta)$, which we will take to be the least, such that any integer greater than r can be partitioned into distinct positive integers greater than β, whose reciprocals sum to α. Little is known about $r(\alpha,\beta)$, except that unpublished work of D. H. Lehmer shows that 77 cannot be partitioned in this way, so that $r(1,1) = 77$.

Graham conjectures that for n sufficiently large (about 10^4 ?) we can similarly partition $n = x_1^2 + x_2^2 + \ldots + x_t^2$ with $\sum 1/x_i = 1$. We can also ask for a decomposition $n = p(x_1) + p(x_2) + \ldots + p(x_t)$ where $p(x)$ is any "reasonable" polynomial; for example $x^2 + x$ is unreasonable since it takes only even values.

In answer to a question of L.-S. Hahn: is there a set of integers, each having an immediate neighbor, the sum of whose reciprocals is an integer, Peter Montgomery gave the examples {1,2,7,8,13,14,39,40,76,77,285,286} and {2,3,4,5,6,7,9,10,17,18,34,35,84,85} whose reciprocals each add to 2.

L.-S. Hahn also asks: if the positive integers are partitioned into a finite number of sets in any way, is there always a set such that *any* positive rational number can be expressed as the sum of the reciprocals of a finite number of distinct members of it? Here it must be possible to choose the set, independent of the rational number. If this is not possible, then given any rational number, can one always choose a set with this property? Now the set can depend on the rational number.

Nagell showed that the sum of the reciprocals of an arithmetic progression is never an integer: see also the paper of Erdős & Niven.

P. J. van Albada & J. H. van Lint, Reciprocal bases for the integers, *Amer. Math. Monthly*, **70**(1963) 170–174.

Michael Anshel & Dorian Goldfeld, Partitions, Egyptian fractions, and free products of finite abelian groups, *Proc. Amer. Math. Soc.*, **111**(1991) 889–899; *MR* **91h**:11104.

E. J. Barbeau, Remarks on an arithmetic derivative, *Canad. Math. Bull.*, **4**(1961) 117–122.

E. J. Barbeau, Computer challenge corner: Problem 477: A brute force program, *J. Recreational Math.*, **9**(1976/77) 30.

E. J. Barbeau, Expressing one as a sum of odd reciprocals: comments and a bibliography, *Crux Mathematicorum* (=*Eureka* (Ottawa)), **3**(1977) 178–181; and see *Math. Mag.*, **49**(1976) 34.

Laurent Beeckmans, The splitting algorithm for Egyptian fractions, *J. Number Theory*, **43**(1993) 173–185.

Leon Bernstein, Zur Lösung der diophantischen Gleichung $m/n = 1/x+1/y+1/z$ insbesondere im Falle $m = 4$, *J. reine angew. Math.*, **211**(1962) 1–10; *MR* **26** #77.

M. N. Bleicher, A new algorithm for the expansion of Egyptian fractions, *J. Number Theory*, **4**(1972) 342–382; *MR* **48** #2052.

M. N. Bleicher and P. Erdős, The number of distinct subsums of $\sum_1^N 1/i$, *Math. Comput.*, **29**(1975) 29–42 (and see *Notices Amer. Math. Soc.*, **20**(1973) A-516).

M. N. Bleicher and P. Erdős, Denominators of Egyptian fractions, *J. Number Theory*, **8**(1976) 157–168; *MR* **53** #7925; II, *Illinois J. Math.*, **20**(1976) 598–613; *MR* **54** #7359.

Robert Breusch, A special case of Egyptian fractions, Solution to Advanced Problem 4512, *Amer. Math. Monthly*, **61**(1954) 200–201.

J. L. Brown, Note on complete sequences of integers, *Amer. Math. Monthly*, **68**(1961) 557–560.

Paul J. Campbell, Bibliography of algorithms for Egyptian fractions (preprint), Beloit Coll., Beloit WI 53511, U.S.A.

Robert Cohen, Egyptian fraction expansions, *Math. Mag.*, **46** (1973) 76–80; *MR* **47** #3300.

D. R. Curtiss, On Kellogg's Diophantine problem, *Amer. Math. Monthly*, **29**(1922) 380–387.

Editor's Note, Odd reciprocals, *Math. Mag.*, **49**(1976) 155–156.

P. Erdős, Egy Kürschák-féle elemi számelméleti tétel áltadanositása, *Mat. es Fys. Lapok* **39**(1932).

P. Erdős, On arithmetical properties of Lambert series, *J. Indian Math. Soc.*, **12**(1948) 63–66.

P. Erdős, On a diophantine equation (Hungarian. Russian and English summaries), *Mat. Lapok*, **1**(1950) 192–210; *MR* **13**, 208.

P. Erdős, On the irrationality of certain series, *Nederl. Akad. Wetensch.* (*Indag. Math.*) **60**(1957) 212–219.

P. Erdős, Sur certaines séries à valeur irrationelle, *Enseignement Math.*, **4**(1958) 93–100.

P. Erdős, *Quelques problèmes de la Théorie des Nombres*, Monographies de l'Enseignement Math. No. 6, Geneva, 1963, problems 72–74.

P. Erdős, Comment on problem E2427, *Amer. Math. Monthly*, **81**(1974) 780–782.

P. Erdős, Some problems and results on the irrationality of the sum of infinite series, *J. Math. Sci.*, **10**(1975) 1–7.

P. Erdős & I. Joó, On the expansion $1 = \sum q^{-n_i}$, *Period. Math. Hungar.*, **23**(1991) 27–30; *MR* **92i**:11030.

P. Erdős & Ivan Niven, Some properties of partial sums of the harmonic series, *Bull. Amer. Math. Soc.*, **52**(1946) 248–251; *MR* **7**, 413.

Paul Erdős & Sherman Stein, Sums of distinct unit fractions, *Proc. Amer. Math. Soc.*, **14**(1963) 126–131.

P. Erdős & E. G. Straus, On the irrationality of certain Ahmes series, *J. Indian Math. Soc.*, **27**(1968) 129–133.

P. Erdős & E. G. Straus, Some number theoretic results, *Pacific J. Math.*, **36**(1971) 635–646.

P. Erdős & E. G. Straus, Solution of problem E2232, *Amer. Math. Monthly*, **78**(1971) 302–303.

P. Erdős & E. G. Straus, On the irrationality of certain series, *Pacific J. Math.*, **55**(1974) 85–92; *MR* **51** #3069.

P. Erdős & E. G. Straus, Solution to problem 387, *Nieuw Arch. Wisk.*, **23**(1975) 183.

Nicola Franceschine, Egyptian Fractions, MA Dissertation, Sonoma State Coll. CA, 1978.

Charles N. Friedman, Sums of divisors and Egyptian fractions, *J. Number Theory*, **44**(1993) 328–339.

S. W. Golomb, An algebraic algorithm for the representation problem of the Ahmes papyrus, *Amer. Math. Monthly*, **69**(1962) 785–786.

S. W. Golomb, On the sums of the reciprocals of the Fermat numbers and related irrationalities, *Canad. J. Math.*, **15**(1963) 475–478.

R. L. Graham, A theorem on partitions, *J. Austral. Math. Soc.*, **3**(1963) 435–441; *MR* **29** #64.

R. L. Graham, On finite sums of unit fractions, *Proc. London Math. Soc.* (3), **14**(1964) 193–207; *MR* **28** #3968.

R. L. Graham, On finite sums of reciprocals of distinct nth powers, *Pacific J. Math.*, **14**(1964) 85–92; *MR* **28** #3004.

H. S. Hahn, Old wine in new bottles: Solution to E2327, *Amer. Math. Monthly*, **79**(1972) 1138 (and see Editor's comment).

Chao Ko, Chi Sun & S. J. Chang, On equations $4/n = 1/x + 1/y + 1/z$, *Acta Sci. Natur. Szechuanensis*, **2**(1964) 21–35.

Li De Lang, On the equation $4/n = 1/x+1/y+1/z$, *J. Number Theory*, **13**(1981) 485–494; *MR* **83e**:10026.

Liang-Shin Hahn, Problem E2689, *Amer. Math. Monthly*, **85**(1978) 47; Solution, Peter Montgomery & Dean Hickerson, **86**(1979) 224.

Liang-Shin Hahn, Problems and solutions (Japanese), *Sûgaku*, **31**(1979) 376.

Liang-Shin Hahn, Egyptian fractions (Chinese), *Mathmedia*, **15**(1980) 8–12.

Gerd Hofmeister & Peter Stoll, Note on Egyptian fractions, *J. reine angew. Math.*, **362**(1985) 141–145; *MR* **87a**:11025.

Allan Wm. Johnson, Letter to the Editor, *Crux Mathematicorum* (= *Eureka* (Ottawa)), **4**(1978) 190.

O. D. Kellogg, On a diophantine problem, *Amer. Math. Monthly*, **28**(1921) 300–303.

F. Mittelbach, Anzahl- und Dichteuntersuchungen bei Stammbruchdarstellungen von Brüchen, Diplomarbeit, Fachbereich Mathematik, Joh. Gutenberg-Univ., Mainz, 1988.

T. Nagell, *Skr. Norske Vid. Akad. Kristiania I*, 1923, no. 13 (1924) 10–15.

D. J. Newman, Problem 76-5: an arithmetic conjecture, *SIAM Rev.*, **18**(1976) 118.

R. Obláth, Sur l'équation diophantienne $4/n = 1/x_1 + 1/x_2 + 1/x_3$, *Mathesis*, **59**(1950) 308–316; *MR* **12**, 481.

J. C. Owings, Another proof of the Egyptian fraction theorem, *Amer. Math. Monthly*, **75**(1968) 777–778.

G. Palamà, Su di una congettura di Sierpiński relativa alla possibilità in numeri naturali della $5/n = 1/x_1 + 1/x_2 + 1/x_3$, *Boll. Un. Mat. Ital.*(3) **13**(1958) 65–72; *MR* **20** #3821.

G. Palamà, Su di una congettura di Schinzel, *Boll. Un. Mat. Ital.*(3) **14**(1959) 82–94; *MR* **22** #7989.

L. A. Rosati, Sull'equazione diofantea $4/n = 1/x_1 + 1/x_2 + 1/x_3$, *Boll. Un. Mat. Ital.*(3), **9**(1954) 59–63; *MR* **15**, 684.

J. W. Sander, On $4/n = 1/x+1/y+1/z$ and Rosser's sieve, *Acta Arith.*, **59**(1991) 183–204; *MR* **92j**:11031.

Andrzej Schinzel, Sur quelques propriétés des nombres $3/n$ et $4/n$, où n est un nombre impair, *Mathesis*, **65**(1956) 219–222; *MR* **18**, 284.

W. Schwarz, *Einführung in Siebmethoden der analytischen Zahlentheorie*, Bibl. Inst., Mannheim-Wien-Zurich, 1974; *MR* **53** #13147.

Jiří Sedláček, Über die Stammbrüche, *Časopis Pěst. Mat.*, **84**(1959) 188–197; *MR* **23** #A829.

W. Sierpiński, Sur les décompositions de nombres rationale en fractions primaires, *Mathesis*, **65**(1956) 16–32; *MR* **17**, 1185.

W. Sierpiński, *On the Decomposition of Rational Numbers into Unit Fractions* (Polish), Pánstwowe Wydawnictwo Naukowe, Warsaw, 1957.

W. Sierpiński, Sur une algorithme pour developper les nombres réels en séries rapidement convergentes, *Bull. Int. Acad. Sci. Cracovie Ser. A Sci. Math.*, **8**(1911) 113–117.

David Singmaster, The number of representations of one as a sum of unit fractions (mimeographed note), 1972.

B. M. Stewart, Sums of distinct divisors, *Amer. J. Math.*, **76**(1954) 779–785; *MR* **16**, 336.

B. M. Stewart & W. A. Webb, Sums of fractions with bounded numerators, *Canad. J. Math.*, **18**(1966) 999–1003; *MR* **33** #7297.

R. J. Stroeker & R. Tijdeman, Diophantine equations, *Computational Methods in Number Theory, Part II*, Math. Centrum Tracts **155**, Amsterdam, 1982; *MR* **84i**:10014.

J. J. Sylvester, On a point in the theory of vulgar fractions, *Amer. J. Math.*, **3**(1880) 332–335, 387–388.

Gérald Tenenbaum & Hisashi Yokota, Length and denominators of Egyptian fractions III, *J. Number Theory*, **35**(1990) 150–156; *MR* **91g**:11028.

R. C. Vaughan, On a problem of Erdős, Straus and Schinzel, *Mathematika*, **17**(1970) 193–198.

H. S. Wilf, Reciprocal bases for the integers, Res. Problem 6, *Bull. Amer. Math. Soc.*, **67**(1961) 456.

Koichi Yamamoto, On the diophantine equation $4/n = 1/x + 1/y + 1/z$, *Mem. Fac. Sci. Kyushū Univ. Ser. A*, **19**(1965) 37–47.

Koichi Yamamoto, On a conjecture of Erdős, *Mem. Fac. Sci. Kyushū Univ. Ser. A*, **18**(1964) 166–167; *MR* **30** #1968.

Hisashi Yokota, On number of integers representable as sums of unit fractions, *Canad. Math. Bull.*, **33**(1990) 235–241; *MR* **909**:11029.

Hisashi Yokota, On a problem of Erdős and Graham, *J. Number Theory*, **39**(1991) 327–338; *MR* **90d**:11104.

D12 Markoff numbers.

A diophantine equation which has excited a great deal of interest is

$$x^2 + y^2 + z^2 = 3xyz.$$

It obviously has what Cassels has called the singular solutions, (1,1,1) and (1,1,2) (with the usual definition, the variety has only the singular solution (0,0,0)). All solutions can be generated from these since the equation is a quadratic in each of the variables, so one integer solution leads to a second, and it can be shown that, apart from the singular solutions, all solutions have distinct values of x, y and z, so that each such solution is a **neighbor** of just three others (Figure 10). The numbers 1, 2, 5, 13, 29, 34, 89, 169, 194, 233, 433, 610, 985, ... are called **Markoff numbers**. To avoid trivialities, assume that $0 < x \le y \le z$ (so that the inequalities are strict if $y \ge 2$). An outstanding problem is whether every Markoff number z defines a unique integer solution (x, y, z). There are occasional claims to have proved that the Markoff numbers are unique in this sense, but so far proofs appear to be fallacious.

If $M(N)$ is the number of triples with $x \le y \le z \le N$, then Zagier has shown that $M(N) = C(\ln N)^2 + O((\ln N)^{1+\epsilon})$ where $C \approx 0.180717105$, and calculations lead him to conjecture that the nth Markoff number, m_n, is

$\left(\frac{1}{3}+O(n^{-1/4+\epsilon})\right) A^{\sqrt{n}}$ where $A{+}e^{1/\sqrt{C}} \approx 10.5101504$. He has no results on distinctness, but can show that the problem is equivalent to the insolvability of a certain system of diophantine equations.

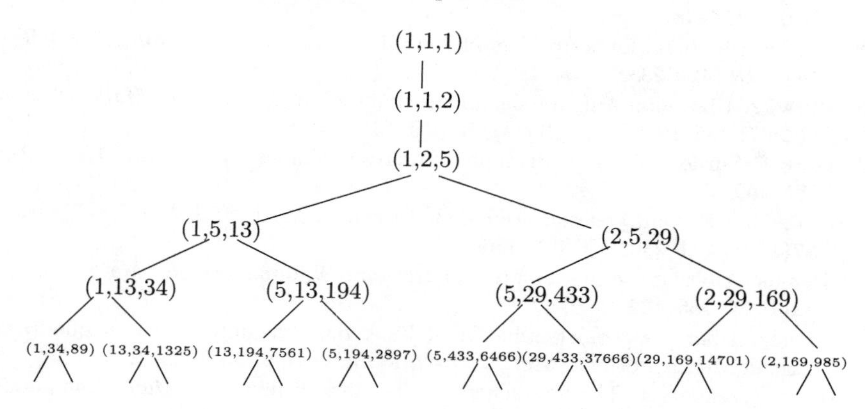

Figure 10. The Tree of Markoff Solutions.

Markoff's equation is a special case of the more general **Hurwitz equation**

$$x_1^2 + x_2^2 + \cdots + x_n^2 = ax_1x_2\cdots x_n$$

for which there are no integer solutions if $a > n$, and for $a = n$ all integer solutions can be generated from (1,1,...,1). For any a, $1 \le a \le n$, there is a finite set of solutions which generates all others. Baragar has shown that, for any g, there are infinitely many pairs (a, n) so that the equation requires at least g generators. Let $M(n, N)$ be the number of solutions of the Hurwitz equation with $a = n$ and each $|x_i| \le N$, then Baragar has also shown that $M(n, N)$ grows like $C(\ln N)^{\alpha(n)+\epsilon}$ for all $\epsilon > 0$ and that $M(n, N) = \Omega((\ln N)^{\alpha(n)-\epsilon}$, where (Zagier) $\alpha(3) = 2$, but $\alpha(4)$ lies between 2.33 and 2.64 (later improved to $2.43 < \alpha(4) < 2.47$). Also that

$$\frac{2\ln n}{\ln 4} \le \alpha(n) \le \frac{3\ln n}{\ln 4}.$$

Arthur Baragar, The Hurwitz equations, Proc. Conf. Markoff Spectrum & Anal. Number Theory, Provo UT, 1991; Lect. Notes Pure Appl. Math., **147** 9–16, Dekker, New York, 1993.

J. W. S. Cassels, An Introduction to Diophantine Approximation, Cambridge, 1957, 27–44.

H. Cohn, Approach to Markoff's minimal forms through modular functions, *Ann. Math. Princeton*(2) **61**(1955) 1–12; *MR* **16**, 801e.

T. W. Cusick, The largest gaps in the lower Markoff spectrum, *Duke Math. J.*, **41**(1974) 453–463; *MR* **57** #5902.

T. W. Cusick, On Perrine's generalized Markoff equations, *Aequationes Math.*, **46**(1993) 203–211.

L. E. Dickson, *Studies in the Theory of Numbers*, Chicago Univ. Press, 1930, Chap. VII.

G. Frobenius, Über die Markoffschen Zahlen, *S.-B. Preuss. Akad. Wiss. Berlin* (1913) 458–487.

Norman P. Herzberg, On a problem of Hurwitz, *Pacific J. Math.*, **50**(1974) 485–493; *MR* **50** #233.

A. Hurwitz, Über eine Aufgabe der unbestimmten Analysis, *Arch. Math. Phys.*, **3**(1907) 185–196; *Math. Werke*, ii, 410–421.

A. Markoff, Sur les formes quadratiques binaires indéfinies, *Math. Ann.*, **15**(1879) 381–409.

Serge Perrine, Sur une généralisation de la théorie de Markoff, *J. Number Theory*, **37**(1991) 211–230; *MR* **92c**:11067.

R. Remak, Über indefinite binäre quadratische Minimalformen, *Math. Ann.*, **92**(1924) 155–182.

R. Remak, Über die geometrische Darstellung der indefiniten binären quadratischen Minimalformen, *Jber. Deutsch Math.-Verein*, **33**(1925) 228–245.

Gerhard Rosenberger, The uniqueness of the Markoff numbers, *Math. Comput.*, **30**(1976) 361–365; but see *MR* **53** #280.

Joseph H. Silverman, The Markoff equation $X^2+Y^2+Z^2 = aXYZ$ over quadratic imaginary fields, *J. Number Theory*, **35**(1990) 72–104; *MR* **91i**:11028.

L. Ja. Vulah, The diophantine equation $p^2 + 2q^2 + 3r^2 = 6pqr$ and the Markoff spectrum (Russian), *Trudy Moskov. Inst. Radiotehn. Èlektron. i Avtomat. Vyp. 67 Mat.*(1973) 105–112, 152; *MR* **58** #21957.

Don B. Zagier, On the number of Markoff numbers below a given bound *Math. Comput.*, **39**(1982) 709–723; *MR* **83k**:10062.

D13 The equation $x^xy^y = z^z$.

Erdős asked for solutions of the equation $x^xy^y = z^z$, apart from the trivial ones $y = 1$, $x = z$. Chao Ko found an infinity of solutions, of which the first three are

x	y	z
12^6	6^8	$2^{11}3^7$
224^{14}	112^{16}	$2^{68}7^{15}$
61440^{30}	30720^{32}	$2^{357}15^{31}$

Did he find them all?

Claude Anderson conjectured that the equation $w^wx^xy^y = z^z$ has no solutions with $1 < w < x < y < z$, but Chao Ko & Sun Qi had earlier found an infinity of counterexamples to a generalization of the conjecture to any number of variables:

$$\begin{aligned} x_1 &= k^{k^n(k^{n+1}-2n-k)+2n}(k^n-1)^{2(k^n-1)} \\ x_2 &= k^{k^n(k^{n+1}-2n-k)}(k^n-1)^{2(k^n-1)+2} \end{aligned}$$

$$\begin{aligned} x_3 &= \cdots = x_k = k^{k^n(k^{n+1}-2n-k)+n}(k^n-1)^{2(k^n-1)+1} \\ z &= k^{k^n(k^{n+1}-2n-k)+n+1} \end{aligned}$$

where, for $k \geq 3$, $n > 0$, and, for $k = 2$, $n > 1$. E.g., $w = 3^{12}2^6$, $x = 3^{13}2^5$, $y = 3^{14}2^4$, $z = 3^{14}2^5$. Ajai Choudhry also found a parametric solution for $k = 3$.

Chao Ko, Note on the diophantine equation $x^xy^y = z^z$, *J. Chinese Math. Soc.*, **2**(1940) 205–207; *MR* **2**, 346.

Chao Ko & Sun Qi, On the equation $\prod_{i=1}^k x_i^{x_i}$, *J. Sichuan Univ.*, **2**(1964) 5–9.

W. H. Mills, An unsolved diophantine equation, *Report Inst. Theory of Numbers*, Boulder CO, 1959, 258–268.

D14 $a_i + b_j$ made squares.

Leo Moser asked for integers a_1, a_2, b_j $(1 \leq j \leq n)$ such that the $2n$ numbers $a_i + b_j$ are all squares. This can be achieved by making $a_2 - a_1$ a sufficiently composite number; for example $a_1 = 0$, $a_2 = 2^{2n+1}$, $b_j = (2^{2n-j} - 2^{j-1})^2$.

John Leech observes that the extension to integers a_1, a_2, a_3, b_j $(1 \leq j \leq n)$ is also solvable for any n. We may take a_1, a_2 to be $(x \pm y)^2$; a_3 can have an arbitrary value $x^2 + \lambda xy + y^2$, which can then be made square by putting $x = u^2 - v^2$, $y = 2uv + \lambda v^2$. Any values of u and v will give triads of squares with differences in this proportion. The problem to find values of u, v so that the scale factor is a rational square reduces to finding rational points on an elliptic curve; arbitrarily many rational values of b_j can then be simultaneously scaled to give integers a_1, a_2, a_3, b_j $(1 \leq j \leq n)$ for any n. A much studied case is $\lambda = 0$, corresponding to sets of rational right-angled triangles of equal area. We can also specialize to fix $a_1 = b_1 = 0$. Provided that a_2, a_3 are squares p^2, q^2 such that q/p has a representation as the product of two *distinct* ratios $(u^2 - v^2)/2uv$, then it is again an elliptic curve problem to find rational squares $b_j = r_j^2$ such that both $p^2 + r_j^2$ and $q^2 + r_j^2$ are squares, and again we can rescale to find integers p, q, r_j $(1 \leq j \leq n)$ for any n such that $p^2 + r_j^2$ and $q^2 + r_j^2$ are integer squares (cf. **D20**). For example, 13/6 has representations $(u_1, v_1, u_2, v_2) = (9, 4, 5, 1)$ and $(8, 5, 9, 1)$ which yield

0^2	351^2	650^2	1728^2	3200^2
720^2	801^2	970^2	1872^2	3280^2
1560^2	1599^2	1690^2	2328^2	3560^2

More generally we seek integers a_i $(1 \leq i \leq m)$, b_j $(1 \leq j \leq n)$. Jean Lagrange has produced the matrix

$$\begin{bmatrix} (54,150,111)^2 & (56,150,79)^2 & (72,234,177)^2 & (72,186,57)^2 \\ (6,78,96)^2 & (16,48,56)^2 & (48,192,168)^2 & (48,120,12)^2 \\ (54,318,384)^2 & (56,312,376)^2 & (72,360,408)^2 & (72,312,372)^2 \\ (6,-50,-96)^2 & (16,0,-56)^2 & (48,176,168)^2 & (48,104,-12)^2 \end{bmatrix}$$

of squares of quadratic forms, $(a,b,c) = au^2 + buv + cv^2$, which yields an infinity of solutions with $m = n = 4$. For example, $u = 2$, $v = 1$ gives

$$\begin{bmatrix} 627^2 & 603^2 & 933^2 & 717^2 \\ 276^2 & 216^2 & 744^2 & 444^2 \\ 1236^2 & 1224^2 & 1416^2 & 1284^2 \\ 172^2 & 8^2 & 712^2 & 388^2 \end{bmatrix}$$

Lagrange, in a letter dated 83-03-13, sends the matrices

$$\begin{bmatrix} 59^2 & 112^2 & 144^2 & 207^2 & 592^2 & 1351^2 & 4077^2 \\ 229^2 & 248^2 & 264^2 & 303^2 & 632^2 & 1369^2 & 4083^2 \\ 499^2 & 508^2 & 516^2 & 537^2 & 772^2 & 1439^2 & 4107^2 \end{bmatrix}$$

and

$$\begin{bmatrix} 18^2 & 234^2 & 346^2 & 514^2 \\ 282^2 & 366^2 & 446^2 & 586^2 \\ 477^2 & 531^2 & 589^2 & 701^2 \\ 1122^2 & 1146^2 & 1174^2 & 1234^2 \end{bmatrix}$$

In these examples, the a_i, b_j are not squares. If the a_i, b_j are themselves squares, then they provide configurations relevant to **D20** (which see) where Lagrange & Leech have made considerable progress. Their triad and tetrad of squares a_i^2 $(i = 1,2,3)$ and b_j^2 $(j = 1,2,3,4)$ with all $a_i^2 + b_j^2$ squares lead to a 4×5 array

$$\begin{bmatrix} 0^2 & 7422030^2 & 8947575^2 & 22276800^2 & 44142336^2 \\ 9282000^2 & 11184530^2 & 12892425^2 & 24132200^2 & 45107664^2 \\ 26822600^2 & 27830530^2 & 28275625^2 & 34867000^2 & 51652664^2 \\ 60386040^2 & 60840450^2 & 61045335^2 & 64364040^2 & 74799864^2 \end{bmatrix}$$

in the present problem.

D15 Numbers whose sums in pairs make squares.

Erdős & Leo Moser (and see earlier references) also asked the analogous question: are there, for every n, n distinct numbers such that the sum of any pair is a square? For $n = 3$ we can take

$$a_1 = \tfrac{1}{2}(q^2 + r^2 - p^2) \quad a_2 = \tfrac{1}{2}(r^2 + p^2 - q^2) \quad a_3 = \tfrac{1}{2}(p^2 + q^2 - r^2)$$

and for $n = 4$ we may augment these by taking s to be any number expressible as the sum of two squares in three distinct ways

$$s = u^2 + p^2 = v^2 + q^2 = w^2 + r^2 \text{ and } a_4 = s - \tfrac{1}{2}(p^2 + q^2 - r^2).$$

Jean Lagrange has given a quite general parametric solution for $n = 5$ and a simplification of it which appears to give a majority of all solutions. He tabulates the first 80 solutions, calculated by J.-L. Nicolas. The smallest is

$$-4878 \quad 4978 \quad 6903 \quad 12978 \quad 31122$$

and the smallest positive solution (at most one number can be negative) is

$$7442 \quad 28658 \quad 148583 \quad 177458 \quad 763442.$$

In a letter dated 72-05-19 he sends the following solution for $n = 6$:

$$-15863902 \quad 17798783 \quad 21126338 \quad 49064546 \quad 82221218 \quad 447422978$$

In fact the problem goes back to T. Baker who found five integers whose sums in pairs were squares, and C. Gill who found five whose sums in threes were squares.

Lagrange also found sets of n squares of which any $n-1$ have their sum square. For $n = 3$, 5 and 8, the smallest such are the squares of (44,117,240), (28,64,259,392,680) and (79,112,204,632,896,916,1828,2092).

Martin LaBar asked for a proof or disproof that a 3×3 magic square can be constructed from nine distinct integer squares. This requires that the nine quantities x^2, y^2, z^2, $y^2 + z^2 - x^2$, $z^2 + x^2 - y^2$, $x^2 + y^2 - z^2$, $2x^2 - y^2$, $2x^2 - z^2$, $3x^2 - y^2 - z^2$ be distinct perfect squares. Not very likely, though it is not difficult to get four of the eight magic sums right.

T. Baker, *The Gentleman's Diary* or *Math. Repository*, London, 1839, 33–35, Question 1385.

C. Gill, *Application of the Angular Analysis to Indeterminate Problems of Degree 2*, New York, 1848, p. 60.

Martin LaBar, Problem 270, *Canad. Math. J.*, **15**(1984) 69.

Jean Lagrange, Cinq nombres dont les sommes deux à deux sont des carrés, *Séminaire Delange-Pisot-Poitou* (*Théorie des nombres*) 12ᵉ *année*, **20**(1970-71) 10pp.

Jean Lagrange, Six entiers dont les sommes deux à deux sont des carrés, *Acta Arith.*, **40**(1981) 91–96.

Jean Lagrange, Sets of n squares of which any $n-1$ have their sum square, *Math. Comput.*, **41**(1983) 675–681.

Jean-Louis Nicolas, 6 nombres dont les sommes deux à deux sont des carrés, *Bull. Soc. Math. France*, Mém. No 49–50 (1977) 141–143; *MR* **58** #482.

A. R. Thatcher, A prize problem, *Math. Gaz.*, **61**(1977) 64.

A. R. Thatcher, Five integers which sum in pairs to squares, *Math. Gaz.*, **62**(1978) 25.

D16 Triples with the same sum and same product.

The problem to find as many different triples of positive integers as possible with the same sum and the same product has been solved by Schinzel: you can have arbitrarily many. In the interim Stephane Vandemergel found 13 triples each with sum 17116 and product $2^{10}3^35^27^211 \cdot 13 \cdot 19$. It may be of interest to ask for the smallest sums or products with each multiplicity. For example, for 4 triples, J. G. Mauldon finds the smallest common sum to be 118: (14,50,54), (15,40,63), (18,30,70), (21,25,72) and the smallest common product to be 25200: (6,56,75), (7,40,90), (9,28,100), (12,20,105).

Lorraine L. Foster & Gabriel Robins, Solution to Problem E2872, *Amer. Math. Monthly*, **89**(1982) 499–500.
J. G. Mauldon, Problem E2872, *Amer. Math. Monthly*, **88**(1981) 148.

D17 Product of blocks of consecutive integers not a power.

Erdős & Selfridge have proved that the product of consecutive integers is never a power, and the binomial coefficient $\binom{n}{k}$ (see **B31**) is never a power for $n \geq 2k \geq 8$. If $k = 2$, then $\binom{n}{k}$ is a square infinitely often, but Tijdeman's methods (see **D9**) will probably show that it is never a nontrivial higher power (for cubes and fourth powers, see Mordell's book), and that $k = 3$ never gives a power, apart from $n = 50$ (see **D3**).

Erdős & Graham ask if the product of two or more disjoint blocks of consecutive integers can be a power. Pomerance has noted that

$$\prod_{i=1}^{4}(a_i - 1)a_i(a_i + 1)$$

is a square if $a_1 = 2^{n-1}$, $a_2 = 2^n$, $a_3 = 2^{2n-1} - 1$, $a_4 = 2^{2n} - 1$, but Erdős & Graham suggest that if $l \geq 4$, then $\prod_{i=1}^{k}\prod_{j=1}^{l}(a_i + j)$ is a square on only a finite number of occasions.

K. R. S. Sastry notes that the product of the blocks $(n-1)n(n+1)$ and $(2n-2)(2n-1)2n$ is a square if $(n+1)(2n-1) = m^2$. This is equivalent to a Pell equation with an infinity of solutions. E.g. $n = 74$ gives

$$(73 \cdot 74 \cdot 75)(146 \cdot 147 \cdot 148) = 73^2 \cdot 74^2 \cdot 210^2$$

Erdős also asks if the product of (more than one) consecutive odd numbers is never a power (higher than the first)? Is the product of 4 consecutive members of an A.P. never a power? Euler showed that it cannot be a square. Fermat had shown that the members cannot be squares individually, while a nonsquare divisor must divide two distinct terms, either (a) 2 divides the first & third or the second & fourth, or (b) 3 divides the first

& fourth, or both. (a) alone is impossible mod 8, (b) alone is impossible mod 3, but we could have $6t^2$, u^2, $2v^2$, $3w^2$. But this implies $w^2 + t^2 = v^2$ and $w^2 + 4t^2 = u^2$, which can be disproved by descent – one Pythagorean ratio can't be twice another. For higher powers Leech notes that we can't have three cubes in A.P.

Sastry asks for which k can the product of four consecutive terms of an A.P. be a k-gonal number, where the r-th **k-gonal number** is

$$\tfrac{1}{2}r((k-2)r-(k-4)).$$

Not for $k = 4$, as Euler showed, but Sastry finds solutions for all other k except 7, 14 and 37. Are these impossible?

P. Erdős, On consecutive integers, *Nieuw Arch. Wisk.*, **3**(1955) 124–128.

P. Erdős & J. L. Selfridge, The product of consecutive integers is never a power, *Illinois J. Math.*, **19**(1975) 292–301.

D18 Is there a perfect cuboid? Four squares whose sums in pairs are square. Four squares whose differences are square.

Is there a rational box? Our treatment of this notorious unsolved problem is owed almost entirely to John Leech. Does there exist a **perfect cuboid**, with integer edges x_i, face diagonals y_i and body diagonal z; are there solutions of the simultaneous diophantine equations

(A) $$x_{i+1}^2 + x_{i+2}^2 = y_i^2,$$

(B) $$\sum x_i^2 = z^2 \qquad ?$$

($i = 1, 2, 3$; and where necessary, subscripts are reduced modulo 3.)

Martin Gardner asked if any six of x_i, y_i, z could be integers. Here there are three problems: just the body diagonal z irrational; just one edge x_3 irrational; just one face diagonal y_1 irrational.

Problem 1. We require solutions to the three equations (A). Suppose such solutions have **generators** a_i, b_i where

$$x_{i+1} : x_{i+2} : y_i = 2a_ib_i : a_i^2 - b_i^2 : a_i^2 + b_i^2$$

Then we want integer solutions of

(C) $$\prod \frac{a_i^2 - b_i^2}{2a_ib_i} = 1.$$

We can assume that the generator pairs have opposite parity and replace (C) by

(D) $$\frac{a_1^2 - b_1^2}{2a_1b_1} \cdot \frac{a_2^2 - b_2^2}{2a_2b_2} = \frac{\alpha^2 - \beta^2}{2\alpha\beta}$$

An example is

$$\text{(E)} \qquad \frac{6^2-5^2}{2\cdot 6\cdot 5}\cdot\frac{11^2-2^2}{2\cdot 11\cdot 2}=\frac{8^2-5^2}{2\cdot 8\cdot 5}$$

Kraitchik gave $241+18-2$ cuboids with odd edge less than a million. Lal & Blundon listed all cuboids obtainable from (D) with a_1, b_1; α, $\beta \le 70$, including the curious pair $(1008, 1100, 1155)$, $(1008, 1100, 12075)$. Leech has deposited a list of all solutions of (D) with two pairs of a_1, b_1; a_2, b_2; α, $\beta \le 376$.

Reversal of the cyclic order of subscripts in (C) leads to the **derived cuboid**: example (E) gives the least solution $(240, 44, 117)$, known to Euler, and the derived cuboid $(429, 2340, 880)$. Note that $240\cdot 429 = 44\cdot 2340 = 117\cdot 880$.

Several parametric solutions are known: the simplest, also known to Euler, is

$$\text{(F)} \qquad \alpha = 2(p^2-q^2), \qquad a_1 = 4pq, \qquad b_1 = \beta = p^2+q^2$$

For a_1, b_1 fixed, (D) is equivalent to the plane cubic curve

$$\frac{a_1^2-b_1^2}{2a_1b_1}=\frac{u^2-1}{2u}\cdot\frac{2v}{v^2-1}$$

whose rational points are finitely generated, so Mordell tells us that one solution leads to an infinity. But not all rationals a_1/b_1 occur in solutions: $a_1/b_1 = 2$ is impossible, so there is no rational cuboid with a pair of edges in the ratio 3:4.

Problem 2. Just an edge irrational. We want $x_1^2+x_2^2=y_3^2$ with $t+x_1^2$, $t+x_2^2$, $t+y_3^2$ all squares. This was proposed by "Mahatma" and readers gave $x_1 = 124$, $x_2 = 957$, $t = 13852800$. Bromhead extended this to a parametric solution. An infinity of solutions is given by

$$\text{(G)} \qquad (x_1, x_2, y_3) = 2\xi\eta\zeta(\xi, \eta, \zeta), \qquad t = \zeta^8 - 6\xi^2\eta^2\zeta^4 + \xi^4\eta^4$$

where (ξ, η, ζ) is a Pythagorean triple. The simplest such is $\xi = 5$, $\eta = 12$; $x_1 = 7800$, $x_2 = 18720$; $t = 211773121$. An earlier solution was given by Flood.

These are not all. We seek solutions of

$$\text{(H)} \qquad x_1^2+x_2^2=y_3^2, \qquad z^2 = x_1^2+y_1^2 = x_2^2+y_2^2$$

other than $z = y_3$ $(t = 0)$. Leech found 100 primitive solutions with $z < 10^5$, 46 of which had $t > 0$. The generators for (H) satisfy

$$\text{(I)} \qquad \frac{\alpha_1^2+\beta_1^2}{2\alpha_1\beta_1}\cdot\frac{2\alpha_2\beta_2}{\alpha_2^2+\beta_2^2}=\frac{a^2-b^2}{2ab}$$

so solutions for fixed x_2/x_1 correspond to rational points on the cubic curve

$$x_1 v(u^2+1) = x_2 u(v^2+1) \tag{J}$$

The trivial solution $t = 0$ corresponds to several **ordinary points** which generate an infinity of solutions for each ratio x_2/x_1. Solutions form cycles of four:

$$\begin{aligned} &\zeta^2 = \xi_1^2 + \eta_1^2 = \xi_2^2 + \eta_2^2 = \xi_3^2 + \eta_3^2 = \xi_4^2 + \eta_4^2, \qquad \xi_1\xi_3 = \xi_2\xi_4, \\ &\xi_1^2 + \xi_2^2,\ \xi_2^2 + \xi_3^2,\ \xi_3^2 + \xi_4^2,\ \xi_4^2 + \xi_1^2 \qquad \text{all squares,} \end{aligned} \tag{K}$$

corresponding to two pairs x_2/x_1 which correspond to two points collinear with the point $t = 0$ on (J). Conversely, such a pair of points corresponds to a cycle of four solutions.

Problem 3. Just one face diagonal irrational. There are two closely related problems: find three integers whose sums and differences are all squares; find three squares whose differences are squares. The cuboid form of the problem asks for integers satisfying

$$x_2^2+y_2^2 = z^2, \qquad x_1^2+x_3^2 = y_2^2, \qquad x_1^2+x_2^2 = y_3^2 \tag{L}$$

whose generators satisfy

$$\frac{\alpha_1^2 - \beta_1^2}{2\alpha_1\beta_1} \cdot \frac{\alpha_3^2 - \beta_3^2}{2\alpha_3\beta_3} = \frac{\alpha_2^2 + \beta_2^2}{2\alpha_2\beta_2}. \tag{M}$$

Write $u_i = (\alpha_i^2 - \beta_i^2)^2/4\alpha_i^2\beta_i^2$ and (M) becomes $u_1u_3 = 1 + u_2$, an equation investigated by many in the contexts of **cycles** and **frieze patterns**. Solutions occur in cycles of five! Leech listed 35 such with $\alpha_1, \beta_1, \alpha_2, \beta_2 \le 50$ and deposited in UMT a list of all cycles with two pairs $\alpha_i, \beta_i \le 376$. There is a close connexion with Napier's rules and the construction of rational spherical triangles.

Solutions to (L) are given by $x_2^2 = z^2 - y_2^2 = (p^2 - q^2)(r^2 - s^2)$, $x_3^2 = z^2 - y_3^2 = 4pqrs$, when the products of the numerators and denominators of $(p^2 - q^2)/2pq$ and $(r^2 - s^2)/2rs$ are each squares. Euler made p, q, r, s squares and found differences of fourth powers, e.g., $3^4 - 2^4$, $9^4 - 7^4$, $11^4 - 2^4$, whose products in pairs are squares. The first two give the second smallest solution (117,520,756) of this type, whose cycle includes the smallest (104,153,672), also known to Euler.

We can also express $z^2 = x_2^2 + y_2^2 = x_3^2 + y_3^2$ as the sum of two squares in two different ways: $x_3/x_1 = (\alpha_2^2 - \beta_2^2)/2\alpha_2\beta_2$, $x_2/x_1 = (\alpha_3^2 - \beta_3^2)/2\alpha_3\beta_3$ give $z^2 = 4(\alpha_2^2\alpha_3^2 + \beta_2^2\beta_3^2)(\alpha_2^2\beta_3^2 + \alpha_3^2\beta_2^2)$. Euler made each factor square and found two rational right triangles of equal area $\frac{1}{2}\alpha_2\alpha_3\beta_2\beta_3$. Diophantus solved this with $\beta_2/\alpha_2 = (s+t)/2r$, $\beta_3/\alpha_3 = s/t$, where $r^2 = s^2 + st + t^2$, $s = l^2 - m^2$, $t = m^2 - n^2$. Put $(l, m, n) = (1, 2, -3)$ and we have a cycle containing the third, fourth and fifth smallest of these cuboids. Leech found 89 with $z < 10^5$.

For fixed α_1/β_1 a nontrivial solution to (M) corresponds to an ordinary point, which generates an infinity of solutions, on the curve

$$(\alpha_1^2 - \beta_1^2)u(v^2 - 1) = 2\alpha_1\beta_1 v(u^2 + 1).$$

The tangent at the point generates a cycle of special interest.

(M), like (D), but unlike (I), does not have nontrivial solutions for all ratios α/β; e.g., there are none with $\alpha/\beta = 2$ or 3 and again no cuboids with edges in the ratio 3:4. Here we do not have "derived" cuboids.

Two other parametric forms for ratios $(p^2 - q^2)/2pq$ whose product and quotient are squares, are

$$p = 2m^2 \pm n^2, \qquad r = m^2 \pm 2n^2, \qquad q = s = m^2 \mp n^2.$$

Four squares whose sums in pairs are square. A solution of (C) gives three such squares; it may be portrayed as a trivalent **vertex** of a graph, the three edges joining it to **nodes** representing generator pairs for rational triangles. If such a pair occurs in one solution it occurs in infinitely many, so the valence of a *node* is infinite. We seek a subgraph homeomorphic to a tetrahedron K_4 whose vertices give four solutions of (C) and whose edges contain nodes corresponding to generator pairs common to pairs of solutions of (C). Lists of solutions have revealed no such subgraph; indeed, not even a closed circuit! Until a circuit is encountered, we need not distinguish between the pairs a, b and $a \pm b$.

So no example of four such squares is known. Construction of four squares with five square sums of pairs is straightforward.

A. R. Thatcher related the problem to the equation $y^2 = -x^8 + 35x^4 - 25$. The only integer solutions are $(\pm 1, \pm 3)$, but there may be a finite number of other rational solutions. Even if not, this does not preclude solutions to the original problem.

Four squares whose differences are square. This problem extends (M) analogously to the above extension of (C). A *vertex* is now *pentavalent*, adjacent to a cycle of five nodes or generator pairs. The cyclic order of the edges is important, but not the sense of rotation. A solution of (M) corresponds to three *consecutive* edges, and here we must distinguish between α, β and $\alpha \pm \beta$: the corresponding nodes are joined with double edges in Figure 11. The nodes are again of infinite valence. Four squares of the required type would correspond to a subgraph homeomorphic to K_6, with six vertices and 15 nodes, one on each edge. The only cycle so far found is shown in Figure 11 and this will *not* serve as part of such a subgraph. Although a solution is unlikely, there do not appear to be any congruence conditions which forbid it.

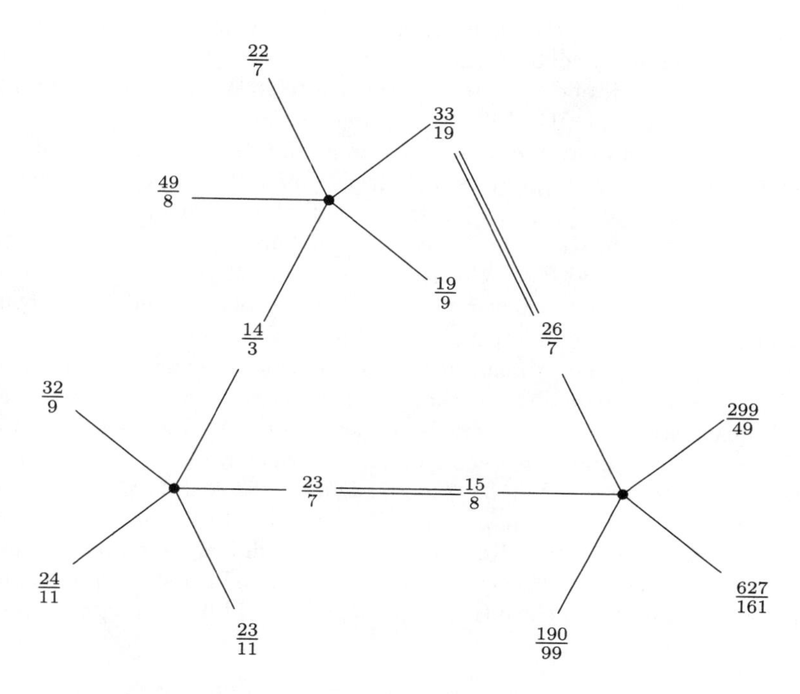

Figure 11. Three Cycles of Five Generator Pairs.

Though there is now no necessary connexion between the generator pairs α, β and $\alpha \pm \beta$, solutions of (M) containing both pairs do occur. Such a solution leads to a sequence of four squares with sums of two or three *consecutive* terms all square. How long can such a sequence be? For more than four we need a sequence of 5-cycles of solutions of (M) each containing adjacent generator pairs α, β; $\bar{\alpha}$, $\bar{\beta}$ where $\alpha \pm \beta$, $\bar{\alpha} \pm \bar{\beta}$ belong one each to the neighboring cycles. Leech found the sequence

$$(56,31)(17,6);(23,11)(23,7);(15,8)(26,7);(33,19)(77,19);(48,29)(35,4);(39,31)(13,9)$$

where $23, 11 = 17 \pm 6$, etc., which gives a sequence of eight such squares. The squares of the edges of a perfect cuboid would form an infinite (periodic) sequence. Four nonzero squares with differences all square would lead to a sequence with terms three apart in constant ratio: an integer ratio would give an infinite sequence. Leech has since given the longer sequence

$$(14,1)(224,37);\ (261,187)(155,132);\ (287,23)(23,7);$$
$$(15,8)(26,7);\ (33,19)(77,19);\ (48,29)(35,4);\ (39,31)(13,9)$$

with "both ends surprisingly small" and asks "are they really ends?" He has no proof that a pair (a, b) can occur while $(a+b, a-b)$ does not. He has some other sequences of the same length, but so far none longer. Randall Rathbun was unable to extend this sequence before (14,1) or after (13,9), but he was able to extend the earlier sequence to seven terms by prefixing (26767,2185)(87,25); or (940,693)(87,25); before (56,31).

The perfect rational cuboid. None of the known numerical solutions to problems 1, 2 and 3 gives a perfect cuboid, and many parametric solutions, for example (G), can be shown not to yield one. Spohn used Pocklington's work to show that *one* of the two mutually derived cuboids of (F) is not perfect and E. Z. Chein and Jean Lagrange have each shown that the derived cuboid is never perfect. On the other hand, no known parametric solution is complete so impossibility can't be proved from these alone. A solution of the problem in the previous section need not lead to a perfect cuboid. Korec showed that the least edge of a perfect rational cuboid must exceed 10^6. Extensive searches, mainly by Randall Rathbun and Torbjorn Granlund, have shown that a perfect rational cuboid must have all its edges greater than 333750000. During the search, the results of which have been deposited in UMT, 6800 + 6380 + 6749 solutions of the three problems were found. Recently Korec has shown that the greatest edge is greater than 10^9. Leech amplifies a result of Horst Bergmann to show that the product of the edges, face diagonals and body diagonal must be divisible by

$$2^8 \times 3^4 \times 5^3 \times 7 \times 11 \times 13 \times 17 \times 19 \times 29 \times 37.$$

Unsolved problems. Do three cycles of solutions of (M) exist whose graph is as in Figure 12? Here we've adopted John Leech's convention of writing the ratio of the sides as a fraction, e.g. $\frac{y_1}{x_1}$, where the same pair of generators belongs to both cycles (as the pair 14, 3 in Figure 11), but writing it as a ratio, e.g. $x_2 : x_3$, where a pair of generators belongs to one cycle, and their sum and difference to the other (as 15, 8 and 23, 7 in Figure 11).

Are there ratios $(p^2-q^2)/2pq$, $(r^2+s^2)/2rs$ with product and quotient both of the form $(m^2-n^2)/2mn$? Is there a nontrivial solution of

$$(a^2c^2 - b^2d^2)(a^2d^2 - b^2c^2) = (a^2b^2 - c^2d^2)^2?$$

Such a solution would lead to a perfect cuboid. Is there a 5-cycle of solutions of (M) with

$$\alpha_1/\beta_1 = (p^2-q^2)/2pq, \quad \alpha_2/\beta_2 = (r^2+s^2)/2rs?$$

What circuits, if any, occur in the graph of solutions of (C)? What circuits occur in the graph of solutions of (M)? Are there cycles of solutions of (I) other than those of form (K)? What ratios besides 3/4 can*not* occur as ratios of edges of cuboids in Problem 1? In Problem 3? Is there a parallelepiped with all edges, face diagonals and body diagonals rational? Rathbun has found 41 pairs of primitive cuboids in which two edges of one equal two of the other. Twenty-one of these are pairs of solutions to Problem 1, with the body diagonal irrational; 13 are pairs of solutions to Problem 2; three are pairs of solutions to Problem 3, with a face diagonal irrational. Three are solutions to Problems 1 and 3, while the last is a solution to Problems 1 and 2.

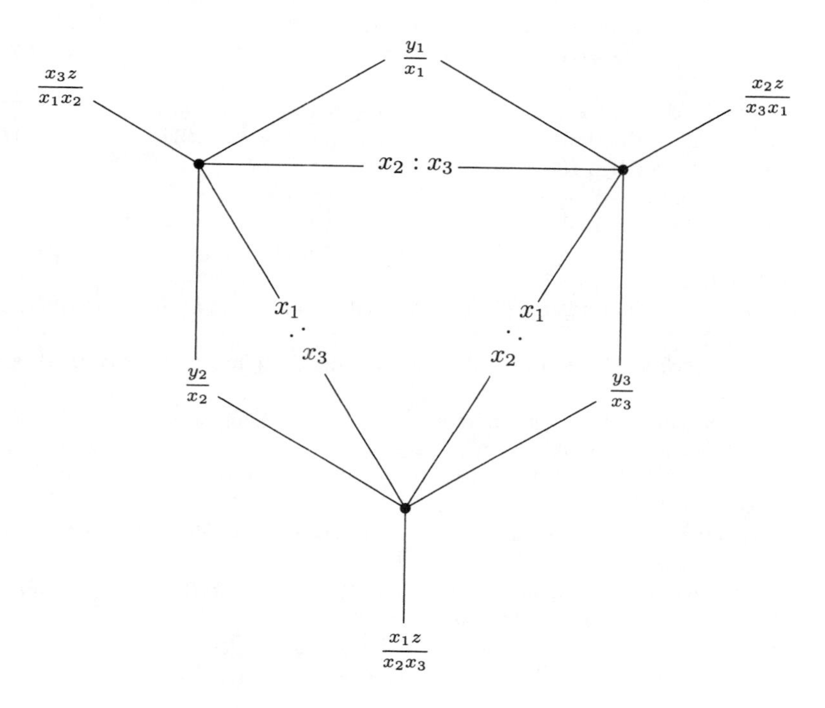

Figure 12. Are there three Cycles of Solutions of (M) like this?

Andrew Bremner, Pythagorean triangles and a quartic surface, *J. reine angew. Math.*, **318**(1980) 120–125.

Andrew Bremner, The rational cuboid and a quartic surface, *Rocky Mountain J. Math.*, **18**(1988) 105–121.

T. Bromhead, On square sums of squares, *Math. Gaz.*, **44**(1960) 219–220; *MR* **23** #A1594.

Ezra Brown, $x^4 + dx^2y^2 + y^4$: some cases with only trivial solutions — and a solution Euler missed, *Glasgow Math. J.*, **31**(1989) 297–307; *MR* **91d**:11026.

W. Burnside, Note on the symmetric group, *Messenger of Math.*, **30**(1900–01) 148–153; *J'buch* **32**, 141–142.

E. Z. Chein, On the derived cuboid of an Eulerian triple, *Canad. Math. Bull.*, **20**(1977) 509–510; *MR* **57** #12375.

W. J. A. Colman, On certain semi-perfect cuboids, *Fibonacci Quart.*, **26** (1988) 54–57 (see also 338–343).

J. H. Conway & H. S. M. Coxeter, Triangulated polygons and frieze patterns, *Math. Gaz.*, **57**(1973) 87–94, 175–183 and refs. on pp. 93–94.

H. S. M. Coxeter, Frieze patterns, *Acta Arith.*, **18**(1971) 297–310; *MR* **44** #3980.

L. E. Dickson, *History of the Theory of Numbers*, Vol. 2, Diophantine Analysis, Washington, 1920: ch. 15, ref. 28, p. 448 and cross-refs. to pp. 446–458; ch. 19, refs 1–30, 40–45, pp. 497–502, 505–507.

Marcus Engel, Numerische Lösung eines Quaderproblems, *Wiss. Z. Pädagog. Hochsch. Erfurt/Mühlhausen Math.-Natur. Reihe*, **22**(1986) 78–86; *MR* **87m**: 11021.

P. W. Flood, *Math. Quest. Educ. Times*, **68**(1898), 53.

Martin Gardner, Mathematical Games, *Scientific Amer.*, **223**#1(Jul.1970) 118; correction, #3(Sep.1970)218.

W. Howard Joint, Cycles, Note 1767, *Math. Gaz.*, **28**(1944) 196–197.

Ivan Korec, Nonexistence of a small perfect rational cuboid, I., II., *Acta Math. Univ. Comenian.*, **42/43**(1983) 73–86, **44/45**(1984) 39–48; *MR* **85i**:11004, **86c**:11013.

Ivan Korec, Lower bounds for perfect rational cuboids, *Math. Slovaca*, **42** (1992) 565–582.

Maurice Kraitchik, On certain rational cuboids, *Scripta Math.*, **11**(1945) 317–326; *MR* **8**, 6.

Maurice Kraitchik, *Théorie des nombres*, t.3, Analyse Diophantine et applications aux cuboides rationnels, Paris, 1947.

Maurice Kraitchik, Sur les cuboides rationnels, in *Proc. Internat. Congr. Math.* 1954, Vol. 2, Amsterdam, 33–34.

Jean Lagrange, Sur le dérivé du cuboïde eulerien, *Canad. Math. Bull.*, **22** (1979) 239–241; *MR* **80h**:10022.

Jean Lagrange, Sets of n squares of which any $n-1$ have their sum square, *Math. Comput.*, **41**(1983) 675–681; *MR* **84j**:10012.

M. Lal & W. J. Blundon, Solutions of the Diophantine equations $x^2+y^2=l^2$, $y^2+z^2=m^2$, $z^2+x^2=n^2$, *Math. Comput.*, **20**(1966) 144–147; *MR* **32** #4082.

J. Leech, The location of four squares in an arithmetical progression with some applications, in *Computers in Number Theory*, Academic Press, London, 1971, 83–98; *MR* **47** #4913.

J. Leech, The rational cuboid revisited, *Amer. Math. Monthly*, **84**(1977) 518–533; corrections (Jean Lagrange) **85**(1978) 473; *MR* **58** #16492.

J. Leech, Five tables related to rational cuboids, *Math. Comput.*, **32**(1978) 657–659.

John Leech, A remark on rational cuboids, *Canad. Math. Bull.*, **24**(1981) 377–378; *MR* **83a**:10022.

John Leech, Four integers whose twelve quotients sum to zero, *Canad. J. Math.*, **38**(1986) 1261–1280; *MR* **88a**:11031; addendum 90-08-18.

R. C. Lyness, Cycles, Note 1581, *Math. Gaz.*, **26**(1942) 62; Note 1847, **29** (1945) 231–233; Note 2952, **45**(1961) 207–209.

"Mahatma", Problem 78, *The A.M.A.* [J. Assist. Masters Assoc. London] **44**(1949) 188; Solutions: J. Hancock, J. Peacock, N. A. Phillips, 225.

Eliakim Hastings Moore, The cross-ratio of $n!$ Cremona-transformations of order $n-3$ in flat space of $n-3$ dimensions, *Amer. J. Math.*, **30**(1900) 279–291; *J'buch* **31**, 655.

L. J. Mordell, On the rational solutions of the indeterminate equations of the third and fourth degrees, *Proc. Cambridge Philos. Soc.*, **21**(1922) 179–192.

H. C. Pocklington, Some diophantine impossibilities, *Proc. Cambridge Philos. Soc.*, **17** (1914) 110–121, esp. p. 116.

Randall L. Rathbun, Table of equal area Pythagorean triangles, from coprimitive sets of integer generator pairs, iii+199pp. Deposited in UMT file; see *Math. Comput.*, **62**(1994) Review #11.

Randall L. Rathbun & Torbjorn Granlund, The classical rational cuboid table of Maurice Kraitchik, revised and enlarged, v(3pp. errata)+135pp. Deposited in UMT file; see *Math. Comput.*, **62**(1994) Review #10.

Randall L. Rathbun & Torbjorn Granlund, The integral cuboid table, with body, edge and face type of solutions, vii+399pp. (2 vols) + The integer cuboid auxiliary table, 100pp. Deposited in UMT file; see *Math. Comput.*, **62**(1994) Review #12.

W. W. Sawyer, Lyness's periodic sequence, Note 2951, *Math. Gaz.*, **45**(1961) 207.

Waclaw Sierpiński, Pythagorean Triangles, *Scripta Math. Studies*, **9**(1962), Yeshiva University, New York, Chap. 15, pp. 97–107.

W. Sierpiński, *A Selection of Problems in the Theory of Numbers*, Pergamon, Oxford, 1964, p. 112.

W. G. Spohn, On the integral cuboid, *Amer. Math. Monthly*, **79**(1972) 57–59; *MR* **46** #7158.

W. G. Spohn, On the derived cuboid, *Canad. Math. Bull.*, **17**(1974) 575–577; *MR* **51** #12693.

D19 Rational distances from the corners of a square.

Is there a point all of whose distances from the corners of the unit square are rational? It was earlier thought that there might not be any nontrivial example (i.e., an example not on the side of the square) of a point with *three* such rational distances, but John Conway & Mike Guy found an infinity of integer solutions of

$$(s^2+b^2-a^2)^2+(s^2+b^2-c^2)^2=(2bs)^2$$

where a, b, c are the distances of a point from three corners of a square of side s. There are relations between such solutions as shown in Figure 13.

For the fourth distance d to be an integer we also need $a^2+c^2=b^2+d^2$. In the three-distance problem, one of s, a, b, c is divisible by 3, one by 4, and one by 5. In the four-distance problem, s is a multiple of 4 and a, b, c, d are odd (assuming that there is no common factor). If s is not a multiple of 3 (respectively 5) then two of a, b, c, d are divisible by 3 (resp. 5).

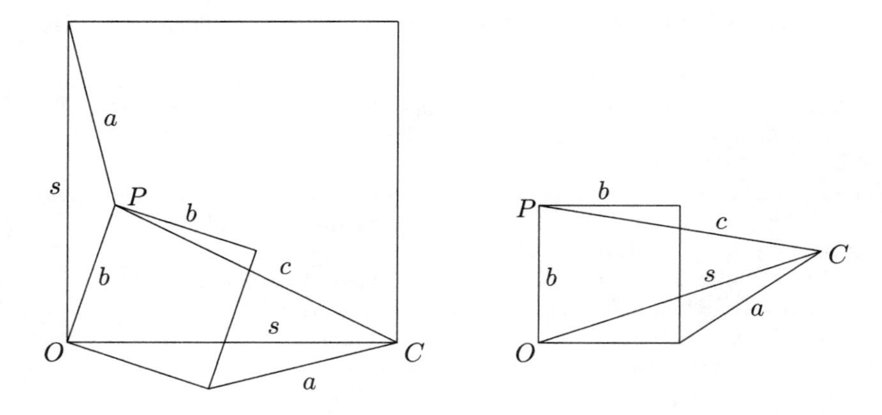

Figure 13. A Solution of the Three-Distance Problem, and its Inverse.

If the problem is generalized to a rational rectangle, then $a^2 + c^2 = b^2 + d^2$ is still required. This is the basis of a Martin Gardner puzzle [Mathematical Games, *Sci. Amer.* **210** #6 (June 1964), Problem 2, p.116]; see also the Dodge reference below. A similar problem with a square with an irrational side and one irrational distance occurs in Dudeney, Canterbury Puzzles, No. 66, pp. 107–109, 212–213. Gardner sends a copy of correspondence between Leslie J. Upton and J. A. H. Hunter. Hunter's 67-03-21 letter gives an infinity of solutions with the three distances in A.P.: $a = m^2 - 2mn + 2n^2$, $b = m^2 + 2n^2$, $c = m^2 + 2mn + 2n^2$ and $s^2 = 2m^2(m^2 + 4n^2)$ where s is an integer if $m = 2(u^2 + 2uv - v^2)$, $n = u^2 - 2uv - v^2$. For example, there is a point at distances 85, 99 and 113 from three consecutive corners of a square of side 140. It can be shown that the fourth distance is never rational in such solutions.

John Leech found points solving the three-distance problem which are dense in the plane. These include the Conway-Guy solutions, and the 'inverses' (in the sense of Fig. 13) of the Hunter solutions. But there are other solutions, and in some sense we now know "all" of them. Consider the more general problem of dissecting a rational square into rational triangles. It is known that at least four triangles are needed and there are just four candidate arrangements, the δ-configuration, the κ-configuration, the ν-configuration and the χ-configuration. The first two turn out to be "duals" and solutions are given by the rational points on an infinite family of elliptic curves. The first few hundred have been investigated by Bremner and Guy, who have also dealt similarly with the ν-configurations. The χ-configuration, i.e., the "four distance" problem, remains as an astonishingly hard nut to crack.

There are infinitely many solutions of the corresponding problem of integer distances a, b, c from the corners of an equilateral triangle of side t. In each of these one of a, b, c, t is divisible by 3, one by 5, one by 7 and one by 8. John Leech has sent us a neat and elementary proof of the fact that the points at rational distances from the vertices of *any* triangle with rational sides are dense in the plane of the triangle. This result was proved earlier by Almering; see reference at **D21**. Arnfried Kemnitz notes that $a = m^2 + n^2$; $b, c = m^2 \pm mn + n^2$ with $m = 2(u^2 - v^2)$, $n = u^2 + 4uv + v^2$ gives $t = 8(u^2 - v^2)(u^2 + uv + v^2)$ and an infinity of solutions in which the points are neither on the sides nor the circumcircle of the triangle. A computer search showed that (57,65,73,112) was the smallest such.

Thomas Berry writes the last displayed equation as

$$2(s^4 + b^4) + a^4 + c^4 = 2(s^2 + b^2)(a^2 + c^2)$$

and notes that this, and the corresponding equation

$$t^4 + a^4 + b^4 + c^4 = t^2a^2 + t^2b^2 + t^2c^2 + b^2c^2 + c^2a^2 + a^2b^2$$

for the equilateral triangle, both represent **Kummer surfaces**, i.e., quartic surfaces with just 16 singular points. They are not isomorphic, but are of the same special type, known as **tetrahedroids**.

There are the following consequences:

- A Kummer surface is not rational: there is no general parametric solution of either problem, in the sense that there are no polynomials (resp. rational functions) giving all integer (resp. rational) solutions.

- One-parameter families of solutions correspond to parametrizable curves on the surface. For example, the 16 conics (which always exist on a Kummer surface) give, in the equilateral triangle problem, points on the sides and circumcircle. The solution given by Arnfried Kemnitz corresponds to the plane section $b + c = 2a$.

- The elliptic curves used by Bremner & Guy to find the delta-lambda configurations form a pencil on the former surface, and since the surfaces are both tetrahedroids, there may be an elliptic pencil on the "equilateral triangle" surface, which allows an analogous attack.

Berry generalizes Almering's result by showing that if the squares on the sides of a triangle are rational and at least one side is rational, then the set of points at rational distances from all three vertices is dense in the plane of the triangle.

Jerry Bergum asks for what integers n do there exist positive integers x, y with $x \perp y$, x even, and $x^2 + y^2 = b^2$ & $x^2 + (y - nx)^2 = c^2$ both perfect squares. If $n = 2m(2m^2 + 1)$, then $x = 4m(4m^2 + 1)$, $y = mx + 1$

is a solution. There are no solutions if $n = \pm 1$, ± 2, ± 4, ± 11 or $\pm p$ where $p \equiv 3 \bmod 4$ and $p^2 + 4$ is prime, e.g., ± 3, ± 7. Bergum has several infinite families of values of n for which there are solutions, e.g., $n = 8t^2 \pm 4t + 2$ with $t > 0$. There are solutions with $n = \pm 5$, ± 6, ± 8, ± 9, ± 14, ± 19. If $n = 8$, the least x for which there is a y is $x = 2996760 = 2^3 \cdot 3 \cdot 5 \cdot 13 \cdot 17 \cdot 113$ and if $n = 19$ the least x is 2410442371920. One solution is $n = 5$, $x = 120$, $y = 391$ as may be seen from Fig. 15(b)! The connexion between this problem and the original one is that (x, y) are the coordinates of P at distances b and c from the origin O and an adjacent corner of the square of side $s = nx$ where n is an integer.

Ron Evans notes that the problem may be stated: which integers n occur as the ratios base/height in integer-sided triangles? The sign of n is $\pm$ according as the triangle is acute or obtuse (e.g., $n = -29$, $x = 120$, $y = 119$ is a solution). He also asks the dual problem: find every integer-sided triangle whose base divides its height. Here the height/base ratios 1 and 2 can't occur, but 3 can (e.g., base 4; sides 13, 15; height 12). If a ratio can occur, are there infinitely many primitive triangles for which it occurs? K. R. S. Sastry gives the triangles (3389, 21029, 24360) and (26921, 42041, 68880) in each of which the ratio base/height is 42, and (25,26,3) and (17,113,120) with ratios 1/8 and 8 (the third member of the triple is the base in each case).

J. H. J. Almering, Rational quadrilaterals, *Nederl. Akad. Wetensch. Proc. Ser. A*, **66** = *Indagationes Math.*, **25**(1963) 192–199; II **68** = **27**(1965) 290–304; *MR* **26** #4963, **31** #3375.

T. G. Berry, Points at rational distance from the corners of a unit square, *Ann. Scuola Norm. Sup. Pisa Cl. Sci.*(4), **17**(1990) 505–529; *MR* **92e**:11021.

T. G. Berry, Points at rational distances from the vertices of a triangle, *Acta Arith.*, **62**(1992) 391–398.

T. G. Berry, Triangle distance problems and Kummer surfaces, (to appear).

Andrew Bremner & Richard K. Guy, The delta-lambda configurations in tiling the square, *J. Number Theory*, **32**(1989) 263–280; *MR* **90g**:11031.

Andrew Bremner & Richard K. Guy, Nu-configurations in tiling the square, *Math. Comput.*, **59**(1992) 195–202, S1–S20; *MR* **93a**:11019.

Clayton W. Dodge, Problem 966, *Math. Mag.*, **49**(1976) 43; partial solution **50**(1977) 166–167; comment **59**(1986) 52.

R. B. Eggleton, Tiling the plane with triangles, *Discrete Math.*, **7**(1974) 53–65.

R. B. Eggleton, Where do all the triangles go? *Amer. Math. Monthly*, **82** (1975) 499–501.

Ronald Evans, Problem E2685, *Amer. Math. Monthly*, **84**(1977) 820.

N. J. Fine, On rational triangles, *Amer. Math. Monthly*, **83**(1976) 517–521.

Richard K. Guy, Tiling the square with rational triangles, in R. A. Mollin (ed.) Number Theory & Applications, *Proc. N.A.T.O. Adv. Study Inst., Banff 1988*, Kluwer, Dordrecht, 1989, 45-101.

W. H. Hudson, Kummer's Quartic Surface, reprinted with a foreword by W. Barth, Cambridge Univ. Press, 1990.

Arnfied Kemnitz, Rational quadrangles, Proc. 21st SE Conf. Combin. Graph Theory Comput., Boca Raton 1990, *Congr. Numer.*, **76**(1990) 193–199; *MR* **92k**:11034.

J. G. Mauldon, An impossible triangle, *Amer. Math. Monthly*, **86**(1979) 785–786.

C. Pomerance, On a tiling problem of R. B. Eggleton, *Discrete Math.*, **18** (1977) 63–70.

D20 Six general points at rational distances.

The first edition of this book asked "are there six points in the plane, no three on a line, no four on a circle, all of whose mutual distances are rational?" Leech pointed out that one such configuration is obtained by fitting six copies of a triangle whose sides and medians are all rational. Such triangles were studied by Euler (see **D21**): the simplest has sides 68, 85, 87 and medians the halves of 158, 131, 127 [Fig. 14(a)]. Harborth & Kemnitz have shown that this triangle leads to the minimal configuration of six points, no three collinear, no four concyclic, at integer mutual distances. A related configuration is obtained by inversion in a concentric circle; the six triangles are then similar but no longer congruent. Can any such configuration be extended? Or are there any sets of more than six such points? Kemnitz has exhibited an unsymmetrical set of six points at integer distances, 13 of them distinct, the largest 319 [Fig. 14(b)].

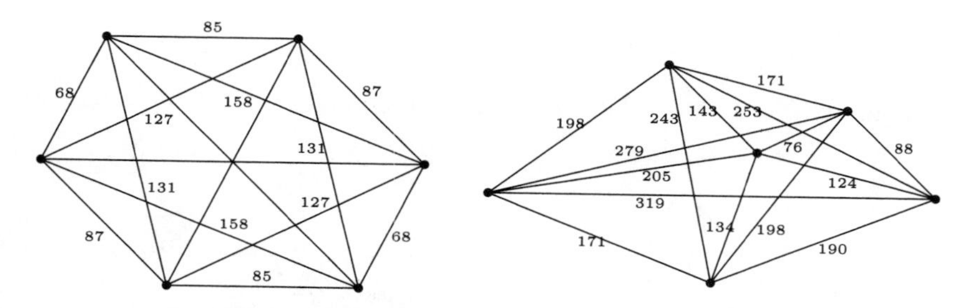

Figure 14(a). A triangle with rational medians, reflected in its centroid.

Figure 14(b). Six points at integer distances, 13 of them distinct.

There are two opposite extreme conjectures: (a) that there is a fixed number c such that any n points in a plane whose mutual distances are rational include at least $n-c$ which are collinear or concyclic, and (b) (ascribed to Besicovitch, but in 1959 he expressed the contrary opinion) that any polygon can be approximated arbitrarily closely by a polygon with all its sides and diagonals rational. If (a) is valid, what is the maximum value of c?

If we have an infinite sequence of points $\{x_i\}$ with all distances rational, can we characterize the set of limit points? It was already known to Euler that they could be dense on a circle. If the sequence is dense in the plane, Ulam conjectured that not all distances could be rational. Does it contain a dense subsequence with all distances irrational?

We can choose a straight line and two points at unit distance from it on a perpendicular line; then points on the first line at distances of the form $(u^2 - v^2)/2uv$ from the intersection form an infinite set of points with rational mutual distances. By inversion in a circle centred at one of the off-line points, we obtain a dense set of points on a circle, together with its centre, with rational mutual distances. This proves Conjecture (b) for *cyclic* polygons. Peeples (who quotes Huff) extended this to a straight line with four points at distances $\pm p$, $\pm q$ from it on a perpendicular line. If q/p has a representation as the product of two *distinct* ratios of the form $(u^2 - v^2)/2uv$, then it has an infinity of them, and there will be an infinity of points on the first line at rational distances from each other and from the four off-line points. Thus c is at least 4 in Conjecture (a). By inversion in a circle centred at one of the four off-line points, we obtain a dense set of points on a circle, together with its centre and a pair of points inverse in the circle, with rational mutual distances. Thus c is at least 3 for the circle in Conjecture (a). Are these the maximum values of c for infinite sets?

What finite sets surpass these values of c? Leech gave an infinite family of sets of nine points, with no more than four on any line or circle, so $c = 5$ for these sets. They are based on solutions of the simultaneous Diophantine equations

$$x^2 + y^2 = \square, \quad x^2 + z^2 = \square, \quad x^2 + (y+z)^2 = \square, \quad x^2 + (y-z)^2 = \square;$$

most simply $x = 120$, $y = 209$, $z = 182$ [Fig. 15(b)].

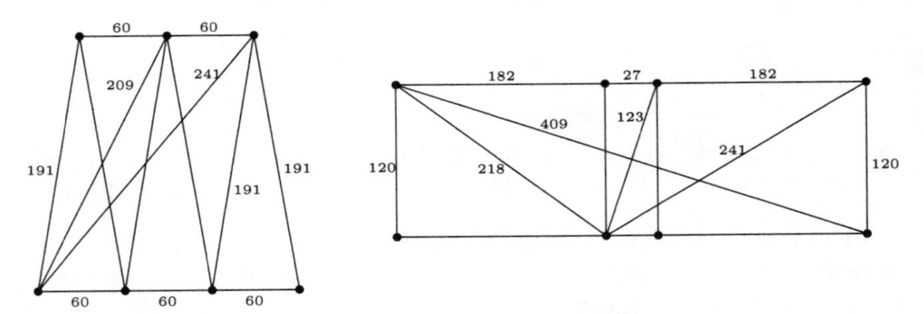

Figure 15 (a) & (b). Leech's Configurations of Points at Rational Distances.

Lagrange and Leech have given infinite families of pairs of triads of integers a_1, a_2, a_3, and b_1, b_2, b_3, such that the nine sums $a_i^2 + b_j^2$ are all squares. These lead to sets of 13 points on two perpendicular lines, comprising their intersection, points $(\pm a_i, 0)$ on one line and points $(0, \pm b_j)$

on the other, with integer mutual distances and no more than seven on a line or four on a circle, so $c = 6$ for these sets. The simplest example has $a_i = 952, 1800, 3536$ and $b_j = 960, 1785, 6630$. They also found an example in which one triad is extended to a tetrad; it has

$$\begin{aligned} a_i &= 9282000, \quad 26822600, \quad 60386040, \\ b_j &= 7422030, \quad 8947575, \quad 22276800, \quad 44142336, \end{aligned}$$

but these still give only $c = 6$. Leech has since found three further examples. Can a pair of tetrads be found with all 16 sums $a_i^2 + b_j^2$ square? If so, we would obtain a set of 17 points with integer mutual distances and $c = 8$.

Noll and Bell search for configurations with no three points collinear and no four concyclic, but using only lattice points. They call such configurations **N-clusters**. They, and independently William Kalsow & Bryan Rosenburg, found the 6-cluster (0,0), (132,−720), (546,−272), (960,−720), (1155,540), (546,1120). They define the **extent** of an N-cluster to be the radius of the smallest circle, centred at one of the points, which contains them all. They find 91 nonequivalent prime 6-clusters of extent less than 20937, but no 7-clusters.

M. Altwegg, Ein Satz über Mengen von Punkten mit ganzzahliger Entfernung, *Elem. Math.*, **7**(1952) 56–58.

D. D. Ang, D. E. Daykin & T. K. Sheng, On Schoenberg's rational polygon problem, *J. Austral. Math. Soc.*, **9**(1969) 337–344; *MR* **39** #6816.

A. S. Besicovitch, Rational polygons, *Mathematika*, **6**(1959) 98; *MR* **22** #1557.

D. E. Daykin, Rational polygons, *Mathematika*, **10**(1963) 125–131; *MR* **30** #63.

D. E. Daykin, Rational triangles and parallelograms, *Math. Mag.*, **38**(1965) 46–47.

H. Harborth, On the problem of P. Erdős concerning points with integral distances, *Annals New York Acad. Sci.*, **175**(1970) 206–207.

H. Harborth, Antwort auf eine Frage von P. Erdős nach fünf Punkten mit ganzzahligen Abständen, *Elem. Math.*, **26**(1971) 112–113.

H. Harborth & A. Kemnitz, Diameters of integral point sets, *Intuitive geometry (Siófok, 1985), Colloq. Math. Soc. János Bolyai*, **48**(1987) 255–266, North-Holland, Amsterdam-New York; *MR* **88k**:52011.

G. B. Huff, Diophantine problems in geometry and elliptic ternary forms, *Duke Math. J.*, **15**(1948) 443–453.

A. Kemnitz, Integral drawings of the complete graph K_6, in Bodendiek & Henn, *Topics in Combinatorics and Graph Theory, Essays in honour of Gerhard Ringel*, Physica-Verlag, Heidelberg, 1990, 421–429.

Jean Lagrange, Points du plan dont les distances mutuelles sont rationnelles, *Séminaire de Théorie des Nombres de Bordeaux*, 1982-1983, Exposé no. 27, Talence 1983.

Jean Lagrange, Sur une quadruple équation. *Analytic and elementary number theory* (Marseille, 1983), 107–113, *Publ. Math. Orsay*, 86-1, *Univ. Paris XI, Orsay*, 1986; *MR* **87g**:11036.

Jean Lagrange & John Leech, Two triads of squares, *Math. Comput.*, **46** (1986) 751–758; *MR* **87d**: 11018.
John Leech, Two diophantine birds with one stone, *Bull. London Math. Soc.*, **13**(1981) 561–563; *MR* **82k**:10017.
D. N. Lehmer, Rational triangles, *Annals Math. Ser. 2*, **1**(1899–1900) 97–102.
L. J. Mordell, Rational quadrilaterals, *J. London Math. Soc.*, **35**(1960) 277–282; *MR* **23** #A1593.
L. C. Noll & D. I. Bell, n-clusters for $1 < n < 7$, *Math. Comput.*, **53** (1989) 439–444.
W. D. Peeples, Elliptic curves and rational distance sets, *Proc. Amer. Math. Soc.*, **5**(1954) 29–33; *MR* **15** 645f.
T. K. Sheng, Rational polygons, *J. Austral. Math. Soc.*, **6**(1966) 452–459; *MR* **35** #137.
T. K. Sheng & D. E. Daykin, On approximating polygons by rational polygons, *Math. Mag.*, **38**(1966) 299–300; *MR* **34** #7463.

D21 Triangles with integer sides, medians and area.

Is there a triangle with integer sides, medians and area? There are, in the literature, incorrect "proofs" of impossibility, but the problem remains open. It may be instructive to would-be solvers to find the fallacies in the arguments of Schubert and of Eggleston, referred to below. For some time it had been suspected that not even two rational medians were possible in a Heron triangle, but discoveries by Randall L. Rathbun, Arnfried Kemnitz and R. H. Buchholz have shown that they can occur. Even more recently, Rathbun has found infinitely many such, and it seems reasonable to conjecture that there are infinitely many infinite families, but this remains undemonstrated, as does the existence or impossibility of three rational medians. If we don't require the area to be rational, there are many solutions. Euler gave a parametric solution of degree five,

$$a = 6\lambda^4 + 20\lambda^2 - 18, \quad b, c = \lambda^5 \pm \lambda^4 - 6\lambda^3 \pm 26\lambda^2 + 9\lambda \pm 9$$

with medians $-2\lambda^5 + 20\lambda^3 + 54\lambda$, $\pm\lambda^5 + 3\lambda^4 \pm 26\lambda^3 - 18\lambda^2 \pm 9\lambda + 27$. Recently George Cole has shown that, up to symmetry, there are just two parametric solutions of degree five, Euler's and a new one. These have not so far yielded a non-degenerate triangle with rational area.

A host of problems arise from Pythagorean triangles. Eckert asked if there are two distinct Pythagorean triples whose products are equal, i.e., is there a solution of

$$xy(x^4 - y^4) = zw(z^4 - w^4)$$

in nonzero integers and Prothro asked if one product could be *twice* the other. More generally, Leech asked what small integers are the ratios of two such products? Trivially $xy(x^4 - y^4) = 8zw(z^4 - w^4)$ when $x,\ y = z \pm w$. Since every product is divisible by $3 \cdot 4 \cdot 5$, many integers are possible, from $13 = \frac{5 \cdot 12 \cdot 13}{3 \cdot 4 \cdot 5}$ upwards. More subtly $11 = \frac{21 \cdot 220 \cdot 221}{13 \cdot 84 \cdot 85}$, and imprimitively

$$6 = \frac{6^3 \cdot 24 \cdot 143 \cdot 145}{135 \cdot 352 \cdot 377}.$$

Are these the smallest?

He also notes that one product is *eight* times the other on setting $x, y = z \pm w$.

How many primitive Pythagorean triangles can have the same area? A triple of such, with generators (77,38), (78,55) and (138,5) was found by Charles L. Shedd in 1945. In 1986 Rathbun found three more:
(1610,869), (2002,1817), (2622,143) (2035,266), (3306,61), (3422,55) and (2201,1166), (2438,2035), (3565,198).
A fifth triple, (7238,2465), (9077,1122), (10434,731), was found independently on consecutive days by Dan Hoey and Rathbun. Is there an infinity of triples? Are there quadruples?

Sastry asks for Pythagorean triangles with a square and a triangle for legs and a pentagonal number, $\frac{1}{2}n(3n-1)$, for hypotenuse. Are there any nontrivial examples besides (3,4,5) and (105,100,145)? Does it help to allow pentagonal numbers of negative rank, $\frac{1}{2}n(3n+1)$?

Despairing of solving the problem of the rational box (**D18**), some people have investigated other polyhedra all of whose distances are integers. There are seven topologically different convex hexahedra, for example, and integer examples have been found by Harborth & Kemnitz (see **D20**) and by Peterson & Jordan. Sastry asked for solutions to the rational box problem, but using triangular numbers instead of squares. Charles Ashbacher gave the triangular numbers 66, 105, 105 whose sums of pairs and whose total are all triangular.

J. H. J. Almering, Heron problems, thesis, Amsterdam, 1950.

Ralph Heiner Buchholz, On triangles with rational altitudes, angle bisectors or medians, PhD thesis, Univ. of Newcastle, Australia, 1989.

George Raymond Cole, Triangles all of whose sides and medians are rational, PhD dissertation, Arizona State University, May 1991.

Ernest J. Eckert, Problem 994, *Crux Mathematicorum*, **10**(1984) 318; comment **12**(1986) 109.

H. G. Eggleston, A proof that there is no triangle the magnitudes of whose sides, area and medians are integers, Note 2204, *Math. Gaz.*, **35**(1951) 114–115.

H. G. Eggleston, Isosceles triangles with integral sides and two integral medians, Note 2347, *Math. Gaz.*, **37**(1953) 208–209.

Albert Fässler, Multiple Pythagorean number triples, *Amer. Math. Monthly*, **98**(1991) 505–517; *MR* **92d**:11021.

Martin Gardner, Mathematical Games: Simple proofs of the Pythagorean theorem, and sundry other matters, *Sci. Amer.* **211**#4 (Oct. 1964) 118–126.

Blake E. Peterson, Integer polyhedra, Ph.D. dissertation, Washington State University, 1993.
Blake E. Peterson & James H. Jordan, Integer polyhedra and the perfect box, (preprint, 93-01-08).
Blake E. Peterson & James H. Jordan, Integer hexahedra equivalent to perfect boxes, *Amer. Math. Monthly*, **101**(1994).
E. T. Prothro, *Amer. Math. Monthly* **95**(1988) 31.
Randall L. Rathbun, Letter to the Editor, *Amer. Math. Monthly*, **99**(1992) 283–284.
K. R. S. Sastry, Problem 1725, *Crux Mathematicorum*, **18**(1992) 75.
K. R. S. Sastry, Problem 1832, *Crux Mathematicorum*, **19**(1993) 112.
H. Schubert, *Die Ganzzahligkeit in der algebraischen Geometrie*, Leipzig, 1905, 1–16.

D22 Simplexes with rational contents.

Are there simplexes in any number of dimensions, all of whose contents (lengths, areas, volumes, hypervolumes) are rational? The answer is "yes" in two dimensions; there are infinitely many **Heron triangles** with rational sides and area. An example is a triangle of sides 13, 14, 15 which has area 84. The answer is also "yes" in three dimensions, but can all tetrahedra be approximated arbitrarily closely by such rational ones?

John Leech notes that four copies of an acute-angled Heron triangle will fit together to form such a tetrahedron, provided that the volume is made rational, and this is not difficult. E.g., three pairs of opposite edges of lengths 148, 195, 203. This is the smallest example: he finds the next few triples to be

$$(533, 875, 888), (1183, 1479, 1804), (2175, 2296, 2431), (1825, 2748, 2873),$$

$$(2180, 2639, 3111), (1887, 5215, 5512), (6409, 6625, 8484), (8619, 10136, 11275).$$

He also suggests examining references on p. 224 of Vol II of Dickson's *History* [**A17**]:

R. Güntsche, *Sitzungsber. Berlin Math. Gesell.*, **6**(1907) 38–53.
R. Güntsche, *Archiv Math. Phys.*(3), **11**(1907) 371.
E. Haentzschel, *Sitzungsber. Berlin Math. Gesell.*, **12**(1913) 101–108 & **17**(1918) 37–39 (& *cf.* **14**(1915) 371).
O. Schultz, Ueber Tetraeder mit rationalen Masszahlen der Kantenlängen und des Volumen, Halle, 1914, 292 pp.

Dickson appealed for a copy of this last. Did he ever get one? Does anyone know of a copy? Would they be willing to donate it, or offer it for sale, to the Strens Collection?

Leech also notes that this problem is answered positively in three dimensions by solutions to Problem 3 in **D18** (find a box which is rational except for one face diagonal). This problem was published as Problem 930 in *Crux Mathematicorum*, **10**(1984) #3, p. 89, and the solution by the COPS (presumably an acronym for the Carleton (Ottawa) Problem Solvers) is:

Take a tetrahedron with a path of three mutually perpendicular edges, $a = p^2q^2 - r^2s^2$, $b = 2pqrs$, $c = p^2r^2 - q^2s^2$. Then $a^2 + b^2$, $b^2 + c^2$ are squares and $a^2 + b^2 + c^2 = (p^4 + s^4)(q^4 + r^4)$ is a square if

$$p^4 + s^4 = q^4 + r^4.$$

[John Leech notes "but not only if" and gives four casual examples,

$$(1^4 + 2^4)(2^4 + 13^4) = 697^2; \quad (1^4 + 2^4)(38^4 + 43^4) = 9673^2;$$

$$(1^4 + 2^4)(314^4 + 863^4) = 1275643^2; \quad (1^4 + 3^4)(9^4 + 437^4) = 1729298^2,$$

which imply further ones of type $(2^4 + 13^4)(38^4 + 43^4)$.]

This equation was solved by Euler. The solution mentioned in **D9** is

$$\begin{aligned} p,\ q &= x^7 + x^5y^2 - 2x^3y^4 \pm 3x^2y^5 + xy^6 \\ r,\ s &= x^6y \pm 3x^5y^2 - 2x^4y^3 + x^2y^5 + y^7 \end{aligned}$$

but this is not in any sense complete.

Buchholz found that the only rational tetrahedron with edge lengths ≤ 156 was that with edge lengths 117, 80, 53, 52, 51, 84, face areas 1800, 1890, 2016, 1170, and volume 18144. He also shows that a *regular* d-dimensional simplex with rational edge has rational d-dimensional volume just if d is of shape $4k(k+1)$ or $2k^2 - 1$.

Dove & Sumner relax the condition that the faces of a tetrahedron have rational area and find two tetrahedra with volume 3 and pairs of opposite edges (32,76) (33,70) (35,44) and (21,58) (32,76) (47,56). They ask if there are infinitely many tetrahedra with integer edges and the same integer volume. Is there such a tetrahedron with volume any given multiple of 3? They have examples from 3 to 99, except for 87.

The tetrahedron with a pair of opposite edges 896 and 990 and the other four edges each 1073, two face areas 436800 and 471240, and volume 62092800, is mentioned by Sierpiński and by Leitzmann.

Ralph Heiner Buchholz, Perfect pyramids, *Bull. Austral. Math. Soc.*, **45**(1991) 353–368.

Kevin L. Dove & John L. Sumner, Tetrahedra with integer edges and integer volume, *Math. Mag.*, **65**(1992) 104–111.

K. È. Kalyamanova, Rational tetrahedra (Russian), *Izv. Vyssh. Uchebn. Zaved. Mat.*, **1990** 73–75; *MR* **92b**:11014.

W. Lietzmann, Der pythagoreisch Lehrsatz, Leipzig, 1965, p. 91 [not in 1930 edition].

W. Sierpiński, *Pythagorean Triangles*, New York, 1962, p. 107.

D23 Some quartic equations.

Another of many unsolved diophantine equations is

$$(x^2-1)(y^2-1)=(z^2-1)^2$$

though Schinzel & Sierpiński have found all solutions for which $x-y=2z$. Cao Zhen-Fu has shown that the only solutions satisfying $x-y=lz$ for other values of $l\le 30$ are $|x|=|y|$ or $|z|=1$, and Wang Yan Bin that these are the only solutions with $x-y=z^2+1$.

Kashihara has shown that all solutions of

$$(x^2-1)(y^2-1)=(z^2-1)$$

can be derived from the trivial solutions $(n,1,1)$ and $(1,n,1)$.

For the equation $x^2-1=y^2(z^2-1)$, Mignotte has shown that if z is large, then the greatest prime factor of y is at least $c\ln\ln y$.

Ron Graham has observed that the diophantine equations

$$2x^2(x^2-1)=3(y^2-1) \quad \text{and} \quad (2x-1)^2=2^n-7$$

each have the solutions $x=0$, 1, 2, 3, 6 and 91. Is this merely an example of the Strong Law of Small Numbers? Evidently so! In a forthcoming paper Stroeker & de Weger find that Graham's equation has one pair and five quadruples of solutions. They note that it is unfair to count the solutions $x=0$ and $x=1$ of the Ramanujan-Nagell equation as separate, while 'forgetting' the solutions $x=-1,\ -2,\ -5,\ -90$.

Baragar has shown that the equation

$$x(x+1)y(y+1)=z(z+1),$$

studied by Katayama, is equivalent to a Markoff type equation (see **D12**)

$$x^2+y^2+z^2=2xyz+5$$

and has counted the number of solutions of size less than N.

Cao Zhen-Fu, A generalization of the Schinzel-Sierpiński system of equations (Chinese; English summary), *J. Harbin Inst. Tech.*, **23**(1991) 9–14; *MR* **93b**:11026.

Kenji Kashihara, The Diophantine equation $x^2-1=(y^2-1)(z^2-1)$ (Japanese; English summary), *Res. Rep. Anan College Tech. No.* **26**(1990) 119–130; *MR* **91d**:11025.

Shin-ichi Katayama & Kenji Kashihara, On the structure of the integer solutions of $z^2=(x^2-1)(y^2-1)-a$, *J. Math. Tokushima Univ.*, **24**(1990) 1–11; *MR* **93c**:11013.

Maurice Mignotte, A note on the equation $x^2-1=y^2(z^2-1)$, *C. R. Math. Rep. Acad. Sci. Canada*, **13**(1991) 157–160; *MR* **92j**:11026.

A. Schinzel & W. Sierpiński, Sur l'equation diophantienne $(x^2-1)(y^2-1) = [((y-x)/2)^2-1]^2$, *Elem. Math.*, **18**(1963) 132–133; *MR* **29** #1180.
Wang Yan-Bin, On the Diophantine equation $(x^2-1)(y^2-1) = (z^2-1)^2$ (Chinese, English summary), *Heilongjiang Daxue Ziran Kexue Xuebao*, **1989** *no.* 4 84–85; *MR* **91e**:11028.

D24 Sum equals product.

For $k > 2$ the equation $a_1 a_2 \cdots a_k = a_1 + a_2 + \cdots + a_k$ has the solution $a_1 = 2$, $a_2 = k$, $a_3 = a_4 = \cdots = a_k = 1$. Schinzel showed that there is no other solution for $k = 6$ or $k = 24$. Misiurewicz has shown that $k = 2$, 3, 4, 6, 24, 114 (misprinted as 144 in *Elem. Math.* and in the first edition of this book), 174 and 444 are the only $k < 1000$ for which there is exactly one solution.

This problem seems first to have been asked by Trost, arising from the solution of $a_1 a_2 \cdots a_k = a_1 + a_2 + \cdots + a_k = 1$ in rationals. For $k = 3$ this is due to Sierpiński; for $k > 3$ to Schinzel. Editorial comment in *Amer. Math. Monthly*, extended the result to $k \le 10000$ and M. L. Brown gave necessary and sufficient conditions on k and extended the search to $k \le 50000$.

M. L. Brown, On the diophantine equation $\sum X_i = \prod X_i$, *Math. Comput.*, **42**(1984) 239–240; *MR* **85d**:11030.
M. Misiurewicz, Ungelöste Probleme, *Elem. Math.*, **21**(1966) 90.
E. P. Starke & others, Solution to Problem E2262 [1970, 1008] & editorial comment, *Amer. Math. Monthly*, **78**(1971) 1021–1022.
E. Trost, Ungelöste Probleme, Nr. **14**, *Elem. Math.*, **11**(1956) 134–135.

D25 Equations involving factorial n.

Are the only solutions of $n! + 1 = x^2$ given by $n = 4$, 5 and 7? Overholt has related this problem to a conjecture of Szpiro. Erdős & Obláth dealt with the equation $n! = x^p \pm y^p$ with $x \perp y$ and $p > 2$. For the case $p = 2$ with the plus sign, see Leech's remark at **D2**; and for the minus sign, split $n!$ into two even factors: $4! = 5^2 - 1^2 = 7^2 - 5^2$; $5! = 11^2 - 1^2 = 13^2 - 7^2 = 17^2 - 13^2 = 31^2 - 29^2$. The number of solutions is $\frac{1}{2}d(n!/4)$.

Simmons notes that $n! = (m-1)m(m+1)$ for $(m, n) = (2, 3)$, (3,4), (5,5) and (9,6) and asks if there are other solutions. More generally he asks if there are any other solutions of $n! + x = x^k$. This is a variation on the question of asking for $n!$ to be the product of k consecutive integers in a nontrivial way ($k \ne n + 1 - j!$). Compare **B23**.

In a 93-05-07 letter to Ron Graham, Nobuhisa Abe states that $x(x+1)\cdots(x+k) = y^2 - 1$ has the unique solution $(x, y) = (2, 71)$ for $k = 5$ and no solutions for $k = 7$ or 11.

Berend & Osgood have shown that the set of n for which the equation $P(x) = n!$ has an integer solution x has density zero if $P(x)$ is a polynomial of degree ≥ 2 with integer coefficients.

Daniel Berend & Charles F. Osgood, On the equation $P(x) = n!$ and a question of Erdős, *J. Number Theory*, **42**(1992) 189–193; *MR* **93e**:11016.

B. Brindza & P. Erdős, On some Diophantine problems involving powers and factorials, *J. Austral. Math. Soc. Ser. A* **51**(1991) 1–7; *MR* **92i**:11036.

H. Brocard, Question 1532, *Nouv. Corresp. Math.*, **2**(1876) 287; *Nouv. Ann. Math.*(3) **4**(1885) 391.

P. Erdős & R. Obláth, Über diophantische Gleichungen der Form $n! = x^p \pm y^p$ und $n! \pm m! = x^p$, *Acta Szeged*, **8**(1937) 241–255.

M. Kraitchik, Recherches sur la Théorie des Nombres, tome 1, Gauthier-Villars, Paris, 1924, 38–41.

Marius Overholt, The Diophantine equation $n! + 1 = m^2$, *Bull. London Math. Soc.*, **25**(1993) 104; *MR* **93m**:11026.

Richard M. Pollack & Harold N. Shapiro, The next to last case of a factorial diophantine equation, *Comm. Pure Appl. Math.*, **26**(1973) 313–325; *MR* **50** #12915.

Gustavus J. Simmons, A factorial conjecture, *J. Recreational Math.*, **1**(1968) 38.

D26 Fibonacci numbers of various shapes.

Stark asks which Fibonacci numbers (see **A3**) are half the difference or sum of two cubes. This is related to the problem of finding all complex quadratic fields of class number 2. Examples: $1 = \frac{1}{2}(1^3+1^3)$, $8 = \frac{1}{2}(2^3+2^3)$, $13 = \frac{1}{2}(3^3-1^3)$. Antoniadis has related all such fields to solutions of certain diophantine equations, and solved them all but two, which were later settled by de Weger.

Cohn showed that the only square Fibonacci numbers are 0, 1 and 144, and Luo Ming confirmed Vern Hoggatt's conjecture that the only triangular ones, i.e. of the form $\frac{1}{2}m(m+1)$, are 0, 1, 3, 21 and 55, and later that the only such Lucas numbers are 1, 3 and 5778.

Jannis A. Antoniadis, Über die Kennzeichnung zweiklassiger imaginär-quadratischer Zahlkörper durch Lösungen diophantisher Gleichungen, *J. reine angew. Math.*, **339**(1983) 27–81; *MR* **85g**:11098.

J. H. E. Cohn, On square Fibonacci numbers, *J. London Math. Soc.*, **39**(1964) 537–540; *MR* **29** #1166.

D. G. Gryte, R. A. Kingsley & H. C. Williams, On certain forms of Fibonacci numbers, *Proc. 2nd Louisiana Conf. Combin. Graph Theory & Comput.*, *Congr. Numer.*, **3**(1971) 339–344.

V. E. Hoggatt, Problem 3, *WA St. Univ. Conf. Number Theory*, 1971, p. 225.

Luo Ming, On triangular Fibonacci numbers, *Fibonacci Quart.*, **27**(1989) 98–108; *MR* **90f**:11013.

Luo Ming, On triangular Lucas numbers, *Applications of Fibonacci numbers, Vol.* 4 (1990), 231–240, Kluwer, Dordrecht, 1991; *MR* **93i**:11016.

Neville Robbins, Fibonacci and Lucas numbers of the forms $w^2 - 1$, $w^3 \pm 1$, *Fibonacci Quart.*, **19**(1981) 369–373.

H. M. Stark, Problem 23, *Summer Institute on Number Theory*, Stony Brook, 1969.

Ray Steiner, On Fibonacci numbers which are one more than a square, *J. reine angew. Math.*, **262/263**(1973) 171–182.

Ray Steiner, On triangular Fibonacci numbers, *Utilitas Math.*, **9**(1976) 319–327.

M. H. Tallman, Problem H23, *Fibonacci Quart.*, **1**(1963) 47.

Charles R. Wall, On triangular Fibonacci numbers, *Fibonacci Quart.*, **23** (1985) 77–79.

B. M. M. de Weger, A diophantine equation of Antoniadis, *Number Theory and Applications* (Banff, 1988), 547–553, *NATO Adv. Sci. Inst. Ser. C: Math. Phys. Sci.*, **265**, Kluwer, 1989; *MR* **92f**:11048.

B. M. M. de Weger, A hyperelliptic Diophantine equation related to imaginary quadratic number fields with class number 2, *J. reine angew. Math.*, **427**(1992) 137–156; *MR* **93d**:11034.

D27 Congruent numbers.

Congruent numbers are perhaps confusingly named; they are related to pythagorean triangles and have an ancient history. Several examples (5, 6, 14, the seventeen entries CA in Table 7, and ten more greater than 1000) are in an Arab manuscript more than a thousand years ago. But it is only for the last ten years, since the work of Tunnell, that we have a reasonably complete understanding of them. They are those integers a for which

$$x^2 + ay^2 = z^2 \quad \text{and} \quad x^2 - ay^2 = t^2$$

have simultaneous integer solutions. Part of the fascination is the often inordinate size of the smallest solutions. For example, $a = 101$ is a congruent number and Bastien gave the smallest solution:

$$x = 20\,1524246294\,9760001961 \qquad y = 1\,1817143185\,2779451900$$

$$z = 23\,3914843530\,6225006961 \qquad t = 16\,2812437072\,7269996961$$

and in spite of improved computing techniques and machines, it may still be some time before some other of the more recalcitrant examples are discovered. Some other large specimens, found by J. A. H. Hunter, M. R. Buckley and K. Gallyas, are given in the first edition of this book.

Congruent numbers are equivalently defined as those a for which there are solutions of the diophantine equation

$$x^4 - a^2y^4 = u^2$$

Table 7. Congruent (C) and Noncongruent (N) Numbers less than 1000. The entry for $a = 40c+r$ is in column c and row r.

r \ c	0	1	2	3	4	5	6	7	8	9	10	11	12	13	14	15	16	17	18	19	20	21	22	23	24	r
1	NB	C1	□	□	CG	N9	N1	N1	N9	□	N1	□	NJ	N1	CG	N1	N1	N9	C&	C1	□	□	N1	N9	□	1
2	NB	NB	N2	NX	□	NX	□	NT	NJ	NX	NJ	CG	NT	□	N2	CG	NJ	NJ	□	NJ	NT	NX	□	NX	NL	2
3	N3	N3	N3	N&	N3	NJ	□	N3	C&	□	NJ	N3	NJ	N3	N3	□	N3	N3	CJ	NJ	NJ	NJ	N3	NJ	□	3
4	□	□	□	□	□	□	□	□	□	□	□	□	□	□	□	□	□	□	□	□	□	□	□	□	□	4
5	C5	□	CG	□	CG	CG	□	C&	□	CJ	□	C&	CJ	□	C&	□	C&	CJ	□	□	C&	□	CJ	□	CL	5
6	C6	C6	C6	□	C6	C6	C&	CA	C6	C&	CJ	C6	□	C6	C6	CJ	CG	□	□	C6	CG	□	C6	C6	CG	6
7	C7	C7	CG	C7	C7	□	CJ	C&	CJ	C7	CJ	CJ	C7	C&	□	C7	C7	CJ	C7	CJ	CJ	□	C7	□	C7	7
8	□	□	□	□	□	□	□	□	□	□	□	□	□	□	□	□	□	□	□	□	□	□	□	□	□	8
9	□	□	N1	N9	□	N9	N9	□	NJ	□	N1	N1	N9	□	N1	C&	N9	C&	□	N1	N1	N9	C&	N1	NJ	9
10	NX	□	□	NL	N&	CA	□	NL	CA	NL	CG	□	□	NL	NJ	NL	□	NJ	NJ	NJ	□	□	CG	NJ	NJ	10
11	N3	NB	NB	N3	□	N3	N3	CG	N3	CG	NJ	NJ	N3	□	N3	NJ	CG	N3	C&	NJ	N3	NJ	□	□	N3	11
12	□	□	□	□	□	□	□	□	□	□	□	□	□	□	□	□	□	□	□	□	□	□	□	□	□	12
13	C5	C5	CG	CJ	C5	CJ	CJ	C5	□	C5	CJ	CJ	CJ	C&	CL	C5	C5	□	C5	C5	C&	C5	CL	CL	CJ	13
14	C6	□	C6	C6	CG	C6	C6	□	C6	CJ	□	C6	CJ	CJ	C&	C6	CJ	C6	C6	□	C&	CJ	CJ	C6	C&	14
15	CA	CG	CG	□	□	C&	CG	CJ	CJ	□	CJ	C&	□	C&	□	C&	CJ	CJ	□	□	CJ	□	CJ	C&	□	15
16	□	□	□	□	□	□	□	□	□	□	□	□	□	□	□	□	□	□	□	□	□	□	□	□	□	16
17	N1	N9	N1	C1	N9	NJ	C1	□	N1	NJ	N9	C1	NJ	N9	N1	N1	□	NJ	N9	C&	N9	N1	NT	N1	N1	17
18	□	NX	□	CG	N2	NX	NJ	NX	□	□	NJ	NX	NJ	NX	□	NJ	C&	NX	□	NX	N2	NJ	NT	NT	NJ	18
19	N3	N3	□	N3	N3	C&	NJ	CG	NJ	N3	N3	□	N3	□	NJ	N3	N3	NJ	N3	NJ	□	N3	NJ	NT	NJ	19
20	□	□	□	□	□	□	□	□	□	□	□	□	□	□	□	□	□	□	□	□	□	□	□	□	□	20
21	CA	C5	C5	C&	C5	CA	□	CJ	CJ	CJ	C5	C5	CJ	C5	CJ	□	C5	C5	CG	CJ	C5	CJ	C&	C5	□	21
22	C6	C6	CG	C6	C&	CJ	C6	C6	□	C6	C6	CG	C6	C6	C&	C6	C6	□	CJ	CJ	CJ	C6	CJ	CJ	C6	22
23	C7	□	C7	C&	CJ	C7	C7	CJ	□	C7	□	C7	C7	CL	CG	CL	C&	CJ	C7	□	C7	C7	C&	C&	C7	23
24	□	□	□	□	□	□	□	□	□	□	□	□	□	□	□	□	□	□	□	□	□	□	□	□	□	24
25	□	CA	NJ	CG	NJ	□	CG	NJ	NJ	NJ	□	CG	C&	NJ	□	□	NJ	NJ	NJ	NJ	□	NJ	C&	□	C&	25
26	NX	NB	NX	N2	N&	C&	NJ	□	NX	C&	C&	N2	NJ	CA	NX	N2	□	NT	NX	NJ	NJ	C&	NJ	NJ	NJ	26
27	□	N3	N3	□	N&	N3	NJ	N3	N3	□	NJ	N3	□	N3	N3	NT	NJ	NJ	□	N3	N3	□	N3	N3	C&	27
28	□	□	□	□	□	□	□	□	□	□	□	□	□	□	□	□	□	□	□	□	□	□	□	□	□	28
29	C5	CG	C5	C5	□	C5	C5	CJ	C5	C5	CA	CJ	C5	□	CJ	C&	C&	C5	CJ	CJ	C5	CJ	□	C&	CJ	29
30	CA	CA	CA	□	CA	CJ	□	CG	□	CA	CJ	CG	CG	□	CJ	□	C&	C&	□	CJ	CJ	C&	C&	□	□	30
31	C7	C7	CG	C7	C7	CA	C7	C7	□	C&	C7	CJ	C&	CJ	CJ	C7	C&	□	C7	C&	CJ	CJ	C7	C&	C7	31
32	□	□	□	□	□	□	□	□	□	□	□	□	□	□	□	□	□	□	□	□	□	□	□	□	□	32
33	N9	N1	N1	□	N1	N1	NJ	C1	C1	N9	N1	N9	□	NJ	N1	N9	N1	NJ	N9	C&	□	□	N9	N1	N9	33
34	CA	NX	N&	CA	C&	□	N2	NX	NJ	NX	CG	NJ	C&	NX	□	NX	C&	NJ	NL	NX	NJ	NJ	N2	□	NJ	34
35	NB	□	N&	NJ	NJ	NJ	□	□	NJ	C&	NJ	□	NJ	NJ	NJ	NJ	□	NJ	NJ	NJ	NJ	□	C&	NJ	C&	35
36	□	□	□	□	□	□	□	□	□	□	□	□	□	□	□	□	□	□	□	□	□	□	□	□	□	36
37	C5	CG	□	C5	C5	CJ	C5	C5	CG	C5	CJ	□	CG	C5	CL	□	C5	CL	C5	C5	□	C5	CL	C&	C5	37
38	C6	CG	C6	C6	□	CJ	C6	C&	C6	C6	CG	C6	C&	□	CJ	CJ	CJ	C6	C6	CG	C6	C6	□	C6	C6	38
39	CG	C7	C&	C&	C7	C7	□	C&	C7	C&	C7	C7	CJ	CJ	C7	□	CJ	C7	CG	C&	C7	C&	C7	C&	□	39
40	□	□	□	□	□	□	□	□	□	□	□	□	□	□	□	□	□	□	□	□	□	□	□	□	□	40
r \ c	0	1	2	3	4	5	6	7	8	9	10	11	12	13	14	15	16	17	18	19	20	21	22	23	24	r

Dickson's *History* gives many early references, including Leonardo of Pisa (Fibonacci); Genocchi; and Gérardin, who gave 7, 22, 41, 69, 77, the twenty Arabic examples and the forty-three entries CG in Table 7. We need consider only squarefree values of a; of the 608 such that are less than 1000, 361 are congruent and 247 are noncongruent. It has long been conjectured that squarefree numbers are congruent if they are $\equiv 5$, 6 or 7 mod 8. This is now known to be true [modulo some widely believed conjectures concerning elliptic curves]. The entries C5, C7 and C6 in Table 7 are for primes $\equiv 5$ or 7 mod 8 and the doubles of primes $\equiv 3$ mod 8. Bastien observed that the following are noncongruent: primes $\equiv 3$ mod 8; products of two such primes; the doubles of primes $\equiv 5$ mod 8; the doubles of the products of two such primes; and the doubles of primes $\equiv 9$ mod 16; these are the respective entries N3, N9, NX, NL and N2 in Table 7. He gave some other noncongruent numbers (entries NB, though $a = 1$ is due to Fermat, and many others were known earlier, e.g. to Genocchi) and stated that a was noncongruent if it was a prime $\equiv 1$ mod 8 with $a = b^2 + c^2$ and $b + c$ a nonresidue (see **F5**) of a. This accounts for several of the entries N1.

Note that the entries 1, 3, 5 and 7 serve as a table of primes in these residue classes mod 8.

Entries C& and N& are from Alter, Curtz & Kubota, and CJ and NJ from Jean Lagrange's thesis.

Ronald Alter, The congruent number problem, *Amer. Math. Monthly*, **87** (1980) 43–45.

R. Alter & T. B. Curtz, A note on congruent numbers, *Math. Comput.*, **28**(1974) 303–305; *MR* **49** #2527 (not #2504 as in *MR* indexes); correction **30**(1976) 198; *MR* **52** #13629.

R. Alter, T. B. Curtz & K. K. Kubota, Remarks and results on congruent numbers, *Proc. 3rd S.E. Conf. Combin. Graph Theory Comput., Congr. Numer.* **6** (1972) 27–35; *MR* **50** #2047.

L. Bastien, Nombres congruents, *Intermédiaire Math.*, **22**(1915) 231–232.

B. J. Birch, Diophantine analysis and modular functions, *Proc. Bombay Colloq. Alg. Geom.*, 1968.

J. W. S. Cassels, Diophantine equations with special reference to elliptic curves, *J. London Math. Soc.*, **41**(1966) 193–291.

L. E. Dickson, *History of the Theory of Numbers*, Vol. 2, Diophantine Analysis, Washington, 1920, 459–472.

A. Genocchi, Note analitiche sopra Tre Sritti, *Annali di Sci. Mat. e Fis.*, **6**(1855) 273–317.

A. Gérardin, Nombres congruents, *Intermédiaire Math.*, **22**(1915) 52–53.

H. J. Godwin, A note on congruent numbers, *Math. Comput.*, **32** (1978) 293–295; **33** (1979) 847; *MR* **58** #495; **80c**:10018.

Jean Lagrange, Thèse d'Etat de l'Université de Reims, 1976.

Jean Lagrange, Construction d'une table de nombres congruents, *Bull. Soc. Math. France Mém.* No. 49–50 (1977) 125–130; *MR* **58** #5498.

Paul Monsky, Mock Heegner points and congruent numbers, *Math. Z.*, **204** (1990) 45–68; *MR* **91e**:11059.

Paul Monsky, Three constructions of rational points on $Y^2 = X^3 \pm NX$, *Math. Z.*, **209**(1992) 445–462; *MR* **93d**:11058.

L. J. Mordell, *Diophantine Equations*, Academic Press, London, 1969, 71–72.

Kazunari Noda & Hideo Wada, All congruent numbers less than 10000, *Proc. Japan Acad. Ser. A Math. Sci.*, **69**(1993) 175–178.

S. Roberts, Note on a problem of Fibonacci's, *Proc. London Math. Soc.*, **11**(1879–80) 35–44.

P. Serf, Congruent numbers and elliptic curves, in *Computational Number Theory* (Proc. Conf. Number Theory, Debrecen, 1989), de Gruyter, 1991, 227–238; *MR* **93g**:11068.

N. M. Stephens, Congruence properties of congruent numbers, *Bull. London Math. Soc.*, **7**(1975) 182–184; *MR* **52** #260.

Jerrold B. Tunnell, A classical diophantine problem and modular forms of weight 3/2, *Invent. Math.* **72**(1983) 323–334; *MR* **85d**:11046.

D28 A reciprocal diophantine equation.

Mordell asked for the integer solutions of

$$\frac{1}{w} + \frac{1}{x} + \frac{1}{y} + \frac{1}{z} + \frac{1}{wxyz} = 0.$$

Several papers have appeared, giving parametric families of solutions. For example, Takahiro Nagashima sends solutions to Mordell's equation: $(w,x,y,z) = (5,3,2,-1)$, $(-7,-3,-2,1)$, $(31,-5,-3,2)$, $(1366,-15,7,-13)$, $(n+1,-n,-1,1)$ and more generally $w = xyz+1$ with

$$x = -2\epsilon h^3 - \delta\epsilon h^2(n-3) + \epsilon h(n-1-2\delta\epsilon) + 1,$$

$$y = 2\delta\epsilon h^2 + \epsilon h(n-3) - \epsilon\delta(n-1) + 1,$$

$$z = -2\delta\epsilon h^2 - \epsilon h(n-1) - 1,$$

where ϵ, $\delta = \pm 1$ independently, but there seems to be no guarantee that these four two-parameter families give all solutions.

Zhang has shown how to obtain all solutions below a given bound, while Clellie Oursler and Judith Longyear have sent extensive analyses which each give a procedure for finding all solutions. That of Longyear extends to the equation $\sum(1/x_i) + \prod(1/x_i) = 0$ with $n(\geq 3)$ variables x_i in place of Mordell's $n = 4$.

Lawrence Brenton & Daniel S. Drucker, On the number of solutions of $\sum_{j=1}^{s}(1/x_j) + 1/(x_1 \cdots x_s) = 1$, *J. Number Theory*, **44**(1993) 25–29.

Cao Zhen-Fu, Mordell's problem on unit fractions. (Chinese. English summary) J. Math. (Wuhan) **7**(1987) 239–244; *MR* **90a**:11032.

Sadao Saito, A diophantine equation proposed by Mordell (Japanese. English summary), *Res. Rep. Miyagi Nat. College Tech. No.* 25 (1988) 101–106; II, *No.* 26 (1990) 159–160; *MR* **91c**:11016–7.

Chan Wah-Keung, Solutions of a Mordell Diophantine equation, *J. Ramanujan Math. Soc.*, **6**(1991) 129-140; *MR* **93d**:11033.

Wen Zhang-Zeng, Investigation of the integer solutions of the Diophantine equation $1/w+1/x+1/y+1/z+1/wxyz = 0$ (Chinese) *J. Chengdu Univ. Natur. Sci.*, **5**(1986) 89–91; *MR* **89c**:11051.

Zhang Ming-Zhi, On the diophantine equation $\frac{1}{x} + \frac{1}{y} + \frac{1}{z} + \frac{1}{w} + \frac{1}{xyzw} = 0$, *Acta Math. Sinica (N.S.)*, **1**(1985) 221–224; *MR* **88a**:11033.

E. Sequences of Integers

Here we are mainly, but not entirely, concerned with infinite sequences; there is some overlap with sections **C** and **A**. An excellent text and source of problems is H. Halberstam & K. F. Roth, *Sequences*, 2nd edition, Springer-Verlag, New York, 1982. Other references are:

P. Erdős, A. Sárközi & E. Szemerédi, On divisibility properties of sequences of integers, in *Number Theory, Colloq. Math. Soc. János Bolyai*, **2**, North-Holland, 1970, 35–49.

H. Ostmann, *Additive Zahlentheorie* I, II, Springer-Verlag, Heidelberg, 1956.

Carl Pomerance & András Sárközi, Combinatorial Number Theory, in R. Graham, M. Grötschel & L. Lovász (editors) *Handbook of Combinatorics*, North-Holland, Amsterdam, 1994.

A. Stöhr, Gelöste und ungelöste Fragen über Basen der natürlichen Zahlenreihe I, II, *J. reine angew. Math.*, **194**(1955) 40–65, 111–140; *MR* **17**, 713.

Paul Turán (editor), *Number Theory and Analysis; a collection of papers in honor of Edmund Landau* (1877–1938), Plenum Press, New York, 1969, contains several papers, by Erdős and others, on sequences of integers.

We will denote by $\mathcal{A} = \{a_i\}$, $i = 1, 2, \ldots$ a possibly infinite strictly increasing sequence of nonnegative integers. The number of a_i which do not exceed x is denoted by $A(x)$. By the **density** of a sequence we will mean $\lim A(x)/x$, if it exists.

E1 A thin sequence with all numbers equal to a member plus a prime.

Erdős offers \$50.00 for a solution of the problem: does there exist a sequence thin enough that $A(x) < c \ln x$, but with every sufficiently large integer expressible in the form $p + a_i$ where p is a prime?

For the analogous problem with squares in place of primes, Leo Moser showed that $A(x) > (1+c)\sqrt{x}$ for some $c > 0$, while Erdős showed that there was a sequence with $A(x) < c\sqrt{x}$. Moser's best value for c was 0.06, but this has been improved to 0.147 by Abbott, to 0.245 by Balasubramanian & Soundarajan, and to 0.273 by Cilleruelo. For the rth powers,

Cilleruelo obtains

$$A(x) > \frac{x^{1-\frac{1}{r}}}{\Gamma(2-\frac{1}{r})\Gamma(1+\frac{1}{r})}$$

For the problem with powers of two in place of primes, Ruzsa obtained the analog of Erdős's result, but it is not known if there is a constant $c > 0$ such that every sequence A for which every positive integer is representable in the form $a + 2^k$ has $A(x) > (1+c)\log_2 x$.

H. L. Abbott, On the additive completion of sets of integers, *J. Number Theory*, **17**(1983) 135–143.

R. Balasubramanian & K. Soundarajan, On the additive completion of squares, II, *J. Number Theory*,**40**(1992) 127–129.

Javier Cilleruelo, The additive completion of kth powers, *J. Number Theory*, **44**(1993) 237–243.

P. Erdős, Problems and results in additive number theory, *Colloque sur la Théorie des Nombres, Bruxelles*, 1955, 127–137, Masson, Paris, 1956.

L. Moser, On the additive completion of sets of integers, *Proc. Symp. Pure Math.*, **8**(1965) Amer. Math. Soc., Providence RI, 175–180.

I. Ruzsa, On a problem of P. Erdős, *Canad. Math. Bull.*, **15**(1972) 309–310.

E2 Density of a sequence with l.c.m. of each pair less than x.

What is the maximum value of $A(x)$ if the least common multiple $[a_i, a_j]$ of each pair of members of the sequence is at most x? It is known that

$$(9x/8)^{1/2} \leq \max A(x) \leq (4x)^{1/2}.$$

The lower bound is obtained by taking all the numbers from 1 up to $\sqrt{x/2}$ and then the even numbers up to $\sqrt{2x}$.

And how many numbers less than x can we find with the greatest common divisor of any pair $< t$ for a given t? If $t < n^{\frac{1}{2}+\epsilon}$, the number $\sim \pi(n)$, while if $t = n^{\frac{1}{2}+c}$, it is $\sim (1+c')\pi(n)$.

Erdős also asks for bounds on $B(x)$, the smallest number so that any subset of $[1,x]$ of cardinality $B(x)$ always contains three members which have pairwise the same least common multiple. Perhaps $B(x) = o(x)$. Again, let $C(x)$ be the corresponding smallest cardinality, so that there are always three numbers with pairwise the same greatest common divisor. No doubt

$$¿ \qquad e^{c_1(\ln x)^{1/2}} < C(x) < e^{c_2(\ln x)^{1/2}} \qquad ?$$

but the best that Erdős has proved is $C(x) < x^{3/4}$.

Given a sequence A, $a_1 < a_2 < \ldots$, Erdős & Szemerédi denote by $F(A,x,k)$ the number of i for which the l.c.m. $[a_{i+1}, a_{i+2}, \ldots, a_{i+k}] < x$, and ask if it is t rue that for every $\epsilon > 0$ there is a k for which $F(A,x,k) < x^{\epsilon}$?

They proved that $F(A, x, 3) < c_1 x^{1/3} \ln x$ for every A, and that there is an A for which $F(A, x, 3) > c_2 x^{1/3} \ln x$ for infinitely many x, but they don't know if there is an A for which this is true for *all* x.

Graham, Spencer & Witsenhausen ask how dense can a sequence of integers be so that $\{n, 2n, 3n\}$ never occur?

There is a good bibliography and many unsolved problems in this area in the paper of Erdős, Sárközy & Szemerédi. See also the references at **B24**.

P. Erdős, Problem, *Mat. Lapok* **2**(1951) 233.

P. Erdős & A. Sárközy, On the divisibility properties of sequences of integers, *Proc. London Math. Soc.* (3), **21**(1970) 97–100; *MR* **42** #222.

P. Erdős, A. Sárközy & E. Szemerédi, On divisibility properties of sequences of integers, in Number Theory *Colloq. Janós Bolyai Math. Soc., Debrecen 1968*, North-Holland, Amsterdam (1970) 35–49; *MR* **43** #4790.

P. Erdős & E. Szemerédi, Remarks on a problem of the *American Mathematical Monthly, Mat. Lapok*, **28**(1980) 121–124; *MR* **82c**:10066.

R. L. Graham, J. H. Spencer & H. S. Witsenhausen, On extremal density theorems for linear forms, in H. Zassenhaus (ed), *Number Theory and Algebra*, Academic Press, New York, 1977, 103–109; *MR* **58** #569.

E3 Density of integers with two comparable divisors.

Is it true that the density of those integers

6, 12, 15, 18, 20, 24, 28, 30, 35, 36, 40, 42, 45, 48, 54, 56, 60, 63, 66, 70, 72, ...

which have two divisors d_1, d_2 such that $d_1 < d_2 < 2d_1$, is one? Erdős has shown that the density exists. There is a connexion with covering congruences (**F13**). Since the first edition this has been solved affirmatively by Maier & Tenenbaum.

P. Erdős, On the density of some sequences of integers, *Bull. Amer. Math. Soc.*, **54**(1948) 685–692; *MR* **10**, 105.

Helmut Maier & G. Tenenbaum, On the set of divisors of an integer, *Invent. Math.*, **76**(1984) 121–128; *MR* **86b**:11057.

E4 Sequence with no member dividing the product of r others.

If no member of the sequence $\{a_i\}$ divides the product of r other terms, Erdős shows that

$$\pi(x) + c_1 x^{2/(r+1)} (\ln x)^{-2} < A(x) < \pi(x) + c_2 x^{2/(r+1)} (\ln x)^{-2}$$

where $\pi(x)$ is the number of primes $\leq x$. If, however, we suppose that the products of any number, not greater than r, of the a_i are distinct, what is $\max A(x)$? For $r \geq 3$, Erdős shows

$$\max A(x) < \pi(x) + O(x^{2/3+\epsilon}).$$

If $r = 1$, so that no term divides any other, the sequence is called **primitive**. Zhang has shown that for a primitive sequence whose members each contain at most four prime factors,

$$\sum_{a_i \leq n} \frac{1}{a_i \ln a_i} \leq \sum_{p \leq n} \frac{1}{p \ln p}$$

(and hence less than 1.64) for $n > 1$, the sums being taken over all members of the sequence up to n and all primes up to n.

P. Erdős, On sequences of integers no one of which divides the product of two others and on some related problems, *Inst. Math. Mec. Tomsk*, **2**(1938) 74–82.

P. Erdős, Extremal problems in number theory V (Hungarian), *Mat. Lapok*, **17**(1966) 135–155.

P. Erdős, On some applications of graph theory to number theory, *Publ. Ramanujan Inst.*, **1**(1969) 131–136.

P. Erdős & Zhang Zhen-Xiang, Upper bound of $\sum 1/(a_i \log a_i)$ for primitive sequences, *Proc. Amer. Math. Soc.*, **117**(1993) 891–895.

Zhang Zhen-Xiang, On a conjecture of Erdős on the sum $\sum_{p \leq n} 1/(p \log p)$, *J. Number Theory*, **39**(1991) 14–17; *MR* **92f**:11131.

Zhang Zhen-Xiang, On a problem of Erdős concerning primitive sequences, *Math. Comput.*, **60**(1993) 827–834; *MR* **93k**:11120.

E5 Sequence with members divisible by at least one of a given set.

Let $D(x)$ be the number of numbers not greater than x which are divisible by at least one a_i where $a_1 < a_2 < \cdots < a_k \leq n$ is a finite sequence. Is $D(x)/x < 2D(n)/n$ for all $x > n$? The number 2 cannot be reduced: for example, $n = 2a_1 - 1$, $x = 2a_1 < a_2$. In the other direction it is known that for each $\epsilon > 0$ there is a sequence which does *not* satisfy the inequality $D(x)/x > \epsilon D(n)/n$.

A. S. Besicovitch, On the density of certain sequences, *Math. Ann.*, **110**(1934) 335–341.

P. Erdős, Note on sequences of integers no one of which is divisible by any other, *J. London Math. Soc.*, **10**(1935) 126–128.

E6 Sequence with sums of pairs not members of a given sequence.

Let $n_1 < n_2 < \cdots$ be a sequence of integers such that $n_{i+1}/n_i \to 1$ as $i \to \infty$, and the $\{n_i\}$ are distributed uniformly mod$d$ for every d; i.e., the number $N(c,d;x)$ of the $n_i \le x$ with $n_i \equiv c \bmod d$ is such that

$$N(c,d;x)/N(1,1;x) \to 1/d \quad \text{as} \quad x \to \infty$$

for each c, $0 \le c < d$, and all d. If $a_1 < a_2 < \cdots$ is an infinite sequence for which $a_j + a_k \ne n_i$ for any i, j, k then Erdős asks: is it true that the density of the a_j is less than $\frac{1}{2}$?

E7 A series and a sequence involving primes.

If p_n is the nth prime, Erdős asks if $\sum(-1)^n n/p_n$ converges. He notes that the series $\sum(-1)^n(n \ln n)/p_n$ diverges.

He also asks if, given three distinct primes and $a_1 < a_2 < a_3 < \ldots$ are all the products of their powers arranged in increasing order, it is true infinitely often that a_i and a_{i+1} are both prime powers. And what if we use k primes or even infinitely many in place of three? Meyer & Tijdeman have asked a similar question for two finite sets S and T of primes with $a_1 < a_2 < a_3 < \ldots$ formed from $S \cup T$. Are there infinitely many i for which a_i is a product of powers of primes from S, while a_{i+1} is a product of powers of primes from T?

E8 Sequence with no sum of a pair a square.

Paul Erdős & David Silverman consider k integers $1 \le a_1 < a_2 < \ldots < a_k \le n$ such that no sum $a_i + a_j$ is a square. Is it true that $k < n(1+\epsilon)/3$, or even that $k < n/3 + O(1)$? The integers $\equiv 1 \bmod 3$ show that if this is true, then it is best possible. They suggest that the same question could be asked for other sequences instead of the squares.

Erdős & Graham added to their book at the proof stage that J. P. Marsias has discovered that the sum of any two integers

$$\equiv 1, 5, 9, 13, 14, 17, 21, 25, 26, 29, 30 \bmod 32$$

is never a square mod 32, so k can be chosen to be at least $11n/32$. This is best possible for the modular version of the problem since Lagarias, Odlyzko & Shearer have shown that if $S \subseteq \mathbb{Z}_n$ and $S+S$ contains no square of $\mathbb{Z}_n$, then $|S| \le 11n/32$.

J. C. Lagarias, A. M. Odlyzko & J. B. Shearer, On the density of sequences of integers the sum of no two of which is a square, I. Arithmetic progressions, *J. Combin. Theory Ser. A*, **33**(1982) 167–185; II. General sequences, **34**(1983) 123–139; *MR* **85d**:11015ab.

E9 Partitioning the integers into classes with numerous sums of pairs.

The conjecture of K. F. Roth, that there exists an absolute constant c so that for every k there is an $n_0 = n_0(k)$ with the following property: For $n > n_0$, partition the integers not exceeding n into k classes $\left\{a_i^{(j)}\right\}$ $(1 \le j \le k)$; then the number of distinct integers not exceeding n which can be written in the form $a_{i_1}^{(j)} + a_{i_2}^{(j)}$ for some j is greater than cn, has been confirmed by Erdős, Sárközy & Sós.

They also investigate the corresponding problem with products in place of sums, where the problem for $k = 2$ remains open.

P. Erdős & A. Sárközy, On a conjecture of Roth and some related problems, II, in R. A. Mollin (ed.) *Number Theory*, Proc. 1st Conf. Canad. Number Theory Assoc., Banff 1988, de Gruyter, 1990, 125–138.

P. Erdős, A. Sárközy & V. T. Sós, On a conjecture of Roth and some related problems, I, *Colloq. Math. Soc. János Bolyai* (1992).

E10 Theorem of van der Waerden. Szemerédi's theorem. Partitioning the integers into classes; at least one contains an A.P.

The well-known theorem of van der Waerden states that for every l there is a number $n(h, l)$ such that if the integers not exceeding $n(h, l)$ are partitioned into h classes, then at least one class contains an arithmetic progression (A.P.) containing $l + 1$ terms. More generally, given l_0, l_1, ..., l_{h-1}, there is always a class V_i $(0 \le i \le h - 1)$ containing an A.P. of $l_i + 1$ terms. Denote by $W(h, l)$, or more generally $W(h; l_0, l_1, \ldots, l_{h-1})$, the least such $n(h, l)$.

Chvátal computed $W(2; 2, 2) = 9$, $W(2; 2, 3) = 18$, $W(2; 2, 4) = 22$, $W(2; 2, 5) = 32$ and $W(2; 2, 6) = 46$ and Beeler & O'Neil give $W(2; 2, 7) = 58$, $W(2; 2, 8) = 77$ and $W(2; 2, 9) = 97$. The values $W(2; 3, 3) = 35$ and $W(2; 3, 4) = 55$ were found by Chvátal and $W(2; 3, 5) = 73$ by Beeler & O'Neil. Stevens & Shantaram found $W(2; 4, 4) = 178$; Chvátal found $W(3; 2, 2, 2) = 27$ and Brown $W(3; 2, 2, 3) = 51$. Beeler & O'Neil also found $W(4; 2, 2, 2, 2) = 76$.

Most proofs of van der Waerden's theorem give poor estimates for $W(h, l)$. Erdős & Rado showed that $W(h, l) > (2lh^l)^{\frac{1}{2}}$ and Moser, Schmidt, and Berlekamp successively improved this to

$$W(h, l) > lh^{c \ln h} \quad \text{and} \quad W(h, l) > h^{l+1-c\sqrt{(l+1)\ln(l+1)}}$$

Moser's bound has been improved for $l \geq 5$ by Abbott & Liu to

$$W(h,l) > h^{c_s (\ln h)^s}$$

where s is defined by $2^s \leq l < 2^{s+1}$, and Everts has shown that $W(h,l) > lh^l/4(l+1)^2$, a result which is sometimes better than Berlekamp's. For $h = 2$ Szabó has recently shown that $W(2,l) > 2^l/l^\epsilon$. All upper bounds were 'ackermanic' in size, until Shelah's proof reduced them to 'wowser' — for an explanation of these terms see the book by Graham, Rothschild & Spencer.

A closely related function, with $l+1 = k$, is the now famous $r_k(n)$, introduced long years ago by Erdős & Turán: the least r such that the sequence $1 \leq a_1 < a_2 < \cdots < a_r \leq n$ of r numbers not exceeding n must contain a k-term A.P. The best bounds when $k = 3$ are due to Behrend, Roth, and Moser:

$$n \exp(-c_1 \sqrt{\ln n}) < r_3(n) < c_2 n / \ln \ln n$$

and for larger k Rankin showed that

$$r_k(n) > n1 - c_s/(\ln n)^{s/(s+1)}$$

where s, much as before, is defined by $2^s < k \leq 2^{s+1}$.

A big breakthrough was Szemerédi's proof that $r_k(n) = o(n)$ for all k, but neither his proof, nor those of Furstenberg and of Katznelson & Ornstein (see Thouvenot) give estimates for $r_k(n)$. Erdős conjectures that

$$¿ \qquad r_k(n) = o(n(\ln n)^{-t}) \quad \text{for every } t \qquad ?$$

This would imply that for every k there are k primes in A.P. See **A5** for a potentially remunerative conjecture of Erdős, which, if true, would imply Szemerédi's theorem.

Another closely related problem was considered by Leo Moser, who wrote the integers in base three, $n = \sum a_i 3^i$ ($a_i = 0$, 1 or 2) and examined the mapping of n into lattice points $(a_1, a_2, a_3, \ldots)$ of infinite-dimensional Euclidean space. He called integers **collinear** if their images are collinear; e.g., $35 \to (2,2,0,1,0,\ldots)$, $41 \to (2,1,1,1,0,\ldots)$ and $47 \to (2,0,2,1,0,\ldots)$ are collinear. He conjectured that every sequence of integers with no three collinear has density zero. If integers are collinear, they are in A.P., but not necessarily conversely (e.g., $16 \to (1,2,1,0,0,\ldots)$, $24 \to (0,2,2,0,0,\ldots)$ and $32 \to (2,1,0,1,0,\ldots)$ are not collinear) so truth of the conjecture would imply Roth's theorem that $r_3(n) = o(n)$.

If $f_3(n)$ is the largest number of lattice points with no three in line in the n-dimensional cube with three points in each edge, then Moser showed that $f_3(n) > c3^n/\sqrt{n}$. It is easy to see that $f_3(n)/3^n$ tends to a limit; is it zero? Chvátal improved the constant in Moser's result to $3/\sqrt{\pi}$ and found the values $f_3(1) = 2$, $f_3(2) = 6$, $f_3(3) = 16$. It is known that $f_3(4) \geq 43$.

More generally, if the n-dimensional cube has k points in each edge, Moser asked for an estimate of $f_k(n)$, the maximum number of lattice points with no k collinear. It is a theorem of Hales & Jewett, with applications to n-dimensional k-in-a-row (tic-tac-toe), that for sufficiently large n, any partition of the k^n lattice points into h classes has a class with k points in line. This implies van der Waerden's theorem on letting the point $(a_0, a_1, \ldots, a_{n-1})$, $(0 \leq a_i \leq k-1)$ correspond to the base k expansion of the integer $\sum a_i k^i$. It is not known whether, for every c and sufficiently large n, it is possible to choose $ck^n/\sqrt{n}$ points without including k in line. It *is* known for *some* c. Inequality (4) in Riddell's second paper quoted below implies that

$$f_k(n) > k^{n+1}/(2\pi e^3(k-1)n)^{\frac{1}{2}}$$

so that one can choose a "line-free" set of $ck^n/\sqrt{n}$ points for some c. In the other direction he obtains $f_3(n) \leq 16 \cdot 3^{n-3}$. He acknowledges Leo Moser's inspiration in obtaining these results.

If you use the **greedy algorithm** to construct sequences not containing an A.P. you don't get a very dense sequence, but you do get some interesting ones. Odlyzko & Stanley construct the sequence $S(m)$ of positive integers with $a_0 = 0$, $a_1 = m$ and each subsequent a_{n+1} is the least number greater than a_n so that a_0, a_1, $\ldots$, a_{n+1} does not contain a three-term A.P. For example

$S(1)$: 0, 1, 3, 4, 9, 10, 12, 13, 27, 28, 30, 31, 36, 37, 39, 40, 81, 82, 84, 85, 90, 91, 93, 94, 108, 109, 111, 112, 117, 118, 120, ...

$S(4)$: 0, 4, 5, 7, 11, 12, 16, 23, 26, 31, 33, 37, 38, 44, 49, 56, 73, 78, 80, 85, 95, 99, 106, 124, 128, 131, 136, 143, ...

If m is a power of three, or twice a power of three, then the members of the sequence are fairly easy to describe (write $S(1)$ in base 3), but for other values the sequences behave quite erratically. Their rates of growth seem to be similar, but this has yet to be proved.

The "simplest" such sequence containing no four-term A.P. is

$$0, 1, 2, 4, 5, 7, 8, 9, 14, 15, 16, 18, 25, 26, 28, 29, 30, 33, 36,$$
$$48, 49, 50, 52, 53, 55, 56, 57, 62, \ldots$$

Is there a simple description of this? How fast does it grow?

If we define the **span** of a set S to be $\max S - \min S$, what is the smallest span $\mathrm{sp}(k,n)$ of a set of n integers containing no k-term A.P.? Zalman Usiskin gives the following values:

n	=	3	4	5	6	7	8	9	10	11	...
$\mathrm{sp}(3,n)$	=	3	4	8	10	12	13	19	24	25	...
$\mathrm{sp}(4,n)$	=		4	5	7	8	9	12	...		

Abbott notes that it follows from Szemerédi's theorem that for each $k \geq 3$, the sequence $\{\mathrm{sp}(k, n+1) - \mathrm{sp}(k, n)\}$ is unbounded, and asks if it contains a bounded subsequence.

A paper of Alfred Brauer, with a magnificent early bibliography, which is relevant to sections **E10** to **E14**, is referred to in **F6**.

H. L. Abbott & D. Hanson, Lower bounds of certain types of van der Waerden numbers, *J. Combin. Theory*, **12**(1972) 143–146.

H. L. Abbott & A. C. Liu, On partitioning integers into progression free sets, *J. Combin. Theory*, **13**(1972) 432–436.

H. L. Abbott, A. C. Liu & J. Riddell, On sets of integers not containing arithmetic progressions of prescribed length, *J. Austral. Math. Soc.*, **18**(1974) 188–193.

Michael D. Beeler & Patrick E. O'Neil, Some new van der Waerden numbers, *Discrete Math.*, **28**(1979) 135–146.

F. A. Behrend, On sets of integers which contain no three terms in arithmetical progression, *Proc. Nat. Acad. Sci. USA* **32**(1946) 331–332; *MR* **8**, 317.

E. R. Berlekamp, A construction for partitions which avoid long arithmetic progressions, *Canad. Math. Bull.*, **11**(1968) 409–414.

E. R. Berlekamp, On sets of ternary vectors whose only linear dependencies involve an odd number of vectors, *Canad. Math. Bull.*, **13**(1970) 363–366.

Thomas C. Brown, Some new Van der Waerden numbers, Abstract 74T-A113, *Notices Amer. Math. Soc.*, **21**(1974) A-432.

T. C. Brown, Behrend's theorem for sequences containing no k-element progression of a certain type, *J. Combin. Theory Ser. A*, **18**(1975) 352–356.

Ashok K. Chandra, On the solution of Moser's problem in four dimensions, *Canad. Math. Bull.*, **16**(1973) 507–511.

V. Chvátal, Some unknown van der Waerden numbers, in *Combinatorial Structures and their Applications*, Gordon and Breach, New York, 1970, 31–33.

J. A. Davis, Roger C. Entringer, Ronald L. Graham & G. J. Simmons, On permutations containing no long arithmetic progressions, *Acta Arith.*, **34**(1977/78) 81–90; *MR* **58** #10705.

P. Erdős, Some recent advances and current problems in number theory, in *Lectures on Modern Mathematics*, Wiley, New York, **3**(1965) 196–244.

P. Erdős & R. Rado, Combinatorial theorems on classifications of subsets of a given set, *Proc. London Math. Soc.*(3), **2**(1952) 417–439; *MR* **16**, 445.

P. Erdős & J. Spencer, *Probabilistic Methods in Combinatorics*, Academic Press, 1974, 37–39.

P. Erdős & P. Turán, On some sequences of integers, *J. London Math. Soc.*, **11**(1936) 261–264.

F. Everts, PhD thesis, Univ. of Colorado, 1977.

H. Furstenberg, Ergodic behaviour of diagonal measures and a theorem of Szemerédi on arithmetic progressions, *J. Analyse Math.*, **31**(1977) 204–256; *MR* **58** #16583.

Joseph L. Gerver & L. Thomas Ramsey, Sets of integers with no long arithmetic progressions generated by the greedy algorithm, *Math. Comput.*, **33**(1979) 1353–1359; *MR* **80k**:10053.

Joseph Gerver, James Propp & Jamie Simpson, Greedily partitioning the natural numbers into sets free of arithmetic progressions, *Proc. Amer. Math. Soc.*, **102**(1988) 765–772

R. L. Graham & B. L. Rothschild, A survey of finite Ramsey theorems, *Proc. 2nd Louisiana Conf. Combin., Graph Theory, Comput., Congr. Numer.*, **3**(1971) 21–40.

R. L. Graham & B. L. Rothschild, A short proof of van der Waerden's theorem on arithmetic progressions, *Proc. Amer. Math. Soc.*, **42**(1974) 385–386.

Ronald L. Graham, Bruce L. Rothschild & Joel H. Spencer, *Ramsey Theory*, 2nd edition, Wiley-Interscience, 1990.

G. Hajós, Über einfache und mehrfache Bedeckungen des n-dimensionalen Raumes mit einem Würfelgitter. *Math. Z.*, **47**(1942) 427–467.

A. W. Hales & R. I. Jewett, Regularity and positional games, *Trans. Amer. Math. Soc.*, **106**(1963) 222–229.

A. Y. Khinchin, *Three Pearls of Number Theory*, Graylock Press, Rochester NY, 1952, 11-17.

Bruce M. Landman & Raymond N. Greenwell, Some new bounds and values for van der Waerden-like numbers, *Graphs Combin.*, **6**(1990) 287–291; *MR* **91k**:11023.

L. Moser, On non-averaging sets of integers, *Canad. J. Math.*, **5**(1953) 245-252; *MR* **14**, 726d, 1278.

Leo Moser, Notes on number theory II. On a theorem of van der Waerden, *Canad. Math. Bull.*, **3**(1960) 23–25; *MR* **22** #5619.

L. Moser, Problem 21, *Proc. Number Theory Conf.*, Univ. of Colorado, Boulder, 1963, 79.

L. Moser, Problem 170, *Canad. Math. Bull.*, **13**(1970) 268.

A. M. Odlyzko & R. P. Stanley, Some curious sequences constructed with the greedy algorithm, Bell Labs. internal memo, 1978.

Carl Pomerance, Collinear subsets of lattice-point sequences – an analog of Szemerédi's theorem, *J. Combin. Theory Ser. A*, **28**(1980) 140–149; *MR* **81m**: 10104.

Jim Propp, What are the laws of greed?, *Amer. Math. Monthly*, **96**(1989) 334–336.

John R. Rabung, On applications of van der Waerden's theorem, *Math. Mag.*, **48**(1975) 142–148.

John R. Rabung, Some progression-free partitions constructed using Folkman's method, *Canad. Math. Bull.*, **22**(1979) 87–91.

R. Rado, Note on combinatorial analysis, *Proc. London Math. Soc.*, **48**(1945) 122–160.

R. A. Rankin, Sets of integers containing not more than a given number of terms in arithmetical progression, *Proc. Roy. Soc. Edinburgh Sect. A*, **65** (1960/61) 332–334; *MR* **26** #95.

J. Riddell, On sets of numbers containing no l terms in arithmetic progression, *Nieuw Arch. Wisk.* (3), **17**(1969) 204–209; *MR* **41** #1678.

J. Riddell, A lattice point problem related to sets containing no l-term arithmetic progression, *Canad. Math. Bull.*, **14**(1971) 535–538; *MR* **48** #265.

K. F. Roth, Sur quelques ensembles d'entiers, *C.R. Acad. Sci. Paris*, **234** (1952) 388–390.

K. F. Roth, On certain sets of integers, *J. London Math. Soc.*, **28**(1953) 104–109 (& see **29**(1954) 20–26); *MR* **14**, 536; *J. Number Theory*, **2**(1970) 125–142; *Period. Math. Hungar.*, **2**(1972) 301–326.

R. Salem & D. C. Spencer, On sets of integers which contain no three terms in arithmetical progession, *Proc. Nat. Acad. Sci.*, **28**(1942) 561–563; *MR* **4**, 131.

R. Salem & D. C. Spencer, On sets which do not contain a given number in arithmetical progession, *Nieuw Arch. Wisk.* (2), **23**(1950) 133–143.

H. Salié, Zur Verteilung natürlicher Zahlen auf elementfremde Klassen, *Ber. Verh. Sächs. Akad. Wiss. Leipzig*, **4**(1954) 2–26.

Wolfgang M. Schmidt, Two combinatorial theorems on arithmetic progressions, *Duke Math. J.*, **29**(1962) 129–140.

S. Shelah, Primitive recursive bounds for van der Waerden numbers, *J. Amer. Math. Soc.*, **1**(1988) 683–697; *MR* **89a**:05017.

G. J. Simmons & H. L. Abbott, How many 3-term arithmetic progressions can there be if there are no longer ones? *Amer. Math. Monthly*, **84**(1977) 633–635; *MR* **57** #3056.

R. S. Stevens & R. Shantaram, Computer generated van der Waerden partitions, *Math. Comput.*, **32**(1978) 635–636.

Zoltán István Szabó, An application of Lovász' local lemma—a new lower bound for the van der Waerden number, *Random Structures Algorithms*, **1**(1990) 343–360; *MR* **92c**:11011.

E. Szemerédi, On sets of integers containing no four terms in arithmetic progression, *Acta Math. Acad. Sci. Hungar.*, **20**(1969) 89–104.

E. Szemerédi, On sets of integers containing no k elements in arithmetic progression, *Acta Arith.*, **27**(1975) 199–245.

J. P. Thouvenot, La démonstration de Furstenberg du théorème de Szemerédi sur les progressions arithmétiques, *Lect. Notes in Math.*, Springer-Verlag Berlin, **710**(1979) 221–232; *MR* **81c**:10072.

B. L. van der Waerden, Beweis einer Baudet'schen Vermutung, *Nieuw Arch. Wisk.* (2), **15**(1927) 212–216.

B. L. van der Waerden, How the proof of Baudet's conjecture was found, in *Studies in Pure Mathematics*, Academic Press, London, 1971, 251–260.

E. Witt, Ein kombinatorische Satz der Elementargeometrie, *Math. Nachr.*, **6**(1952) 261–262.

E11 Schur's problem. Partitioning integers into sum-free classes.

Schur proved that if the integers less than $n!e$ are partitioned into n classes in any way, then $x + y = z$ can be solved in integers within one class. Let $s(n)$ be the largest integer such that there exists a partition of the integers $[1, s(n)]$ into n classes with no solutions in any class. Abbott & Moser obtained the lower bound $s(n) > (89)^{n/4 - c\ln n}$ for some c and all sufficiently large n and Abbott & Hanson obtained $s(n) > c(89)^{n/4}$, improving Schur's own estimate of $s(n) \geq (3^n + 1)/2$. This last result is in fact sharp for $n = 1$, 2 and 3, but it is too low for larger values of n. The value $s(4) = 44$ was computed by Baumert: for example, the first 44 numbers may be split into four sum-free classes

$$\{1,3,5,15,17,19,26,28,40,42,44\}, \quad \{2,7,8,18,21,24,27,33,37,38,43\},$$
$$\{4,6,13,20,22,23,25,30,32,39,41\}, \quad \{9,10,11,12,14,16,29,31,34,35,36\}.$$

Later Fredricksen showed that $s(5) \geq 157$ (see **E12** for his example) and this improves the lower bound for all subsequent Schur numbers: $s(n) \geq c(315)^{n/5}$ $(n > 5)$.

Robert Irving has slightly improved Schur's upper bound from $\lfloor n!e \rfloor$ to $\lfloor n!(e - \frac{1}{24}) \rfloor$. This result also appears in O'Sullivan's Ph.D. thesis (see **E28**). Eugene Levine says that this seems to be the best that can be deduced from Jon Folkman's result that the Ramsey number $R(3,3,3,3) \leq 65$. Also, Schinzel notes that the result ascribed to Irving was attributed by the latter to Earl Glen Whitehead.

Denote by $v = \sigma(m,n)$ the least integer v such that any partition of $\{1,2,\ldots,v\}$ into n subsets has a part containing $a_1, \ldots, a_m$ (not necessarily distinct) which satisfy $a_1 + \ldots + a_{m-1} = a_m$, i.e., $s(n) = \sigma(3,n)$. Beutelspacher & Brestovansky note that $\sigma(m,1) = m-1$ and $\sigma(2,n) = 1$ and prove that $\sigma(m,2) = m^2 - m - 1$. They exhibit 3-sumfree 6- and 7-partitions that show that $\sigma(3,6) \geq 476$ & $\sigma(3,7) \geq 1430$. Hence $\sigma(3,n) \geq \frac{1}{2}(2859 \cdot 3^{n-7} + 1)$ for $n \geq 7$. They also define and investigate Schur numbers of arithmetic progressions.

E. & G. Szekeres and Schönheim have considered what Bill Sands calls an un-Schur problem. Call a partition of the integers $[1,n]$ into three classes **admissible** if there is *no* solution to $x+y=z$ with x, y, z in *distinct* classes. There is no admissible partition with the size of each class $> \frac{1}{4}n$.

Is it true that if the integers are split into r classes, then some class contains three distinct integers x, y, z satisfying $\frac{1}{x} + \frac{1}{y} = \frac{1}{z}$? T. C. Brown has verified this for $r = 2$.

Harvey L. Abbott, PhD thesis, Univ. of Alberta, 1965.

H. L. Abbott & D. Hanson, A problem of Schur and its generalizations, *Acta Arith.*, **20**(1972) 175–187.

H. L. Abbott & L. Moser, Sum-free sets of integers, *Acta Arith.*, **11**(1966) 393–396; *MR* **34** #69.

L. D. Baumert, Sum-free sets, *Jet Propulsion Lab. Res. Summary, No. 36-10*, **1**(1961) 16–18.

Albrecht Beutelspacher & Walter Brestovansky, Generalized Schur numbers, in Combinatorial Theory, *Springer Lecture Notes in Math.*, **969**(1982) 30–38.

S. L. G. Choi, The largest sum-free subsequence from a sequence of n numbers, *Proc. Amer. Math. Soc.*, **39**(1973) 42–44; *MR* **47** #1771.

S. L. G. Choi, J. Komlós & E. Szemerédi, On sum-free subsequences, *Trans. Amer. Math. Soc.*, **212**(1975) 307–313; *MR* **51** #12769.

Paul Erdős, Some problems and results in number theory, in *Number Theory and Combinatorics* (Japan, 1984) World Sci. Publishing, Singapore, 1985, 65–87; *MR* **87g**:11003.

H. Fredricksen, Five sum-free sets, *Proc. 6th SE Conf. Graph Theory, Combin. & Comput., Congressus Numerantium* **14** Utilitas Math., 1975, 309–314.

R. W. Irving, An extension of Schur's theorem on sum-free partitions, *Acta Arith.*, **25**(1973) 55–63.

J. Komlós, M. Sulyok & E. Szemerédi, Linear problems in combinatorial number theory, *Acta Math. Acad. Sci. Hungar.*, **26**(1975) 113–121; *MR* **51** #342.

L. Mirsky, The combinatorics of arbitrary partitions, *Bull. Inst. Math. Appl.*, **11**(1975) 6–9.

J. Schönheim, On partitions of the positive integers with no x, y, z belonging to distinct classes satisfying $x+y=z$, in R. A. Mollin (ed.) Number Theory (*Proc. 1st Conf. Canad. Number Theory Assoc., Banff 1988*, de Gruyter, 1990, 515–528; *MR* **92d**:11018.

I. Schur, Über die Kongruenz $x^m + y^m \equiv z^m \bmod p$, *Jahresb. Deutsche Math.-Verein.*, **25**(1916) 114–117.

Esther & George Szekeres, Adding numbers, *James Cook Math. Notes*, **4** no. 35 (1984) 4073-4075.

W. D. Wallis, A. P. Street & J. S. Wallis, *Combinatorics: Room Squares, Sum-free Sets, Hadamard Matrices*, Springer-Verlag, 1972.

Earl Glen Whitehead, The Ramsey number $N(3,3,3,3;2)$, *Discrete Math.*, **4**(1973) 389–396; *MR* **47** #3229.

Š. Znám, Generalisation of a number-theoretic result, *Mat.-Fyz. Časopis*, **16**(1966) 357–361.

Š. Znám, On k-thin sets and n-extensive graphs, *Math. Časopis*, **17**(1967) 297–307.

E12 The modular version of Schur's problem.

A similar problem to Schur's was considered by Abbott & Wang. Let $t(n)$ be the largest integer m so that there is a partition of the integers from 1 to m into n classes, with no solution to the congruence

$$x + y \equiv z \bmod (m+1)$$

in any class. Clearly $t(n) \leq s(n)$, where $s(n)$ is as in Schur's problem (**E11**), but for $n = 1$, 2 or 3, we have equality, $t(1) = s(1) = 1$, $t(2) = s(2) = 4$, $t(3) = s(3) = 13$. Indeed, the only three partitions of [1,13] into three sets satisfying the sum-free condition,

$$\{1,4,10,13\} \quad \{2,3,11,12\} \quad \{5,6,8,9\}$$

(with 7 in any of the three sets) all satisfy the seemingly more restrictive congruence-free condition, modulo 14, while Baumert's example (**E11**) shows only one failure: $33 + 33 \equiv 21 \bmod 45$ in the second set. In fact

Baumert found 112 ways of partitioning [1,44] into four sum-free sets, and some of these are sum-free mod 45, so $t(4) = 44$. An example is

$$\{\pm1, \pm3, \pm5, 15, \pm17, \pm19\}, \quad \{\pm2, \pm7, \pm8, \pm18, \pm21\}$$

$$\{\pm4, \pm6, \pm13, \pm20, \pm22, 30\}, \quad \{\pm9, \pm10, \pm11, \pm12, \pm14, \pm16\}.$$

Abbott & Wang obtained the inequality

$$f(n_1 + n_2) \geq 2f(n_1)f(n_2)$$

which holds for $f(n) = s(n) - \frac{1}{2}$ and leads to the same lower bound that Schur obtained for his problem, $t(n) \geq (3^n + 1)/2$. Indeed, they obtain evidence that $t(n) = s(n)$. Moreover, the example of Fredricksen

$$\pm\{1, 4, 10, 16, 21, 23, 28, 34, 40, 43, 45, 48, 54, 60\},$$

$$\pm\{2, 3, 8, 9, 14, 19, 20, 24, 25, 30, 31, 37, 42, 47, 52, 65, 70\},$$

$$\pm\{5, 11, 12, 13, 15, 29, 32, 33, 35, 36, 39, 53, 55, 56, 57, 59, 77, 79\},$$

$$\pm\{6, 7, 17, 18, 22, 26, 27, 38, 41, 46, 50, 51, 75\},$$

$$\pm\{44, 49, 58, 61, 62, 63, 64, 66, 67, 68, 69, 71, 72, 73, 74, 76, 78\},$$

which shows that $s(5) \geq 157$ is also sum-free mod 158 so that $t(5) \geq 157$ and $t(n) > c(315)^{n/5}$ as well.

Alon & Kleitman call a subset A of a commutative group **sum-free** if no sum of two elements of A is in A, $(A + A) \cap A = \varnothing$, and they show that every set of n nonzero elements of such a group contains a sum-free subset of cardinality $> \frac{2}{7}n$. That $\frac{2}{7}$ is best possible follows from a result of Rhemtulla & Street, though it can be improved for particular groups. They also show that any set of n nonzero integers contains a sum-free subset of cardinality $> \frac{1}{3}n$, where $\frac{1}{3}$ cannot be replaced by $\frac{12}{29}$. Füredi notes that the set {1,2,3,4,5,6,8,9,10,18} shows that it cannot be replaced by $\frac{2}{5}$: is $\frac{1}{3}$ best possible?

Erdős lets $f(n)$ be the smallest integer for which the integers less than n can be partitioned into $f(n)$ classes so that n is not the sum of distinct members of the same class. For example, $f(11) = 2$, because of the partition $\{1, 3, 4, 5, 9\}$, $\{2, 6, 7, 8, 10\}$, but $f(12) = 3$. Erdős can prove $f(n) < n^{1/3}/\ln n$ but is unable to show that $f(n) > n^{1/3-\epsilon}$.

H. L. Abbott & E. T. H. Wang, Sum-free sets of integers, *Proc. Amer. Math. Soc.*, **67**(1977) 11-16; *MR* **58** #5571.

Noga Alon & Daniel J. Kleitman, Sum-free subsets, *A tribute to Paul Erdős,* Cambridge Univ. Press, Cambridge, 1990, 13–26; *MR* **92f**:11020.

H. Fredricksen, Schur numbers and the Ramsey number $N(3, 3, \ldots, 3; 2)$, *J. Combin. Theory Ser. A*, **27**(1979), 376–377.

A. H. Rhemtulla & Anne Penfold Street, Maximum sum-free sets in elementary Abelian p-groups, *Canad. Math. Bull.*, **14**(1971) 73–80.

E13 Partitioning into strongly sum-free classes.

Turán has shown that if the integers $[m, 5m+3]$ are partitioned into two classes in any way, then in at least one of them the equation $x + y = z$ is solvable with $x \neq y$, and that this is not true for the integers $[m, 5m+2]$. The uniqueness of the partition of $[m, 5m+2]$ into two sum-free sets has been demonstrated by Znám.

Turán also considered the problem where x, y are not necessarily distinct. Define $s(m,n)$ as the least integer s such that however the interval $[m, m+s]$ is partitioned into n classes, one of them contains a solution of $x + y = z$. His result corresponding to the first problem is $s(m,2) = 4m$. Clearly $s(1,n) = s(n) - 1$, where $s(n)$ is as in **E11**, and Irving's result implies that $s(m,n) \leq m\lfloor n!(e-\frac{1}{24})-1\rfloor$. Abbott & Znám (see **E11**) independently noted that $s(m,n) \geq 3s(m,n-1)+m$ so that $s(m,n) \geq m(3^n-1)/2$.

Abbott & Hanson call a class **strongly sum-free** if it contains no solution to either of the equations $x + y = z$ or $x + y + 1 = z$. They show that if $r(n)$ is the least r such that however $[1, r]$ is partitioned into n classes, one of them contains such a solution, then

$$r(m+n) \geq 2r(n)s(m) - r(n) - s(m) + 1.$$

They used this to improve the lower bound for $s(m,n)$; their method, with Fredericksen's example, now gives $s(m,n) > cm(315)^{n/5}$.

Š. Znám, Megjegyzések Turán Pál egy publikálatlan eremén yéhez, *Mat. Lapok*, **14** (1963) 307–310.

E14 Rado's generalizations of van der Waerden's and Schur's problems.

Rado has considered a number of generalizations of van der Waerden's and Schur's problems. For example he shows that for any natural numbers a, b, c, there is a number u so that however the numbers $[1, u]$ are partitioned into two classes, there is a solution of $ax + by = cz$ in at least one of the classes. He gives a value for u, but, as in Schur's original problem, this is not best possible. For instance, with $2x+y = 5z$, the theorem gives $u = 20$, whereas it is true even for $u = 15$, though not for any smaller value of u: neither of the sets

$$\{1,4,5,6,9,11,14\} \qquad \{2,3,7,8,10,12,13\}$$

contains a solution of $2x + y = 5z$. If we are allowed *three* sets, then 45 is the least value for u, since the three sets

$$\{1,4,5,6,9,11,14,16,19,20,21,24,26,29,31,34,36,39,41,44\},$$

$$\{2,3,7,8,10,12,13,15,17,18,22,23,27,28,32,33,37,38,42,43\},$$
$$\{6,7,8,9,25,30,35,40\}$$

contain all the numbers [1,44], even with 6, 7, 8 and 9 duplicated.

Rado called the equation $\sum a_i x_i = 0$, where the a_i are nonzero integers, **n-fold regular** if there is a number $u(n)$, which we can assume to be minimal, such that however the interval $[1, u(n)]$ is partitioned into n classes, at least one class contains a solution to the equation. He called it **regular** if it was n-fold regular for all n, and showed that an equation was regular just if $\sum a_j = 0$ for some subset of the a_i. For example, if $a_1 = a_2 = 1$ and $a_3 = -1$, we have Schur's original problem with $u(n) = s(n)$. Salié and Abbott considered the problem of finding lower bounds for $u(n)$; see **E10** and **E11** for references.

The example with $a_1 = 2$, $a_2 = 1$, $a_3 = -5$ is *not* regular, since, although we have seen that it is both 2-fold and 3-fold regular, it is not 4-fold regular. For, put every number $5^k l$, where $5 \nmid l$, into just one of four classes, according as k is even or odd, and l is $\equiv \pm 1$ or ± 2 mod 5. It can be verified that none of these four classes contains a solution of $2x + y = 5z$.

Rado asked if there exist, for every k, equations which are k-regular, but not $(k+1)$-regular.

For the equations $2x_1 + x_2 = 2x_3$ and $x_1 + x_2 + x_3 = 2x_4$, Salié, Abbott, and Abbott & Hanson obtained successively better lower bounds, culminating in $u(n) > c(12)^{n/3}$ and $c(10)^{n/3}$ respectively.

Vera Sós asks for the maximum size of subset of $[1, n]$ such that Rado's equation has no solution in the subset. For example if $a_1 = a_2 = 1$ and $a_3 = -2$, the answer is in the interval $[n\exp(-\sqrt{\ln n}), n/(\ln n)^\alpha]$. If $a_1 = a_2 = 1$ and $a_3 = a_4 = -1$, we have a Sidon set (compare **C9**) and the answer is $\approx \sqrt{n}$. If $a_1 = a_2 = 1$ and $a_3 = -1$, the answer is $n/2$. It is known more generally that the answer is $o(n)$ just if $x_1 = x_2 = \ldots = 1$ is a solution of Rado's equation. Can the answer ever be comparable to n_α with $\frac{1}{2} < \alpha < 1$?

Compare problems **E10–14** with **C14–16**.

Walter Deuber, Partitionen und lineare Gleichungssysteme, *Math. Z.*, **133** (1973) 109–123.

R. Rado, Studien zur Kombinatorik, *Math. Z.*, **36**(1933) 424–480.

E. R. Williams, M.Sc. thesis, Memorial University, 1967.

E15 A recursion of Göbel.

F. Göbel has remarked that the recursion $x_0 = 1$,

$$x_n = (1 + x_0^2 + x_1^2 + \ldots + x_{n-1}^2)/n \qquad n = 1, 2, \ldots$$

[or, for $n > 0$, $(n+1)x_{n+1} = x_n(x_n + n)$] yields integers

$$x_1 = 2, 3, 5, 10, 28, 154, 3520, 1551880, 267593772160, \ldots$$

for a long time, but Hendrik Lenstra found that x_{43} was not an integer!

The corresponding sequence with cubes in place of squares holds out as far as x_{89}. Henry Ibstedt has made extensive calculations, for various powers k and various initial values a_0. The table shows the rank of the first noninteger member of the sequence

k	2	3	4	5	6	7	8	9	10	11
$x_1 = 2$	43	89	97	214	19	239	37	79	83	239
$x_1 = 3$	7	89	17	43	83	191	7	127	31	389
$x_1 = 4$	17	89	23	139	13	359	23	158	41	239
$x_1 = 5$	34	89	97	107	19	419	37	79	83	137
$x_1 = 6$	17	31	149	269	13	127	23	103	71	239
$x_1 = 7$	17	151	13	107	37	127	37	103	83	239
$x_1 = 8$	51	79	13	214	13	239	17	163	71	239
$x_1 = 9$	17	89	83	139	37	191	23	103	23	169
$x_1 = 10$	7	79	23	251	347	239	7	163	41	239
$x_1 = 11$	34	601	13	107	19	478	37	79	31	389

Raphael Robinson has observed that, in contrast to Göbel's sequence, the recurrence

$$x_n x_{n-k} = a x_{n-p} x_{n-k+p} + b x_{n-q} x_{n-k+q} + c x_{n-r} x_{n-k+r}$$

appears to generate integers from the starting values $x_0 = x_1 = \ldots = x_k = 1$ for any integers $a \geq 0$, $b \geq 0$, $c \geq 0$, $p \geq 1$, $q \geq 1$, $r \geq 1$, k such that $p + q + r = k$.

David Gale, Mathematical Entertainments, *Math. Intelligencer*, **13**(1991) No. 1, 40–43.

Henry Ibstedt, Some sequences of large integers, *Fibonacci Quart.*, **28**(1990) 200–203; *MR* **91h**:11011.

Janice L. Malouf, An integer sequence from a rational recursion, *Discrete Math.*, **110**(1992) 257-261.

Raphael M. Robinson, Periodicity of Somos sequences, *Proc. Amer. Math. Soc.*, **116**(1992) 613–619; *MR* **93a**:11012.

Michael Somos, Problem 1470, *Crux Mathematicorum*, **15**(1989) 208.

E16 Collatz's sequence.

When he was a student, L. Collatz asked if the sequence defined by $a_{n+1} = a_n/2$ (a_n even), $a_{n+1} = 3a_n + 1$ (a_n odd) is tree-like in structure, apart from the cycle 4, 2, 1, 4, ... (Figure 16) in the sense that, starting from any integer a_1, there is a value of n for which $a_n = 1$. This has been verified for all $a_1 \leq 2 \cdot 10^{12}$ and for many larger numbers. Eliahou has shown that any nontrivial cycle has period at least 17087915.

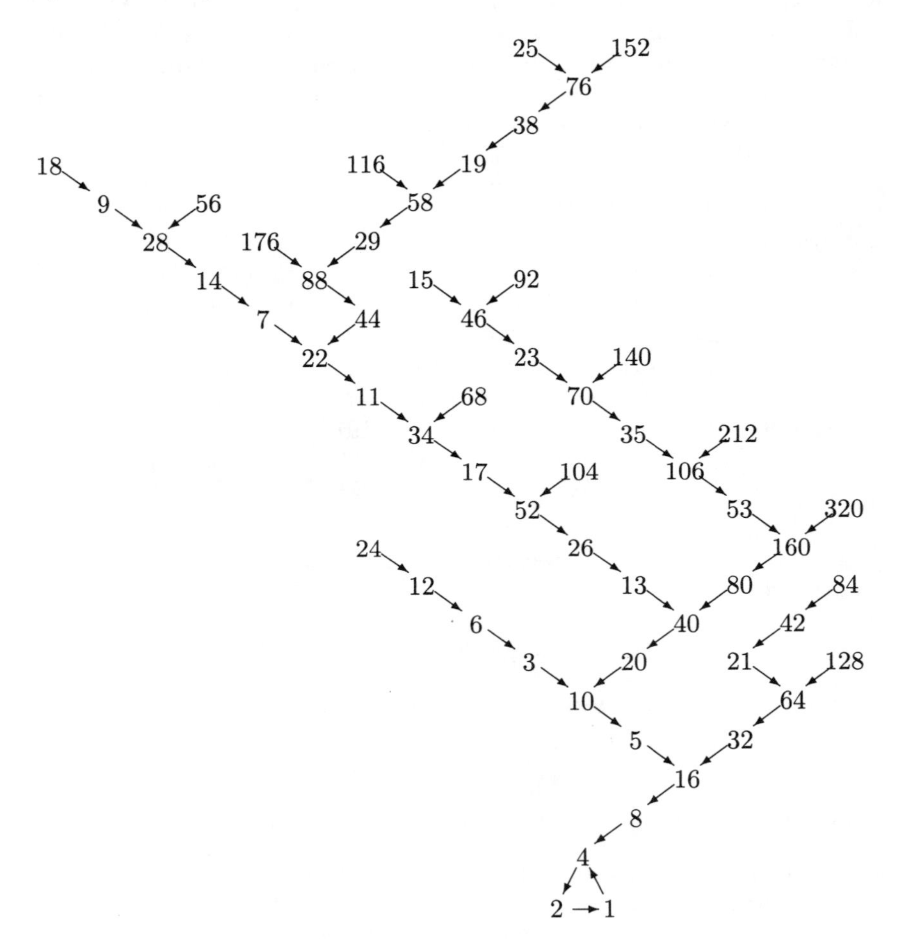

Figure 16. Is the Collatz Sequence Tree-like?

If $3a_n + 1$ is replaced by $3a_n - 1$ (or if we allow negative integers) then it seems likely that any sequence concludes with one of the cycles {1,2}, {5,14,7,20,10} or

{17,50,25,74,37,110,55,164,82,41,122,61,182,91,271,136,68,34}.

This is true for all $a_1 \leq 10^8$.

David Kay and others define the sequence more generally by $a_{n+1} = a_n/p$ if $p|n$ and $a_{n+1} = a_n q + r$ if $p \nmid a_n$ and asks if there are numbers p,

q, r for which the problem can be settled. For $(p,q,r) = (2,5,1)$ or $(2,7,1)$ it certainly seems very plausible that any sequence will increase rapidly, if erratically, but it seems to be just as hard to prove anything as it is for the original problem. The literature is enormous; would-be solvers are urged to study carefully the writings of Lagarias.

Define $f(n)$ to be the largest odd divisor of $3n+1$. Zimian asked if

$$\prod_{i=1}^{m} n_i = \prod_{i=1}^{m} f(n_i)$$

holds for any (multi)set $\{n_i\}$ of integers $n_i > 1$. Erdős found that

$$65 \cdot 7 \cdot 7 \cdot 11 \cdot 11 \cdot 17 \cdot 17 \cdot 13 = 49 \cdot 11 \cdot 11 \cdot 17 \cdot 17 \cdot 13 \cdot 13 \cdot 5.$$

Call an integer n **self-contained** if n divides $f^k(n)$ for some $k \geq 1$. If this happens and if the Collatz sequence $n^* = f^k(n)/n$ reaches 1, then the set

$$\{n, f(n), \ldots, f^{k-1}(n), n^*, f(n^*), \ldots, 1\}$$

is a set such as the above. A computer search for $n \leq 10^4$ yielded five self-contained integers: 31, 83, 293, 347 and 671.

These mappings are not one-to-one; you can't retrace the history of a sequence, since there is often no unique inverse.

J.-P. Allouche, Sur la conjecture de "Syracuse-Kakutani-Collatz," *Séminaire de Théorie des Nombres*, 1978/79, Exp. No. 9, Talence, 1979; *MR* **81g**:10014.

David Applegate & Jeffery C. Lagarias, Density bounds for the $3x+1$ problem, Abstract 882-11-10, *Abstracts Amer. Math. Soc.*, **14**(1993) 414.

Michael Beeler, William Gosper & Rich Schroeppel, Hakmem, Memo 239, Artificial Intelligence Laboratory, M.I.T., 1972, p. 64.

Daniel J. Bernstein, A noniterative 2-adic statement of the $3N+1$ conjecture, *Proc. Amer. Math. Soc.*, (1993).

David Boyd, Which rationals are ratios of Pisot sequences? *Canad. Math. Bull.*, **28**(1985) 343–349; *MR* **86j**:11078.

R. E. Crandall, On the "$3x+1$" problem, *Math. Comput.*, **32**(1978) 1281–1292; *MR* **58** #494.

J. L. Davidson, Some comments on an iteration problem, Proc. 6th Manitoba Conf. Numerical Math., 1976, *Congressus Numerantium*, **18**(1977) 155–159.

S. Eliahou, The $3x+1$ problem: new lower bounds on nontrivial cycle lengths, *Discrete Math.*, **118**(1993) 45–56.

C. J. Everett, Iteration of the number-theoretic function $f(2n) = n$, $f(2n+1) = 3n+2$, *Advances in Math.*, **25**(1977) 42–45; *MR* **56** #15552.

P. Filipponi, On the $3n+1$ problem: something old, something new, *Rend. Mat. Appl.*(7) **11**(1991) 85–103; *MR* **92i**:11031.

L. E. Garner, On the Collatz $3n+1$ algorithm, *Proc. Amer. Math. Soc.*, **82**(1981) 19–22; *MR* **82j**:10090.

Lynn E. Garner, On heights in the Collatz $3n+1$ problem, *Discrete Math.*, **55**(1985) 57–64; *MR* **86j**:11005.

E. Heppner, Eine Bemerkung zum Hasse-Syracuse-Algorithmus, *Arch. Math. (Basel)*, **31**(1977/79) 317–320; *MR* **80d**:10007.

I. N. Herstein & I. Kaplansky, *Matters Mathematical*, 2nd ed., Chelsea, 1978, pp. 44–45.

David C. Kay, *Pi Mu Epsilon J.*, **5**(1972) 338.

I. Korec & Š. Znám, A note on the $3x+1$ problem, *Amer. Math. Monthly*, **94**(1987) 771–772.

I. Krasikov, How many numbers satisfy the $3x+1$ conjecture? *Internat. J. Math. Math. Sci.*, **12**(1989) 791–796; *MR* **90k**:11013.

Jeffrey C. Lagarias, The $3x+1$ problem and its generalizations, *Amer. Math. Monthly*, **92**(1985) 3–23, *MR* **86i**:11043.

Jeffrey C. Lagarias, The set of rational cycles for the $3x+1$ problem, *Acta Arith.*, **56**(1990) 33–53, *MR* **91i**:11024.

J. C. Lagarias, H. A. Porta & K. B. Stolarsky, Asymmetric tent map expansions I: eventually periodic points, *J. London Math. Soc.*, **47**(1993) 542–556.

Jeffrey C. Lagarias & A. Weiss, The $3x+1$ problem: two stochastic models, *Ann. Appl. Probab.*, **2**(1992) 229–261.

K. R. Matthews & A. M. Watts, A generalization of Hasse's generalization of the Syracuse algorithm, *Acta Arith.*, **43**(1984) 167–175; *MR* **85i**:11068.

K. R. Matthews & A. M. Watts, A Markov approach to the generalized Syracuse algorithm, *Acta Arith.*, **45**(1985) 29–42; *MR* **87c**:11071.

Herbert Möller, Über Hasses Verallgemeinerung der Syracuse-Algorithmus (Kakutani's problem), *Acta Arith.*, **34**(1978) 219–226; *MR* **57** #16246.

Helmut Müller, Das "$3n+1$"-Problem, *Mitt. Math. Ges. Hamburg*, **12**(1991) 231–251.

Daniel A. Rawsthorne, Imitation of an iteration, *Math. Mag.*, **58**(1985) 172–176; *MR* **86i**:40001.

J. W. Sander, On the $(3N+1)$-conjecture, *Acta Arith.*, **55**(1990) 241–248; *MR* **91m**:11052.

J. Shallit, The "$3x+1$" problem and finite automata, *Bull. Europ. Assoc. Theor. Comput. Sci.*, **46**(1991) 182–185.

Ray P. Steiner, On the "$Qx+1$ problem", Q odd, *Fibonacci Quart.*, **19**(1981) 285–288; II, 293–296.

Riho Terras, A stopping time problem on the positive integers, *Acta Arith.*, **30**(1976) 241–252; *MR* **58** #27879 (and see **35**(1979) 100–102; *MR* **80h**:10066).

Ilan Vardi, Computational Recreations in *Mathematica*©, Addison-Wesley, Redwood City CA, 1991, Chap. 7.

G. Venturini, Iterates of number-theoretic functions with periodic rational coefficients (generalization of the $3x+1$ problem), *Stud. Appl. Math.* **86**(1992) 185–218; *MR* **93b**:11102.

Stan Wagon, The Collatz problem, *Math. Intelligencer*, **7**(1985) 72–76.

Masaji Yamada, A convergence proof about an integral sequence, *Fibonacci Quart.*, **18**(1980) 231–242; see *MR* **82d**:10026 for errors.

E17 Permutation sequences.

The situation is different, though no more clear, in the case of **permutation sequences**. A simple example, probably the inverse of Collatz's original problem (see **E16** and Lagarias's article cited there), is

$$a_{n+1} = 3a_n/2 \quad (a_n \text{ even}), \qquad a_{n+1} = \lfloor (3a_n+1)/4 \rfloor \quad (a_n \text{ odd}),$$

or, perhaps more perspicuously,

$$2m \to 3m \qquad 4m-1 \to 3m-1 \qquad 4m+1 \to 3m+1$$

from which it is clear that the inverse operation works just as well. So the resulting structure consists only of disjoint cycles and doubly infinite chains. It is not known whether there is a finite or infinite number of each of these, nor even whether an infinite chain exists. It is conjectured that the only cycles are {1}, {2,3}, {4,6,9,7,5} and

$$\{44, 66, 99, 74, 111, 83, 62, 93, 70, 105, 79, 59\}.$$

Mike Guy, with the help of TITAN, showed that any other cycles have period greater than 320. What is the status of the sequence containing the number 8?

…, 97, 73, 55, 41, 31, 23, 17, 13, 10, 15, 11, **8**,
12, 18, 27, 20, 30, 45, 34, 51, 38, 57, 43, 32, 48, 72, …

Do the numbers 8, 14, 40, 64, 80, 82, 104, 136, 172, 184, 188, 242, 256, 274, 280, 296, 352, 368, 382, 386, 424, 472, 496, 526 530, 608, 622, 638, 640, 652, 670, 688, 692, 712, 716, 752, 760, 782, 784, 800, 814, 824, 832, 860, 878, 904, 910, 932, 964, 980, … each belong to a separate sequence?

There are some intriguing paradoxes: as you go "forward" you multiply by 3/2 if the current number is even, and by about 3/4 if it's odd — and get an erratic "pseudo-GP" of common ratio $3/\sqrt{8} \approx 1.060660172$. On the other hand, as you go "backward" you multiply by 2/3 if the current number is a multiple of 3, and by about 4/3 otherwise — a "pseudo-GP" of common ratio $32^{1/3}/3 \approx 1.058267368$. These two numbers should be reciprocal! We have a sort of discrete analog of an everywhere non-differentiable function. The "derivative" on the right is positive; that on the left is negative. Note that, when going "forwards" each successor of an even number is a multiple of 3 — half the numbers are multiples of three!

J. H. Conway, Unpredictable iterations, in *Proc. Number Theory Conf.*, Boulder CO, 1972, 49–52; *MR* **52** #13717.

David Gale, Mathematical Entertainments, *Math. Intelligencer*, **13**(1991) No. 3, 53–55.

G. Venturini, Iterates of number theoretic functions with periodic rational coefficients (generalization of the $3x+1$ problem), *Stud. Appl. Math.*, **86**(1992) 185–218.

E18 Mahler's Z-numbers.

Mahler considered the following problem: given any real number α, let r_n be the fractional part of $\alpha(3/2)^n$. Do there exist **Z-numbers**, for which $0 \le r_n < \frac{1}{2}$ for all n? Probably not. Mahler shows that there is at most one between each pair of consecutive integers, and that, for x large enough, at most $x^{0.7}$ less than x. Flatto has improved on Mahler's results, but the problem remains unsolved.

A similar question is: is there a rational number r/s $(s \ne 1)$ such that $\lfloor (r/s)^n \rfloor$ is odd for all n? Tijdeman proved that for every odd integer $r > 3$, there are real numbers α such that the fractional part of $\alpha(r/2)^n$ is in $[0, \frac{1}{2})$ for all n.

Littlewood once remarked that it was not known that the fractional part of e^n did not tend to 0 as $n \to \infty$.

Leopold Flatto, Z-numbers and β-transformations, *Symbolic Dynamics and its Applications, Contemporary Math.*, **135**, Amer. Math. Soc., 1992, 181–201.

K. Mahler, An unsolved problem on the powers of 3/2, *J. Austral. Math. Soc.*, **8**(1968) 313–321; *MR* **37** #2694.

R. Tijdeman, Note on Mahler's $\frac{3}{2}$-problem, *Kongel. Norske Vidensk. Selsk. Skr.*, **16**(1972) 1–4.

E19 Are the integer parts of the powers of a fraction infinitely often prime?

Forman & Shapiro have proved that infinitely many integers of the form $\lfloor (4/3)^n \rfloor$ and also of the form $\lfloor (3/2)^n \rfloor$ are composite. A. L. Whiteman conjectures that these two sequences also each contain infinitely many primes. The method appears not to work for other rationals.

W. Forman & H. N. Shapiro, An arithmetic property of certain rational powers, *Comm. Pure Appl. Math.*, **20**(1967) 561–573; *MR* **35** #2852.

E20 Davenport-Schinzel sequences.

Form sequences from an alphabet $[1, n]$ of n letters such that there are no immediate repetitions $\ldots aa \ldots$ and no alternating subsequences

$$\ldots a \ldots b \ldots a \ldots b \ldots$$

of length greater than d. Denote by $N_d(n)$ the maximal length of any such sequence; then a sequence of this length is a **Davenport-Schinzel sequence**. The problem is to determine all D-S sequences, and in particular to find $N_d(n)$. We need only consider **normal** sequences in which the first appearance of an integer of the alphabet comes after the first appearance of every smaller one.

The sequences 12131323, 12121213131313232323, and

$$1\ 2\ 1\ 3\ 1\ 4\ 1 \ldots 1\ \overline{n-1}\ 1\ \overline{n-1}\ \overline{n-2} \ldots 3\ 2\ n\ 2\ n\ 3\ n \ldots n\ \overline{n-1}\ n$$

show that $N_4(3) \geq 8$, $N_8(3) \geq 20$, and $N_4(n) \geq 5n-8$. Davenport & Schinzel showed that $N_1(n) = 1$, $N_2(n) = n$, $N_3(n) = 2n-1$; that $N_4(n) = O(n \ln n / \ln \ln n)$, $\lim N_4(n)/n \geq 8$ and, with J. H. Conway, that $N_4(lm+1) \geq 6lm - m - 5l + 2$, so that $N_4(n) = 5n - 8$ $(4 \leq n \leq 10)$. Z. Kolba showed that $N_4(2m) \geq 11m - 13$ and Mills obtained the values of $N_4(n)$ for $n \leq 21$. For example, the sequence

$$abacadaeafafedcbgbhbhgcicigdjdjgekekgkjihflflhliljlkl$$

(which, Günter Rote notes, was misprinted in the first edition) is part of the proof that $N_4(12) = 53$.

Roselle & Stanton fixed n rather than d and obtained $N_d(2) = d$, $N_d(3) = 2\lfloor 3d/2 \rfloor - 4$ $(d > 3)$, $N_d(4) = 2\lfloor 3d/2 \rfloor + 3d - 13$ $(d > 4)$ and $N_d(5) = 4\lfloor 3d/2 \rfloor + 4d - 27$ $(d > 5)$, though Peterkin observed that this last parenthesis should be $(d > 6)$ since $N_6(5) = 34$. Roselle & Stanton also showed that normal D-S sequences of length $N_{2d+1}(5)$ are unique and that there are just two of length $N_{2d+1}(4)$ and $N_{2d}(5)$. Peterkin exhibited the 56 D-S sequences of length $N_5(6) = 29$ and showed that $N_5(n) \geq 7n - 13$ $(n > 5)$ and $N_6(n) \geq 13n - 32$ $(n > 5)$.

Table 8. Values of $N_d(n)$.

$d \backslash n$	1	2	3	4	5	6	7	8	9	10	11	12	13	14	15	16	17	18	19	20	21
1	1	1	1	1	1	1	1	1	1	1	1	1	1	1	1	1	1	1	1	1	1
2	1	2	3	4	5	6	7	8	9	10	11	12	13	14	15	16	17	18	19	20	21
3	1	3	5	7	9	11	13	15	17	19	21	23	25	27	29	31	33	35	37	39	41
4	1	4	8	12	17	22	27	32	37	42	47	53	58	64	69	75	81	86	92	98	104
5	1	5	10	16	22	29															
6	1	6	14	23	34																
7	1	7	16	28	41																
8	1	8	20	35	53																
9	1	9	22	40	61																
10	1	10	26	47	73																

Rennie & Dobson gave an upper bound for $N_d(n)$ in the form

$$(nd - 3n - 2d + 7)N_d(n) \leq n(d-3)N_d(n-1) + 2n - d + 2 \qquad (d > 3)$$

thus generalizing the result of Roselle & Stanton for $d = 4$.

Szemerédi showed that $N_d(n) < c_d n \log^* n$, where $\log^* n$ is a slow-growing function, the least number of iterations of the exponential function needed to exceed n. More recent work, mainly by Sharir, has shown that the order of $N_d(n)$ is $\Theta(n\alpha(n))$, where $\alpha(n)$ is the incredibly slow-growing inverse of the Akermann function. See the paper of Agarwal, Sharir & Shor for precise details.

P. K. Agarwal, M. Sharir & P. Shor, Sharp upper and lower bounds on the length of general Davenport-Schinzel sequences, *J. Combin. Theory Ser. A*, **52**(1989) 228–274; *MR* **90m**:11034.

H. Davenport & A. Schinzel, A combinatorial problem connected with differential equations, *Amer. J. Math.*, **87**(1965) 684–694; II, *Acta Arith.*, **17**(1970/71) 363–372; *MR* **32** #7426; **44** #2619.

Annette J. Dobson & Shiela Oates Macdonald, Lower bounds for the lengths of Davenport-Schinzel sequences, *Utilitas Math.*, **6**(1974) 251–257; *MR* **50** #9781.

Z. Füredi & P. Hajnal, Davenport-Schinzel theory of matrices, *Discrete Math.*, **103**(1992) 233–251.

D. Gardy & D. Gouyou-Beauchamps, Enumerating Davenport-Schinzel sequences, *RAIRO Inform. Théor. Appl.* **26**(1992) 387–402.

S. Hart & M. Sharir, Nonlinearity of Davenport-Schinzel sequences and of generalized path compression schemes, *Combinatorica*, **6**(1986) 151–177.

P. Komjáth, A simplified construction of nonlinear Davenport-Schinzel sequences, *J. Combin. Theory Ser. A*, **49**(1988) 262–267.

W. H. Mills, Some Davenport-Schinzel sequences, *Congress. Numer.* **9**, Proc. 3rd Manitoba Conf. Numer. Math., 1973, pp. 307–313; *MR* **50** #135.

W. H. Mills, On Davenport-Schinzel sequences, *Utilitas. Math.* **9**(1976) 87–112; *MR* **53** #10703.

R. C. Mullin & R. G. Stanton, A map-theoretic approach to Davenport-Schinzel sequences, *Pacific J. Math.*, **40**(1972) 167–172; *MR* **46** #1745.

C. R. Peterkin, Some results on Davenport-Schinzel sequences, *Congress. Numer.* **9**, Proc. 3rd Manitoba Conf. Numer. Math., 1973, pp. 337–344; *MR* **50** #136.

B. C. Rennie & Annette J. Dobson, Upper bounds for the lengths of Davenport-Schinzel sequences, *Utilitas Math.*, **8**(1975) 181–185; *MR* **52** #13624.

D. P. Roselle, An algorithmic approach to Davenport-Schinzel sequences, *Utilitas Math.*, **6**(1974) 91–93; *MR* **50** #9780.

D. P. Roselle & R. G. Stanton, Results on Davenport-Schinzel sequences, *Congress. Numer.* **1** Proc. Louisiana Conf. Combin. Graph Theory, Comput., (1970) 249–267; *MR* **43** #68.

D. P. Roselle & R. G. Stanton, Some properties of Davenport-Schinzel sequences, *Acta Arith.*, **17**(1970/71) 355–362; *MR* **44** #1641.

M. Sharir, Almost linear upper bounds on the length of generalized Davenport-Schinzel sequences, *Combinatorica*, **7**(1987) 131–143.

Micha Sharir, Improved lower bounds on the length of Davenport-Schinzel sequences, *Combinatorica*, **8**(1988) 117–124; *MR* **89j**:11024.

R. G. Stanton & D. P. Roselle, A result on Davenport-Schinzel sequences, *Combinatorial Theory and its Applications*, Proc. Colloq., Balatonfüred, 1969, North-Holland, 1970, pp. 1023–1027; *MR* **46** #3324.

R. G. Stanton & P. H. Dirksen, Davenport-Schinzel sequences, *Ars Combin.*, **1**(1976) 43–51; *MR* **53** #13106.

E. Szemerédi, On a problem of Davenport and Schinzel, *Acta Arith.*, **25** (1973/74) 213–224; *MR* **49** #244.

E21 Thue sequences.

Thue showed that there are infinite sequences on 3 symbols which contain no two identically equal consecutive segments, and sequences on 2 symbols which contain no three identically equal consecutive segments, and many others have rediscovered these results.

If, instead of identically equal segments, we ask to avoid consecutive segments which are *permutations* of one another, Justin constructed a sequence on 2 symbols without five consecutive segments which are permutations of each other, and Pleasants constructed a sequence on 5 symbols without two such consecutive segments. Dekking has solved the (2,4) and (3,3) problems, but describes the (4,2) case as an interesting open problem. Is there a sequence on 4 symbols without consecutive segments which are permutations of each other?

S. Arshon, Démonstration de l'éxistence des suites asymétriques infinies (Russian. French summary), *Mat. Sb.*, **2**(44)(1937) 769–779.

C. H. Braunholtz, Solution to Problem 5030 [1962,439], *Amer. Math. Monthly*, **70**(1963) 675–676.

T. C. Brown, Is there a sequence on four symbols in which no two adjacent segments are permutations of one another? *Amer. Math. Monthly*, **78**(1971) 886–888.

Richard A. Dean, A sequence without repeats on x, x^{-1}, y, y^{-1}, *Amer. Math. Monthly*, **72**(1965) 383–385.

F. M. Dekking, On repetitions of blocks in binary sequences, *J. Combin. Theory Ser. A*, **20**(1976) 292–299.

F. M. Dekking, Strongly non-repetitive sequences and progression-free sets, *J. Combin. Theory Ser. A*, **27**(1979) 181–185.

R. C. Entringer, D. E. Jackson & J. A. Schatz, On non-repetitive sequences, *J. Combin. Theory Ser. A*, **16**(1974) 159–164.

P. Erdős, Some unsolved problems, *Magyar Tud. Akad. Mat. Kutató Int. Közl.*, **6**(1961) 221–254, esp. p. 240.

A. A. Evdokimov, Strongly asymmetric sequences generated by a finite number of symbols, *Dokl. Akad. Nauk SSSR*, **179**(1968) 1268–1271; *Soviet Math. Dokl.*, **7**(1968) 536–539.

Earl Dennet Fife, Binary sequences which contain no BBb, PhD thesis, Wesleyan Univ., Middletown CT, 1976.

D. Hawkins & W. E. Mientka, On sequences which contain no repetitions, *Math. Student*, **24**(1956) 185–187; *MR* **19**, 241.

G. A. Hedlund, Remarks on the work of Axel Thue on sequences, *Nordisk Mat. Tidskr.*, **15**(1967) 147–150; *MR* **37** #4454.

G. A. Hedlund & W. H. Gottschalk, A characterization of the Morse minimal set, *Proc. Amer. Math. Soc.*, **16**(1964) 70–74.

J. Justin, Généralisation du théorème de van de Waerden sur les semi-groupes répétitifs, *J. Combin. Theory Ser. A*, **12**(1972) 357–367.

J. Justin, Semi-groupes répétitifs, *Sém. IRIA, Log. Automat.*, 1971, 101–105, 108; *Zbl.* 274.20092.

J. Justin, Characterization of the repetitive commutative semigroups, *J. Algebra*, **21**(1972) 87–90; *MR* **46** #277. **12**(1972) 357–367.

John Leech, A problem on strings of beads, *Math. Gaz.*, **41**(1957) 277–278.

W. F. Lunnon & P. A.B. Pleasants, Characterization of two-distance sequences, *J. Austral. Math. Soc. Ser. A*, **53**(1992) 198–218; *MR* **93h**:11027.

Marston Morse, A solution of the problem of infinite play in chess, *Bull. Amer. Math. Soc.*, **44**(1938) 632.

Marston Morse & Gustav A. Hedlund, Unending chess, symbolic dynamics and a problem in semigroups, *Duke Math. J.*, **11**(1944) 202.

P. A. B. Pleasants, Non-repetitive sequences, *Proc. Cambridge Philos. Soc.*, **68**(1970) 267–274.

Helmut Prodinger & Friedrich J. Urbanek, Infinite 0-1 sequences without long adjacent identical blocks, *Discrete Math.*, **28**(1979) 277–289.

H. E. Robbins, On a class of recurrent sequences, *Bull. Amer. Math. Soc.*, **43**(1937) 413–417.

A. Thue, Über unendliche Zeichenreihen, *Norske Vid. Selsk. Skr. I Mat.-Nat. Kl. Christiana*, 1906, No. 7, 1–22.

A. Thue, Über die gegensetige Lage gleicher Teile gewisser Zeichenreihen, *Norske Vid. Selsk. Skr. I Mat.-Nat. Kl. Christiana*, 1912, No. 1, 1–67.

E22 Cycles and sequences containing all permutations as subsequences.

Hansraj Gupta asked, for $n \geq 2$, to find the least positive integer $m = m(n)$ for which a cycle $a_1, a_2, \ldots, a_m$ of positive integers, each $\leq n$, exists such that any given permutation of the first n natural numbers appears as a subsequence (not necessarily consecutive) of

$$a_j, a_{j+1}, \ldots, a_1, a_2, \ldots, a_{j-1}$$

for at least one j, $1 \leq j \leq m$. For example, for $n = 5$, such a cycle is 1, 2, 3, 4, 5, 4, 3, 2, 1, 5, 4, 5, so that $m(5) \leq 12$. He conjectures that $m(n) \leq \lfloor n^2/2 \rfloor$.

Motzkin & Straus used the **ruler function** (exponent of the highest power of 2 which divides k), e.g., $n = 5$, $1 \leq k \leq 31$,

$$1, 2, 1, 3, 1, 2, 1, 4, 1, 2, 1, 3, 1, 2, 1, 5, 1, 2, 1, 3, 1, 2, 1, 4, 1, 2, 1, 3, 1, 2, 1,$$

but this doesn't make use of the cyclic options and gives only $m(n) \leq 2^n - 1$.

E23 Covering the integers with A.P.s.

If S is the union of n arithmetic progressions, each with common difference $\geq k$, where $k \leq n$, Crittenden & Vanden Eynden conjecture that S contains all positive integers whenever it contains those $\leq k2^{n-k+1}$. If this is true, it's best possible. They have proved it for $k = 1$ and 2, and Simpson has proved it for $k = 3$.

R. B. Crittenden & C. L. Vanden Eynden, Any n arithmetic progressions covering the first 2^n integers covers all integers, *Proc. Amer. Math. Soc.*, **24**(1970) 475–481.

R. B. Crittenden & C. L. Vanden Eynden, The union of arithmetic progressions with differences not less than k, *Amer. Math. Monthly*, **79**(1972) 630.

R. Jamie Simpson, Ph.D. thesis, Univ. of Adelaide, 1985.

E24 Irrationality sequences.

Erdős & Straus called a sequence of positive integers $\{a_n\}$ an **irrationality sequence** if $\sum 1/a_n b_n$ is irrational for all integer sequences $\{b_n\}$. What are the irrationality sequences? Find some interesting ones. If $\limsup(\log_2 \ln a_n)/n > 1$, where the log is to base 2, then $\{a_n\}$ is an irrationality sequence. Notice that $\{n!\}$ is not an irrationality sequence, because $\sum 1/n!(n+2) = \frac{1}{2}$. Erdős has shown that $\{2^{2^n}\}$ is an irrationality sequence. The sequence 2, 3, 7, 43, 1807, ..., where $a_{n+1} = a_n^2 - a_n + 1$, is *not* an irrationality sequence, since we may take $b_n = 1$ and the sum of the reciprocals is 1, but what about the sequence of alternate terms, 2, 7, 1807, ... ?

P. Erdős, On the irrationality of certain series, *Nederl. Akad. Wetensch. Proc. Ser. A = Indagationes Math.*, **19**(1957) 212–219; *MR* **19**, 252.

P. Erdős, On the irrationality of certain series, *Math. Student*, **36**(1968) 222–226 (1969); *MR* **41** #6787.

P. Erdős, Some problems and results on the irrationality of the sum of infinite series, *J. Math. Sci.*, **10**(1975) 1–7.

P. Erdős & E. G. Straus, On the irrationality of certain Ahmes series, *J. Indian Math. Soc.*, **27**(1963) 129–133 (1969); *MR* **41** #6787.

E25 Silverman's sequence.

The sequence

$$1, 2, 2, 3, 3, 4, 4, 4, 5, 5, 5, 6, 6, 6, 6, 7, 7, 7, 7, 8, 8, 8, 8, 9, 9, 9, 9, 9, \ldots$$

defined by $f(1) = 1$ and $f(n)$ as the number of occurrences of n in a nondecreasing sequence of integers was attributed to David Silverman in the first edition. It was given as a problem by Golomb, and solved by him, by van Lint, and by Marcus and Fine (see reference below): the asymptotic expression for the n-th term is indeed $\tau^{2-\tau} n^{\tau-1}$ where τ is the golden number $(1+\sqrt{5})/2$. The error term $E(n)$ has been investigated by

Ilan Vardi, who conjectures that

$$E(n) = \Omega_{\pm}\left(\frac{n^{\tau-1}}{\ln n}\right)$$

where $E(n) = \Omega_{\pm}(g(n))$ means that there are constants c_1, c_2 such that $E(n) > c_1 g(n)$ and $E(n) < -c_2 g(n)$ are each true for infinitely many n, but he was unable to prove even that $|E(n)|$ is unbounded. He states some related unsolved problems.

Marshall Hall proved the existence of a sequence such that every positive integer occurs uniquely as the difference of two members of the sequence. For example,

$$1, 2, 4, 8, 16, 21, 42, 51, 102, 112, 224, 235, 470, 486, 972, 990, 1980, 2001, \ldots$$

defined by $a_1 = 1$, $a_2 = 2$, $a_{2n+1} = 2a_{2n}$, $a_{2n+2} = a_{2n+1} + r_n$, where r_n is the least natural number which cannot be represented in the form $a_j - a_i$ with $1 \le i < j \le 2n+1$. There are other possible sequences, but the problem of finding the ones with smallest asymptotic growth remains open.

J. Browkin, Solution of a certain problem of A. Schinzel (Polish), *Prace Mat.*, **3**(1959) 205–207.

R. L. Graham, Problem E1910, *Amer. Math. Monthly*, **73**(1966) 775; remark by C. B. A. Peck, **75**(1968) 80–81.

M. Hall, Cyclic projective planes, *Duke Math. J.*, **4**(1947) 1079–1090.

Daniel Marcus & N. J. Fine, Solutions to Problem 5407, *Amer. Math. Monthly*, **74**(1967) 740–743.

Themistocles M. Rassias, A solution to a problem of R. K. Guy & D. Silverman in number theory, 84T-10-336, *Abstracts Amer. Math. Soc.*, **5**(1984) 330.

W. Sierpiński, *Elementary Theory of Numbers*, 2nd English edition (A. Schinzel) PWN, Warsaw, 1987, chap. 12,4 p. 444.

Ilan Vardi, The error term in Golomb's sequence, *J. Number Theory*, **40**(1992) 1–11; *MR* **93d**:11103.

E26 Epstein's Put-or-Take-a-Square game.

Richard Epstein's Put-or-Take-a-Square game is played with one heap of beans. Two players play alternately. A move is to add or subtract the largest perfect square number of beans that is in the heap. That is, the two players alternately name nonnegative integers a_n, where

$$a_{n+1} = a_n \pm \lfloor\sqrt{a_n}\rfloor^2,$$

the winner being the first to name zero. This is a loopy game and many numbers lead to a draw. For example, from 2 the next player will not take

1, allowing his opponent to win, so she goes to 3. Now to add 1 is a bad move, so her opponent goes back to 2. Similarly 6 leads to a draw with best play: 6, 10, 19!, 35, 60, 109!, 209!, 13!, 22!, 6, ..., where ! means a good move, not factorial! For example, 405 is a bad move after 209, since the next player can go to 5 which is a **$\mathcal{P}$-position** (previous player winning). Similarly, from 60, it is bad to go to 11, an **$\mathcal{N}$-position** , one in which the next player can win (by going to 20).

Do either the $\mathcal{P}$-positions

0, 5, 20, 29, 45, 80, 101, 116, 135, 145, 165, 173, 236, 257, 397, 404, 445, 477, 540, 565, 580, 629, 666, 836, 845, 885, 909, 944, 949, 954, 975, 1125, 1177, ...

or the $\mathcal{N}$-positions

1, 4, 9, 11, 14, 16, 21, 25, 30, 36, 41, 44, 49, 52, 54, 64, 69, 71, 81, 84, 86, 92, 100, 105, 120, 121, 126, 136, 141, 144, 149, 164, 169, 174, 189, 196, 201, 208, 216, 225, 230, 245, 252, 254, 256, 261, ...

have positive density?

E. R. Berlekamp, J. H. Conway & R. K. Guy, *Winning Ways for your Mathematical Plays*, Academic Press, London, 1982, Chapter 15.

E27 Max and mex sequences.

In his master's thesis, Roger Eggleton discussed **max sequences**, in which a given finite sequence a_0, a_1, ..., a_n is extended by defining $a_{n+1} = \max_i(a_i + a_{n-i})$. One of the main results is that the first differences are ultimately periodic. For example, starting from 1, 4, 3, 2 we get 7, 8, 11, 12, 15, 16, ... with differences 3, −1, −1, 5, 1, 3, 1, 3, 1, ... What happens to **mex sequences**, where the **mex** of a set of nonnegative integers is the minimum excluded number, or least nonnegative integer which does not appear in the set? Now the sequence 1, 4, 3, 2 continues

0, 0, 0, 0, 0, 5, 1, 1, 1, 1, 1, 6, 2, 2, 0, 0, 0, 0, 0, 5, 1, 1, 1, 1, 1, 6, ...

Are such sequences ultimately periodic?

A. S. Fraenkel is reminded of the sequence $a_i = \lfloor i\alpha \rfloor$, where α is any real number, which satisfies the inequalities

$$\max(a_{n-i}+a_i) \le a_n \le 1+\min(a_{n-i}+a_i), \qquad 1 \le i < n, \qquad n = 2, 3, \ldots .$$

E.g., $\alpha = \frac{1}{2}(1+\sqrt{5})$ generates the sequence 1, 3, 4, 6, 8, 9, 11, 12, 14, 16, 17, 19, 21, ... which complements that for $b_i = \lfloor i\beta \rfloor$, where $1/\alpha + 1/\beta = 1$. These **Beatty sequences** combine to form **Wythoff pairs**.

Motivation for this problem comes from the analysis of **octal games** using the Sprague-Grundy theory, where ordinary addition is replaced by **nim addition**, i.e., addition in base 2 without carry, or XOR. Now 1, 4, 3, 2 leads to

0, 0, 0, 0, 0, 5, 1, 4, 1, 1, 1, 3, 6, 6, 6, 3, 0, 2, 2, 2, 7, 2, 4, 1, ...

The behavior of such sequences remains a considerable mystery, clarification of which would lead to results about nim-like games.

S. Beatty, Problem 3173, *Amer. Math. Monthly*, **33** (1926) 159 (and see J. Lambek & L. Moser, **61** (1954) 454.

E. R. Berlekamp, J. H. Conway & R. K. Guy, *Winning Ways for your Mathematical Plays*, Academic Press, London, 1982, Chapter 4.

Michael Boshernitzan & Aviezri S. Fraenkel, Nonhomogeneous spectra of numbers, *Discrete Math.*, **34**(1981) 325–327; *MR* **82d**: 10077.

Michael Boshernitzan & Aviezri S. Fraenkel, A linear algorithm for nonhomogeneous spectra of numbers, *J. Algorithms*, **5**(1984) 187–198; *MR* **85j**:11183.

R. B. Eggleton, Generalized integers, M.A. thesis, Univ. of Melbourne, 1969.

Ronald L. Graham, Lin Chio-Shih & Lin Shen, Spectra of numbers, *Math. Mag.*, **51**(1978) 174–176; *MR* **58** #10808.

E28 B_2-sequences.

Call an infinite sequence $1 \le a_1 < a_2 < \dots$ an **A-sequence** if no a_i is the sum of distinct members of the sequence other than a_i. Erdős proved that for every A-sequence, $\sum 1/a_i < 103$, and Levine and O'Sullivan improved this to 4. They also gave an A-sequence whose sum of reciprocals is > 2.035. Abbott, and later Zhang, have given the example

$$\{1, 2, 4, 8, 1 + 24k, 35950 + 24t : 1 \le k \le 55, 0 \le t \le 44\}$$

which improves this to 2.0648. A further block of terms will push this past 2.0649, but not as far as 2.065.

If $1 \le a_1 < a_2 < \dots$ is a B_2**-sequence** (compare **C9**), i.e., a sequence where all the sums of pairs $a_i + a_j$ are different, what is the maximum of $\sum 1/a_i$? There are two problems, according as $i = j$ is permitted or not, but Erdős was unable to solve either of them.

The most obvious B_2-sequence is that obtained by the greedy algorithm (compare **E10**). Each term is the least integer greater than earlier terms which does not violate the distinctness of sums condition; $i = j$ is permitted:

1, 2, 4, 8, 13, 21, 31, 45, 66, 81, 97, 123, 148, 182, ...

Mian & Chowla used this to show the existence of a B_2-sequence with $a_k \ll k^3$. If M is the maximum of $\sum 1/a_i$ over all B_2-sequences and S^* is the sum of the reciprocals of the Mian–Chowla sequence, then $M \ge S^* > 2.156$. But

Levine observes that if $t_n = n(n+1)/2$, then $M \le \sum 1/(t_n+1) < 2.374$, and asks if $M = S^*$? Zhang disproves this by showing that $S^* < 2.1596$ and $M > 2.1597$. The latter result is obtained by replacing the next term, 204, in the Mian-Chowla sequence, by 229, and then continuing with the greedy algorithm.

Let $a_1 < a_2 < \cdots$ be an infinite sequence of integers for which all the triple sums $a_i + a_j + a_k$ are distinct. Erdős offers \$500.00 for a proof or disproof of an old conjecture of his, that $\lim a_n/n^3 = \infty$.

H. L. Abbott, On sum-free sequences, *Acta Arith.*, **48**(1987) 93–96.

P. Erdős, Problems and results in combinatorial analysis and combinatorial number theory, *Graph Theory, Combinatorics and Applications*, Vol. 1 (Kalamazoo MI, 1988) 397–406, Wiley, New York, 1991.

Eugene Levine, An extremal result for sum-free sequences, *J. Number Theory*, **12**(1980) 251–257.

Eugene Levine & Joseph O'Sullivan, An upper estimate for the reciprocal sum of a sum-free sequence, *Acta Arith.*, **34**(1977) 9–24; *MR* **57** #5900.

Abdul Majid Mian & S. D. Chowla, On the B$_2$-sequences of Sidon, *Proc. Nat. Acad. Sci. India Sect. A*, **14**(1944) 3–4; *MR* **7**, 243.

J. O'Sullivan, On reciprocal sums of sum-free sequences, PhD thesis, Adelphi University, 1973.

Zhang Zhen-Xiang, A sum-free sequence with larger reciprocal sum, *Discrete Math.*, (1992)

Zhang Zhen-Xiang, A B_2-sequence with larger reciprocal sum, *Math. Comput.*, **60**(1993) 835–839.

Zhang Zhen-Xiang, Finding finite B_2-sequences with larger $m - a_m^{1/2}$, *Math. Comput.*, **61**(1993); *MR* **93m**:11012.

E29 Sequence with sums and products all in one of two classes.

Partition the integers into two classes. Is it true that there is always a sequence $\{a_i\}$ so that all the sums $\sum \epsilon_i a_i$ and all the products $\prod a_i^{\epsilon_i}$ where the ϵ_i are 0 or 1 with all but a finite number zero, are in the same class? Hindman answered this question of Erdős negatively.

Is there a sequence $a_1 < a_2 < \cdots$ so that all the sums $a_i + a_j$ and products $a_i a_j$ are in the same class? Graham proved that if we partition the integers [1,252] into two classes, there are four distinct numbers x, y, $x+y$ and xy all in the same class. Moreover, 252 is best possible. Hindman proved that if we partition the integers [2,990] into two classes, then one class always contains four distinct numbers x, y, $x+y$ and xy. No corresponding result is known for the integers ≥ 3.

Hindman also proved that if we partition the integers into two classes, there is always an infinite sequence $\{a_i\}$ so that all the sums $a_i + a_j$

($i = j$ permitted) are in the same class. On the other hand he found a decomposition into three classes so that no such infinite sequence exists.

J. Baumgartner, A short proof of Hindman's theorem, *J. Combin. Theory Ser. A*, **17**(1974) 384–386.
Neil Hindman, Finite sums with sequences within cells of a partition of n, *J. Combin. Theory Ser. A*, **17**(1974) 1–11.
Neil Hindman, Partitions and sums and products of integers, *Trans. Amer. Math. Soc.*, **247**(1979) 227–245; *MR* **80b**:10022.
Neil Hindman, Partitions and sums and products — two counterexamples, *J. Combin. Theory Ser. A*, **29**(1980) 113–120.

E30 MacMahon's prime numbers of measurement.

MacMahon's "prime numbers of measurement,"

$$1, 2, 4, 5, 8, 10, 14, 15, 16, 21, 22, 25, 26, 28, 33, 34, 35, 36, 38, 40, 42, \dots$$

are generated by excluding all the sums of two or more consecutive earlier members of the sequence.

If m_n is the nth member of the sequence, and M_n is the sum of the first n members, then George Andrews conjectures that

$$¿ \quad m_n \sim n(\ln n)/\ln\ln n \quad ? \qquad \text{and} \qquad ¿ \quad M_n \sim n^2(\ln n)/\ln(\ln n)^2 \quad ?$$

and poses the following, presumably easier, problems: prove $\lim n^{-\Delta} m_n = 0$ for some $\Delta < 2$; prove $\lim m_n/n = \infty$; prove $m_n < p_n$ for every n, where p_n is the nth prime.

Jeff Lagarias suggests excluding only the sums of *two* or *three* consecutive earlier members, and asked if the resulting sequence

$$1, 2, 4, 5, 8, 10, 12, 14, 15, 16, 19, 20, 21, 24, 25, 27, 28, 32, 33, 34,$$
$$37, 38, 40, 42, 43, 44, 46, 47, 48, 51, 53, 54, 56, 57, 58, 59, 61, \dots$$

has density $\frac{3}{5}$. Don Coppersmith has a better heuristic, suggesting that the answer is 'no.'

More generally, if $1 \le a_1 < a_2 < \cdots < a_k \le n$ is a sequence in which no a is the sum of consecutive earlier members, then Pomerance found that $\max k \ge \lfloor \frac{n+3}{2} \rfloor$ and Róbert Freud later showed that $\max k \ge \frac{19}{36}n$. They notice, with Erdős, that $\max k \le \frac{2}{3}n$, even if we only forbid sums of *two* consecutive earlier members. Coppersmith & Phillips have since shown that $\max k \ge \frac{13}{24}n - O(1)$ and they lower the upper bound to

$$\max k \le (\frac{2}{3} - \epsilon)n + O(\ln n) \quad \text{with} \quad \epsilon = \frac{1}{896}.$$

Erdős asks if the lower density of the sequence is zero; perhaps

$$¿ \qquad \frac{1}{\ln x} \sum_{a_i < x} \frac{1}{a_i} \to 0 \qquad ?$$

G. E. Andrews, MacMahon's prime numbers of measurement, *Amer. Math. Monthly*, **82**(1975) 922–923.

Don Coppersmith & Steven Phillips, On a question of Erdős on subsequence sums, (preprint, Nov. 1992).

Róbert Freud, *James Cook Math. Notes*, Jan. 1993.

R. L. Graham, Problem 1910, *Amer. Math. Monthly*, **73**(1966) 775; solution **75**(1968) 80–81.

Jeff Lagarias, Problem 17, W. Coast Number Theory Conf., Asilomar, 1975.

P. A. MacMahon, The prime numbers of measurement on a scale, *Proc. Cambridge Philos. Soc.*, **21**(1923) 651–654.

Štefan Porubský, On MacMahon's segmented numbers and related sequences, *Nieuw Arch. Wisk.*(3) **25**(1977) 403–408; *MR* **58** #5575.

N. J. A. Sloane, *A Handbook of Integer Sequences*, Academic Press, New York, 1973; sequences 363, 416, 1044.

E31 Three sequences of Hofstadter.

Doug Hofstadter has defined three intriguing sequences.

(a) $a_1 = a_2 = 1$ and $a_n = a_{n-a_{n-1}} + a_{n-a_{n-2}}$ for $n \geq 3$. What is the general behavior of this sequence?

1, 1, 2, 3, 3, 4, 5, 5, 6, 6, 6, 8, 8, 8, 10, 9, 10, 11, 11,
12, 12, 12, 12, 16, 14, 14, 16, 16, 16, 16, 20, 17, 17, ...

Are there infinitely many integers 7, 13, 15, 18, ... that get missed out?

(b) $b_1 = 1$, $b_2 = 2$ and for $n \geq 3$, b_n is the least integer greater than b_{n-1} which can be expressed as the sum of two or more *consecutive* terms of the sequence, so it goes

1, 2, 3, 5, 6, 8, 10, 11, 14, 16, 17, 18, 19, 21, 22, 24, 25, 29,
30, 32, 33, 34, 35, 37, 40, 41, 43, 45, 46, 47, 49, 51, ...

This is a sort of dual of MacMahon's prime numbers of measurement (**E30**). How does the sequence grow?

(c) $c_1 = 2$, $c_2 = 3$, and when $c_1, \ldots, c_n$ are defined, form all possible expressions

$c_i c_j - 1 \quad (1 \leq i < j \leq n)$ and append them to the sequence:

2, 3, 5, 9, 14, 17, 26, 27, 33, 41, 44, 50, 51, 53, 65, 69, 77,
80, 81, 84, 87, 98, 99, 101, 105, 122, 125, 129, ...

Does the result include almost all of the integers?

A similar sequence to the first of these three was given by Conway:

1, 1, 2, 2, 3, 4, 4, 4, 5, 6, 7, 7, 8, 8, 8, 8, 9, ...

defined, for $n \geq 3$, by

$$a(n) = a(a(n-1)) + a(n - a(n-1)).$$

The difficult questions were answered in an entertaining paper by Mallows. Several identities have been obtained by Zeitlin. A variation is to define

$$b(n) = b(b(n-1)) + b(n-1-b(n-1)),$$

but this is related to the previous sequence by $b(n-1) = n - a(n)$. But if we write

$$c(n) = c(c(n-2)) + c(n - c(n-2))$$

then the increments become very irregular, and it is not even clear that $c(n)/n$ tends to a limit.

W. A. Beyer, R. G. Schrandt & S. M. Ulam, Computer studies of some history-dependent random processes, LA-4246, Los Alamos Nat. Lab., 1969.

J. H. Conway, Some crazy sequences, videotaped talk at A.T. & T. Bell Laboratories, 88-07-15.

Peter J. Downey & Ralph E. Griswold, On a family of nested recurrences, *Fibonacci Quart.*, 22(1984) 310–317; *MR* **86e**:11013.

P. Erdős & R. L. Graham, Old and New Problems and Results in Combinatorial Number Theory, *Monographie de L'Enseignement Mathématique, Genève*, **28**(1980) 83–84.

Douglas R. Hofstadter, *Gödel, Escher, Bach*, Vintage Books, New York, 1980, p. 137.

Mark Kac, A history-dependent random sequence defined by Ulam, *Adv. in Appl. Math.*, **10**(1989) 270–277; *MR* **91c:**11042.

Péter Kiss & Béla Zay, On a generalization of a recursive sequence, *Fibonacci Quart.*, **30**(1992) 103–109; *MR* **90e**:11022.

Colin L. Mallows, Conway's challenge sequence, *Amer. Math. Monthly*, **98** (1991) 5–20.

David Newman, Problem E3274, *Amer. Math. Monthly*, **95**(1988) 555.

Stephen M. Tanny, A well-behaved cousin of the Hofstadter sequence, *Discrete Math.*, **105**(1992) 227–239; *MR* **93i**:11029.

David Zeitlin, Explicit solutions and identities for Conway's iterated sequence, *Abstracts Amer. Math. Soc.*, **12**(1991).

E32 B_2-sequences formed by the greedy algorithm.

An old problem of Dickson is still unsolved. Given a set of k integers, $a_1 < a_2 < \ldots < a_k$, define a_{n+1} for $n \geq k$ as the least integer greater than a_n which is *not* of the form $a_i + a_j$, $i, j \leq n$. Except for the prescribed section at the beginning of the sequence, these are sum-free sequences formed by the greedy algorithm (compare **C9, C14, E10, E28**).

Is the sequence of differences $a_{n+1} - a_n$ ultimately periodic?

Such sequences may take a long time before the periodicity appears. For example, even for $k = 2$, if we take $a_1 = 1$, $a_2 = 6$, the sequence is

$$1,6,8,10,13,15,17,22,24,29,31,33,36,38,40,45,47,52,54,56,59,61,63,68,\ldots$$

and one can be forgiven for not immediately recognizing the pattern. Try starting with the set {1,4,9,16,25}; after 82 irregular differences, it settles down to a period of length 224.

Queneau (see reference at **C4**) considered the similar problem with $i < j \le n$ in place of $i, j \le n$. Such **0-additive sequences** have also been conjectured to have ultimately periodic differences. Steven Finch has calculated $1\frac{1}{2}$ million terms of the sequence whose first 6 terms are given as $\{3, 4, 6, 9, 10, 17\}$ without detecting any sign of ultimate periodicity of the differences.

Selmer tells me that Dickson's problem is the particular case $h = 2$ of **Stöhr sequences** : let $a_1 = 1$ and define a_{n+1} for $n \ge k$ to be the least integer greater than a_n which can *not* be written as the sum of at most h addends among a_1, a_2, …, a_n. Compare the h-bases of **C12**. In the great majority of cases, the sequence of differences $a_{n+1} - a_n$ turns out to be ultimately periodic, but there are a few of the examined cases where periodicity has not been established.

Neil J. Calkin, Sum-free sets and measure spaces, PhD thesis, Univ. of Waterloo, 1988.

Neil J. Calkin & Steven R. Finch, Necessary and sufficient conditions for a sum-free set to be ultimately periodic (preprint, 1993).

Peter J. Cameron, Portrait of a typical sum-free set, Surveys in Combinatorics 1987, *London Math. Soc. Lecture Notes*, **123**(1987) Cambridge Univ. Press, 13–42.

L. E. Dickson, The converse of Waring's problem, *Bull. Amer. Math. Soc.*, **40** (1934) 711–714.

Steven R. Finch, Are 0-additive sequences always regular? *Amer. Math. Monthly*, **99**(1992) 671–673.

Ernst S. Selmer, On Stöhr's recurrent h-bases for N, *Kgl. Norske Vid. Selsk. Skrifter*, **3**(1986) 1–15.

Ernst S. Selmer & Svein Mossige, Stöhr sequences in the postage stamp problem, No. **32**(Dec. 1984) Dept. Pure Math., Univ. Bergen, ISSN 0332-5407.

E33 Sequences containing no monotone A.P.s.

Erdős & Graham say that a sequence $\{a_i\}$ has a **monotone** A.P. of length k if there are subscripts $i_1 < i_2 < \cdots < i_k$ such that the subsequence a_{i_j}, $1 \le j \le k$ is either an increasing or a decreasing A.P. If $M(n)$ is the number of permutations of $[1, n]$ which have no monotone 3-term A.P., then Davis et. al have shown that

$$M(n) \geq 2^{n-1} \qquad M(2n-1) \leq (n!)^2 \qquad M(2n) \leq (n+1)(n!)^2$$

They ask if $M(n)^{1/n}$ is bounded.

Davis et al have also shown that any permutation of (all) the positive integers must contain an increasing 3-term A.P., but there are permutations with no monotone 5-term A.P. It is not known whether a monotone 4-term A.P. must always occur.

If the positive integers are arranged as a doubly-infinite sequence then a monotone 3-term A.P. must still occur, but it's possible to prevent the occurrence of 4-term ones.

If *all* the integers are to be permuted then Tom Odda has shown that no 7-term A.P. need occur in the singly-infinite case, but little else is known.

J. A. Davis, R. C. Entringer, R. L. Graham & G. J. Simmons, On permutations containing no long arithmetic progressions, *Acta Arith.*, **34**(1977) 81–90; *MR* **58** #10705.

Tom Odda, Solution to Problem E2440, *Amer. Math. Monthly*, **82**(1975) 74.

E34 Happy numbers.

Reg Allenby's daughter came home from school in Britain with the concept of **happy numbers**. If you iterate the process of summing the squares of the decimal digits of a number, then it's easy to see that you either reach the cycle

$$4 \to 16 \to 37 \to 58 \to 89 \to 145 \to 42 \to 20 \to 4$$

or arrive at 1. In the latter case you started from a happy number. The first hundred happy numbers are

1 7 10 13 19 23 28 31 32 44 49 68 70 79 82 86 91 94 97 100
103 109 129 130 133 139 167 176 188 190 192 193 203 208 219 226 230 236 239 262
263 280 291 293 301 302 310 313 319 320 326 329 331 338 356 362 365 367 368 376
379 383 386 391 392 397 404 409 440 446 464 469 478 487 490 496 536 556 563 565
566 608 617 622 623 632 635 637 638 644 649 653 655 656 665 671 673 680 683 694

It seems that about 1/7 of all numbers are happy, but what bounds on the density can be proved? How many consecutive happy numbers can you have? Can there be arbitrarily many? The first pair is 31, 32; the first triple is 1880, 1881, 1882 and an example of five consecutive happy numbers is 44488, 44489, 44490, 44491, 44492. Give bounds on the length of such a consecutive set in terms of the size of its members. How large can the gaps be in the sequence of happy numbers? We can define the **height** of a happy number to be the number of iterations needed to reach 1. For example, the least happy numbers of

height	0	1	2	3	4	5	6
are	1	10	13	23	19	7	356.

Is 78999 the least happy number of height 7? Give bounds for the size of the least happy number of height h.

If we replace squares by cubes, then the situation is dominated, at least in base 10, by the fact that perfect cubes are congruent to 0 or ± 1 mod 9. The density of the corresponding numbers (1, 10, 100, 112, 121, 211, 778, ...) may be zero. Numbers $\equiv 0 \bmod 3$ converge to 153 and numbers $\equiv 2 \bmod 3$ converge to 371 or 407, so that interest is confined to numbers $\equiv 1 \bmod 3$. These converge to 370, or to one of the 3-cycles (55, 250, 133) or (160, 217, 352), or to one of the 2-cycles (919, 1459) or (136, 244), or, just occasionally, to 1. What proportion goes to each?

And what of higher powers? And different bases?

Henry Ernest Dudeney, *536 Puzzles & Curious Problems* (edited Martin Gardner), Scribner's, New York, 1967, Problem 143, pp. 43, 258–259.

Joseph S. Madachy, *Mathematics on Vacation*, Scribner's, New York, 1966, pp. 163–165.

E35 The Kimberling shuffle.

Clark Kimberling considered the array:

$$\begin{array}{ccccccccccccccccccc}
\boxed{1} & 2 & 3 & 4 & 5 & 6 & 7 & 8 & 9 & 10 & 11 & 12 & 13 & 14 & 15 & 16 & 17 & 18 & 19 \\
2 & \boxed{3} & 4 & 5 & 6 & 7 & 8 & 9 & 10 & 11 & 12 & 13 & 14 & 15 & 16 & 17 & 18 & 19 & 20 \\
4 & 2 & \boxed{5} & 6 & 7 & 8 & 9 & 10 & 11 & 12 & 13 & 14 & 15 & 16 & 17 & 18 & 19 & 20 & 21 \\
6 & 2 & 7 & \boxed{4} & 8 & 9 & 10 & 11 & 12 & 13 & 14 & 15 & 16 & 17 & 18 & 19 & 20 & 21 & 22 \\
8 & 7 & 9 & 2 & \boxed{10} & 6 & 11 & 12 & 13 & 14 & 15 & 16 & 17 & 18 & 19 & 20 & 21 & 22 & 23 \\
6 & 2 & 11 & 9 & 12 & \boxed{7} & 13 & 8 & 14 & 15 & 16 & 17 & 18 & 19 & 20 & 21 & 22 & 23 & 24 \\
13 & 12 & 8 & 9 & 14 & 11 & \boxed{15} & 2 & 16 & 6 & 17 & 18 & 19 & 20 & 21 & 22 & 23 & 24 & 25 \\
2 & 11 & 16 & 14 & 6 & 9 & 17 & \boxed{8} & 18 & 12 & 19 & 13 & 20 & 21 & 22 & 23 & 24 & 25 & 26 \\
18 & 17 & 12 & 9 & 19 & 6 & 13 & 14 & \boxed{20} & 16 & 21 & 11 & 22 & 2 & 23 & 24 & 25 & 26 & 27 \\
\cdots & \cdots & \cdots & \cdots & \cdots & \cdots & \cdots & \cdots & \cdots & \cdots & \cdots & \cdots & \cdots & \cdots & \cdots & \cdots & \cdots & \cdots & \cdots
\end{array}$$

in which each row is obtained from the previous by boxing (and expelling) the main diagonal element, and then reading the first number after the box, the first before the box, the second after the box, the second before the box, and so on until all the initial numbers are read off, and then continuing with all the remaining numbers (still in numerical order). Is every number eventually expelled?

The numbers	1	2	3	4	5	6	7	8	9	10
are expelled on rows	1	25	2	4	3	22	6	8	10	5
and the numbers	11	12	13	14	15	16	17	18	19	20
are expelled on rows	32	83	44	14	7	66	169	11	49595	9

and the numbers	40	68	106	147
are expelled on rows	93167	181393	270186	8765242
and the numbers	242	322	502	669
are expelled on rows	16509502	38293016	118850522	653494691

In the alternate reading from the right and left of the box, one could instead start from the left first, leading to the array

$\boxed{1}$	2	3	4	5	6	7	8	9	10	11	12	13	14	15	16	17	18	19
2	$\boxed{3}$	4	5	6	7	8	9	10	11	12	13	14	15	16	17	18	19	20
2	4	$\boxed{5}$	6	7	8	9	10	11	12	13	14	15	16	17	18	19	20	21
4	6	2	$\boxed{7}$	8	9	10	11	12	13	14	15	16	17	18	19	20	21	22
2	8	6	9	$\boxed{4}$	10	11	12	13	14	15	16	17	18	19	20	21	22	23
9	10	6	11	8	$\boxed{12}$	2	13	14	15	16	17	18	19	20	21	22	23	24
8	2	11	13	6	14	$\boxed{10}$	15	9	16	17	18	19	20	21	22	23	24	25
14	15	6	9	13	16	11	$\boxed{17}$	2	18	8	19	20	21	22	23	24	25	26
11	2	16	18	13	8	9	19	$\boxed{6}$	20	15	21	14	22	23	24	25	26	27
.	.	.	.	.	.	.	.	.	.	.	.	.	.	.	.	.	.	.

which displays a similar chaotic behavior.

In each array there is a small amount of pattern: observe knight's move arithmetic progressions of common difference 3. For example, in the original array, for each $y \geq 0$, the number $n + 3y$ is in position $2y + 1$ on row $x + y$, where, for each $t \geq 0$,

$$\begin{array}{llll} & n = 9 \cdot 2^t - 3t - 7, & 12 \cdot 2^t - 3t - 8, & \text{resp.} \quad 15 \cdot 2^t - 3t - 9 \\ \text{when} & x = 3 \cdot 2^t - 1, & 4 \cdot 2^t - 1, & \text{resp.} \quad 5 \cdot 2^t - 1 \end{array}$$

so that $n + 3y$ is expelled on row $2x - 1 = 2y + 1$. I.e.,

$$\begin{array}{llll} n = & 18 \cdot 2^t - 3t - 13, & 24 \cdot 2^t - 3t - 14, & \text{resp.} \; 30 \cdot 2^t - 3t - 15 \\ \text{is expelled on row} & 6 \cdot 2^t - 3, & 8 \cdot 2^t - 3, & \text{resp.} \quad 10 \cdot 2^t - 3 \end{array}$$

for $t = -1, 0, 1, \ldots$.

Clark Kimberling, Problem 1615, *Crux Mathematicorum*, **17**#2(Feb 1991) 44.

E36 Klarner-Rado sequences.

The sequence

1, 2, 4, 5, 8, 9, 10, 14, 15, 16, 17, 18, 20, 26, 27, 28, 29, 30, 32, 33, 34, 36, 40, 44, 47, 50, 51, 52, 53, 54, 56, 57, 58, 60, 62, 63, 64, 66, 68, 72, 80, 83, 86, 87, 88, 89, 92, 93, 94, 98, 99, 100, 101, 102, 104, 105, 106, 108, 110, 111, 112, 114, 116, 120, 122, 123, 124, 126, 128, 132, 134, 136, ...

is the thinnest which contains 1, and whenever it contains x, also contains $2x$, $3x+2$ and $6x+3$. Does it have positive density?

Several questions of this type were asked in the paper:

David A. Klarner & Richard Rado, Arithmetic properties of certain recursively defined sets, *Pacific J. Math.*, **53**(1974) 445–463.

In the review, *MR* **50** #9784, it was stated that a subsequent paper ("Sets generated by a linear operation", same J., to appear) settles many of the conjectures stated in this paper. Did it ever appear? See also:

David A. Klarner & Richard Rado, Linear combinations of sets of consecutive integers, *Amer. Math. Monthly*, **80**(1973) 985–989.

David A. Klarner & Karel Post, Some fascinating integer sequences, *Discrete Math.*, **106/107**(1992) 303–309; *MR* **93i**:11031.

E37 Mousetrap.

Cayley introduced a permutation problem he called **Mousetrap** which is loosely based on the card game Treize. Suppose that the numbers $1, 2, \dots, n$ are written on cards, one to a card. After shuffling (permuting) the cards, start counting the deck from the top card down. If the number on the card does not equal the count, transfer the card to the bottom of the deck and continue counting. If the two are equal then set the card aside and start counting again from 1. The game is **won** if all the cards have been set aside, but lost if the count reaches $n+1$. Cayley proposed two questions.

1. For each n find all the winning permutations of $1, 2, \dots, n$.

2. For each n find the number of permutations that eliminate precisely i cards for each i, $1 \le i \le n$.

A third question arose during our investigations. Consider a permutation for which every number is set aside. The list of numbers in the order that they were set aside is another permutation. Any permutation obtained in this way we call a **reformed** permutation.

3. Characterize the reformed permutations.

The permutation 4213 is a winning permutation which gives rise to the permutation 2134; this in turn gives the reformed permutation 3214 which is not a winning permutation.

4. For a given n, what is the longest sequence of reformed permutations?

5. Are there sequences of arbitrary length? Are there any cycles other than

$$1 \to 1 \to 1 \to 1 \ldots \quad \text{and} \quad 12 \to 12 \to 12 \to 12 \ldots ?$$

Modular Mousetrap. We can play Mousetrap, but instead of counting n, $n+1$, ... , we can start again, ... , n, 1, 2, ... Now at least as many cards get set aside. In fact if n is prime, then either the initial deck is a derangement, or all cards get set aside, so every sequence cycles or terminates in a derangement. The identity permutation $123\ldots n$ will always form a 1-cycle and now there are also examples of nontrivial cycles.

6. Are there k-cycles for every k ? What is the least value of n which yields a k-cycle?

A. Cayley, A Problem in Permutations, *Quart. Math. J.*, **I**(1857), 79.

A. Cayley, On the Game of Mousetrap, *Quart. J. Pure Appl. Math.*, **XV** (1877), 8-10.

A. Cayley, A Problem on Arrangements, *Proc. Roy. Soc. Edinburgh*, **9**(1878) 338–342.

A. Cayley, Note on Mr. Muir's Solution of a Problem of Arrangement, *Proc. Roy. Soc. Edinburgh*, **9**(1878) 388–391.

Richard K. Guy & Richard J. Nowakowski, Mousetrap, Proc. Erdős80 Keszthely Combin. Conf., 1993. [see also *Amer. Math. Monthly*, **101**(1994).]

T. Muir, On Professor Tait's Problem of Arrangement, *Proc. Royal Soc. Edinburgh*, **9**(1878) 382–387.

T. Muir, Additional Note on a Problem of Arrangement, *Proc. Royal Soc. Edinburgh*, **11**(1882) 187–190.

Adolf Steen, Some Formulae Respecting the Game of Mousetrap, *Quart. J. Pure Appl. Math.*, **XV**(1878), 230-241.

Peter Guthrie Tait, *Scientific Papers*, vol. 1, Cambridge, 1898, 287.

E38 Odd sequences.

Call a sequence of n zeros and ones, $\{a_1, \ldots, a_n\}$, **odd** if each of the n sums $\sum_{i=1}^{n-k} a_i a_{i+k}$, $(k = 0, 1, \ldots, n-1)$ is odd. For example, 1101 is odd. Pelikán conjectured that there were no odd sequences if $n \geq 5$, but Peter Alles showed that there are infinitely many: if **o** is an odd sequence of length n and **x** and **z** are sequences of $n-1$ and $3n-2$ zeros, then **oxozo** and **ozoxo** are odd sequences of length $7n-3$. For

$n =$	1	4	12	16	24	25	36	37	40	45
he finds	1	2	2	8	2	4	2	16	2	16

odd sequences and no others with $n \leq 50$. For example, 101011100011 and its reversal. He asks if n is always congruent to 0 or 1 modulo 4, and if, when there are odd sequences, they are always a power of two in number.

Peter Alles, On a conjecture of J. Pelikán, *J. Combin. Theory Ser. A*, **60** (1992) 312–313; *MR* **93i**:11028.

J. Pelikán, Problem, in *Infinite and Finite Sets, Vol. III* (*Keszthely*, 1973), 1549, *Colloq. Math. Soc. János Bolyai*, **10**, North-Holland, Amsterdam, 1975.

F. None of the Above

The first few problems in this miscellaneous section are about **lattice points**, whose Euclidean coordinates are integers. Most of them are two-dimensional problems, but some can be formulated in higher dimensions as well. Some interesting books are

J. W. S. Cassels, *Introduction to the Geometry of Numbers*, Springer-Verlag, New York, 1972.

L. Fejes Tóth, *Lagerungen in der Ebene, auf der Kugel und in Raum*, Springer-Verlag, Berlin, 1953.

J. Hammer, *Unsolved Problems Concerning Lattice Points*, Pitman, 1977.

O.-H. Keller, *Geometrie der Zahlen*, Enzyklopedia der Math. Wissenschaften **12**, B. G. Teubner, Leipzig, 1954.

C. G. Lekkerkerker, *Geometry of Numbers*, Bibliotheca Mathematica **8**, Walters-Noordhoff, Groningen; North-Holland, Amsterdam, 1969.

C. A. Rogers, *Packing and Covering*, Cambridge Univ. Press, 1964.

F1 Gauß's lattice point problem.

A very difficult unsolved problem is **Gauß's problem**. How many lattice points are there inside the circle with centre at the origin and radius r? If the answer is $\pi r^2 + h(r)$, then Hardy & Landau showed that $h(r)$ is *not* $o(r^{1/2}(\ln r)^{1/4})$. It is conjectured that $h(r) = O(r^{1/2+\epsilon})$. Iwaniec & Mozzochi have shown that $h(r) = O(r^{7/11+\epsilon})$, and the best that is known is $h(r) = O(r^{46/73+\epsilon})$, by Huxley.

One can ask analogous questions in three dimensions for the sphere and regular tetrahedron. For the rectangular tetrahedron of **F22** see the paper of Lehmer and also that of Xu & Yau, but their suggested counterexample to Overhagen's upper bound for arbitrary convex bodies is incorrect.

Chen Jing-Run, The lattice points in a circle, *Sci. Sinica*, **12**(1963) 633–649; *MR* **27** #4799.

Javier Cilleruello, The distribution of the lattice points on circles, *J. Number Theory*, **43**(1993) 198–202.

Andrew Granville, The lattice points of an n-dimensional tetrahedron, *Aequationes Math.*, **41**(1991) 234–241; *MR* **92b**:11070.

Martin Huxley, *Proc. London Math. Soc.* (to appear).

Aleksandar Ivić, Large values of the error term in divisor problems and the mean square of the zeta-function, *Invent. Math.*, **71**(1983) 513–520; *MR* **84i**:10046.
H. Iwaniec & C. J. Mozzochi, On the divisor and circle problems, *J. Number Theory*, **29**(1988) 60–93; *MR* **89g**:11091.
D. H. Lehmer, The lattice points of an n-dimensional tetrahedron, *Duke Math. J.*, **7**(1940) 341–353.
T. Overhagen, Zur Gitterpunktanzahl konvexer Körper im 3-dimensionalen euklidischen Raum, *Math. Ann.*, **216**(1975) 217–224; *MR* **57** #281.
Xu Yijing & Stephen Yau S.-T., A sharp estimate of the number of integral points in a tetrahedron, *J. reine angew. Math.*, **423**(1992) 199–219; *MR* **93d**:11067.

F2 Lattice points with distinct distances.

What is the largest number k of lattice points (x, y), $1 \le x, y \le n$, which can be chosen so that their $\binom{k}{2}$ mutual distances are all distinct? It is easy to see that $k \le n$. This bound can be attained for $n \le 7$, for example for $n = 7$ with the points $(1,1)$, $(1,2)$, $(2,3)$, $(3,7)$, $(4,1)$, $(6,6)$ and $(7,7)$, but not for any larger value of n. Erdős & Guy showed that

$$n^{2/3-\epsilon} < k < cn/(\ln n)^{1/4}$$

and they conjecture that

$$¿ \qquad k < cn^{2/3}(\ln n)^{1/6} \qquad ?$$

One can also ask for "saturated" configurations, containing a *minimum* number of points which determine distinct distances, but such that *no* lattice point may be added without duplicating a distance. Erdős observes that this needs at least $n^{2/3-\epsilon}$ lattice points. In one dimension he cannot improve on $O(n^{1/3})$ and suspects that $O(n^{1/2+\epsilon})$ is best possible.

P. Erdős & R. K. Guy, Distinct distances between lattice points, *Elem. Math.*, **25**(1970) 121–123; *MR* **43** #7406.

F3 Lattice points, no four on a circle.

Erdős & Purdy ask how many of the n^2 lattice points (x, y), $1 \le x, y \le n$ can you choose with no four of them on a circle. It is easy to show $n^{2/3-\epsilon}$, but more should be possible.

What is the smallest t so that you can choose t of the lattice points so that the $\binom{t}{2}$ lines that they determine contain all the n^2 lattice points? It is not hard to show $t > cn^{2/3}$ and Noga Alon has since obtained the bounds

$$cn^{d(d-1)/(2d-1)} \le t(n,d) \le Cn^{d(d-1)/(2d-1)} \ln n$$

for the problem in any number, d, of dimensions.

Noga Alon, Economical coverings of sets of lattice points, *Geom. Funct. Anal*, **1**(1991) 224–230; *MR* **92g**:52017.

F4 The no-three-in-line problem.

Can $2n$ lattice points (x, y) $(1 \leq x, y \leq n)$ be selected with no three in a straight line? This has been achieved for $2 \leq n \leq 32$ and for several larger even values of n. Guy & Kelly make four conjectures.

1. There are no configurations with the symmetry of a rectangle which do not have the full symmetry of the square.

2. The only configurations having the full symmetry of the square are those in Figure 17. The $n = 10$ configuration was first found by Acland-Hood. This conjecture has been verified by Flammenkamp for $n \leq 60$.

3. For large enough n, the answer to the initial question is "no," i.e., there are only finitely many solutions to the problem. The total number of configurations, not counting reflexions and rotations, is:

n	2	3	4	5	6	7	8	9	10	11	12	13
#	1	1	4	5	11	22	57	51	156	158	566	499

   and the configurations with specific symmetries have been enumerated for larger values of n.

4. For large n we may select at most $(c+\epsilon)n$ lattice points with no three in line, where $3c^3 = 2\pi^2$, i.e. $c \approx 1.85$.

In the opposite direction, Erdős showed that if n is prime, it is possible to choose n points with no three in line, and Hall, Jackson, Sudbery & Wild have shown that for n large, $(\frac{3}{2} - \epsilon)n$ such points can be found.

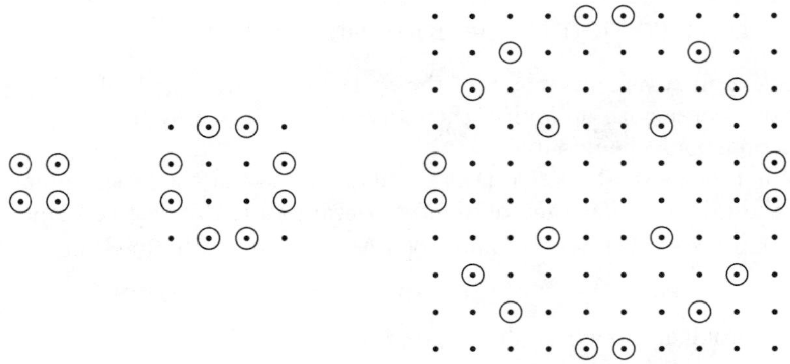

Figure 17. $2n$ Lattice Points, No Three in Line, $n = 2$, 4, 10.

T. Thiele modifies Erdős's construction to show that one can find $(\frac{1}{4} - \epsilon)n$ points with no 3 in line and no 4 on a circle.

The no-three-in-line problem is a discrete analog of an old problem of Heilbronn. Place n (≥ 3) points in a disk (or square, or equilateral triangle) of unit area so as to maximize the smallest area of a triangle formed by three of the points. If we denote this maximum area by $\Delta(n)$, then Heilbronn originally conjectured that $\Delta(n) < c/n^2$, but Komlós, Pintz & Szemerédi disproved this by showing that $\Delta(n) > (\ln n)/n^2$. Roth showed that $\Delta(n) \ll 1/n(\ln \ln n)^{1/2}$; Schmidt improved this to $\Delta(n) \ll 1/n(\ln n)^{1/2}$ and Roth subsequently made the further improvement $\Delta(n) \ll 1/n^{\mu-\epsilon}$, first with $\mu = 2 - 2/\sqrt{5} > 1.1055$ and later with $\mu = (17 - \sqrt{65})/8 > 1.1172$.

Given $3n$ points in the unit square, $n \geq 2$, they determine n triangles in many ways. Choose the partition so as to minimize the sum of the areas, and let $a^*(n)$ be the maximum value of this minimum sum, taken over all configurations of $3n$ points. Then Odlyzko & Stolarsky show that $n^{-1/2} \ll a^*(n) \ll n^{-1/24}$. If the n triangles are required to be *area disjoint* it is not even clear that the sum of their areas tends to zero.

Acland-Hood, *Bull. Malayan Math. Soc.*, **0**(1952-53) E11–12.

Michael A. Adena, Derek A. Holton & Patrick A. Kelly, Some thoughts on the no-three-in-line problem, Proc. 2nd Austral. Conf. Combin. Math., *Springer Lecture Notes*, **403**(1974) 6–17; *MR* **50** #1890.

David Brent Anderson, Update on the no-three-in-line problem, *J. Combin. Theory Ser. A*, **27**(1979) 365–366.

W. W. Rouse Ball & H. S. M. Coxeter, *Mathematical Recreations & Essays*, 12th edition, University of Toronto, 1974, p. 189.

C. E. Corzatt, Some extremal problems of number theory and geometry, PhD dissertation, Univ. of Illinois, Urbana, 1976.

D. Craggs & R. Hughes-Jones, On the no-three-in-line problem, *J. Combin. Theory Ser. A*, **20**(1976) 363–364; *MR* **53** #10590.

Hallard T. Croft, Kenneth J. Falconer & Richard K. Guy, Unsolved Problems in Geometry, Springer-Verlag, New York, 1991, §**E5**.

H. E. Dudeney, *The Tribune*, 1906-11-07.

H. E. Dudeney, *Amusements in Mathematics*, Nelson, Edinburgh, 1917, pp. 94, 222.

Achim Flammenkamp, Progress in the no-three-in-line problem, *J. Combin. Theory Ser. A* (submitted).

Martin Gardner, Mathematical Games, *Sci. Amer.*, **226** #5 (May 1972) 113–114; **235** #4 (Oct 1976) 133–134; **236** #3 (Mar 1977) 139–140.

Michael Goldberg, Maximizing the smallest triangle made by N points in a square, *Math. Mag.*, **45**(1972) 135–144.

R. Goldstein, K. W. Heuer & D. Winter, Partition of S into n triples; solution to Problem 6316, *Amer. Math. Monthly*, **89**(1982) 705–706.

Richard K. Guy, *Bull. Malayan Math. Soc.*, **0**(1952-53) E22.

Richard K. Guy & Patrick A. Kelly, The no-three-in-line problem, *Canad. Math. Bull.*, **11**(1968) 527–531.

R. R. Hall, T. H. Jackson, A. Sudbery & K. Wild, Some advances in the no-three-in-line problem, *J. Combin. Theory Ser. A*, **18**(1975) 336–341.

Heiko Harborth, Philipp Oertel & Thomas Prellberg, No-three-in-line for seventeen and nineteen, *Discrete Math.*, **73**(1989) 89–90; *MR* **90f**:05041.

P. A. Kelly, The use of the computer in game theory, M.Sc. thesis, Univ. of Calgary, 1967.

Torliev Kløve, On the no-three-in-line problem II, III, *J. Combin. Theory Ser. A*, **24**(1978) 126–127; **26**(1979) 82–83; *MR* **57** #2962; **80d**:05020.

J. Komlós, J. Pintz & E. Szemerédi, A lower bound for Heilbronn's problem, *J. London Math. Soc.*(2) **25**(1982) 13–24; *MR* **83i**:10042.

Andrew M. Odlyzko, J. Pintz & Kenneth B. Stolarsky, Partitions of planar sets into small triangles, *Discrete Math.*, **57**(1985) 89–97; *MR* **87e**:52007.

Carl Pomerance, Collinear subsets of lattice point sequences – an analog of Szemerédi's theorem, *J. Combin. Theory Ser. A*, **28**(1980) 140–149.

K. F. Roth, On a problem of Heilbronn, *J. London Math. Soc.*, **25**(1951) 198–204, esp. p. 204; II, III, *Proc. London Math. Soc.*, **25**(1972) 193–212; 543–549.

K. F. Roth, Developments in Heilbronn's triangle problem, *Advances in Math.*, **22**(1976) 364–385; *MR* **55** #2771.

Wolfgang M. Schmidt, On a problem of Heilbronn, *J. London Math. Soc.*, **4**(1971/72) 545–550.

T. Thiele, *J. Combin. Theory Ser A* (submitted).

F5 Quadratic residues. Schur's conjecture.

The **quadratic residues** of a prime p are the nonzero numbers r for which the congruence $r \equiv x^2 \bmod p$ has solutions. There are $\frac{1}{2}(p-1)$ of them in the interval $[1, p-1]$ and they are symmetrically distributed if p is of shape $4k+1$. If $p = 4k-1$, there are more quadratic residues in the interval $[1, 2k-1]$ than in $[2k, 4k-2]$, but all known proofs use Dirichlet's class-number formula. Is there an elementary proof?

For the first few values of d it is easy to remember which primes have d as a quadratic residue:

$d=-1$	$p=4k+1$		
$d=-2$	$p=8k+1,3$	$d=2$	$p=8k\pm1$
$d=-3$	$p=6k+1$	$d=3$	$p=12k\pm1$
$d=-5$	$p=20k+1,3,7,9$	$d=5$	$p=10k\pm1$
$d=-6$	$p=24k+1,5,7,11$	$d=6$	$p=24k\pm1,5$

However, it's just an example of the Strong Law of Small Numbers that in these small cases the residues are just those in the first half or the end quarters of the period, according to the sign of d. The **Legendre symbol**, $\left(\frac{a}{p}\right)$, is often used to indicate the quadratic character of a number a, $a \perp p$, relative to the prime p. Its value is ±1 according as a is, or is not, a quadratic residue of p. For example, $\left(\frac{-1}{p}\right) = \pm1$ according as

$p = 4k \pm 1$. The important properties of this symbol are that $\left(\frac{a}{p}\right) = \left(\frac{c}{p}\right)$ if $a \equiv c \bmod p$; and Gauß's famous **quadratic reciprocity law**, that for odd primes p and q, $\left(\frac{p}{q}\right) = \left(\frac{q}{p}\right)$ unless p and q are both $\equiv -1 \bmod 4$, in which case $\left(\frac{p}{q}\right) = -\left(\frac{q}{p}\right)$. These can be used for making quick verifications of the quadratic character of quite large numbers. For example,

$$\left(\frac{173}{211}\right) = \left(\frac{211}{173}\right) = \left(\frac{38}{173}\right) = \left(\frac{2}{173}\right)\left(\frac{19}{173}\right) = -\left(\frac{19}{173}\right) = -\left(\frac{173}{19}\right) = -\left(\frac{2}{19}\right) = +1$$

in fact $173 \equiv 54^2 \bmod 211$.

A useful generalization of the Legendre symbol is the **Jacobi symbol** $\left(\frac{a}{b}\right)$ which is defined for $a \perp b$ and b any positive odd number, by the product

$$\prod \left(\frac{a}{p_i}\right)$$

of Legendre symbols, where $b = \prod p_i$ is the prime factorization of b, *with repetitions counted.* It has similar properties to the Legendre symbol, but note that, if b is not prime, then $\left(\frac{a}{b}\right) = +1$ does not necessarily imply that a is a quadratic residue of b.

If R (respectively N) is the maximum number of consecutive quadratic residues (respectively nonresidues) modulo an odd prime p, then A. Brauer showed that for $p \equiv 3 \bmod 4$, $R = N < \sqrt{p}$. On the other hand, if $p = 13$, then $N = 4 > \sqrt{13}$, since 5, 6, 7, 8 are all nonresidues of 13. Schur conjectured that $N < \sqrt{p}$ if p is large enough. Hudson proved Schur's conjecture; moreover, he believes that $p = 13$ is the only exception.

A. Brauer, Über die Verteilung der Potenzreste, *Math. Z.*, **35**(1932) 39–50; *Zbl.* **3**, 339.

H. Davenport, *The Higher Arithmetic*, Fifth edition, Cambridge University Press, 1982, pp.74–77.

Richard H. Hudson, On sequences of quadratic nonresidues, *J. Number Theory*, **3**(1971) 178–181; *MR* **43** #150.

Richard H. Hudson, On a conjecture of Issai Schur, *J. reine angew. Math.*, **289**(1977) 215–220; *MR* **58** #16481.

F6 Patterns of quadratic residues.

What patterns of quadratic residues are sure to occur? It is easy to see that a pair of neighboring ones always do, since at least one of 2, 5 and 10 is a residue, so that (1,2), (4,5) or (9,10) is such a pair. In the same way at least one of (1,3), (2,4) or (4,6) is a pair of residues differing by 2; (1,4) is a pair differing by 3; (1,5), (4,8), (6,10) or (12,16) is a pair differing by 4; and so on.

Suppose that each of r, $r+a$, $r+b$ is a quadratic residue modulo p. Emma Lehmer asks: for which pairs (a,b) will such a triplet occur for *all* sufficiently large p? Denote by $\Omega(a,b)$ the least number such that a triplet is assured with $r \le \Omega(a,b)$ for all $p > p(a,b)$, and write $\Omega(a,b) = \infty$ if there is no such finite number. For example, Emma Lehmer showed that $\Omega(1,2) = \infty$, and more generally that $\Omega(a,b) = \infty$ if $(a,b) \equiv (1,2) \bmod 3$; or if $(a,b) \equiv (1,3)$, $(2,3)$ or $(2,4) \bmod 5$; or if $(a,b) \equiv (1,5)$, $(2,3)$ or $(4,6) \bmod 7$. Is $\Omega(a,b)$ finite in all other cases? Emma Lehmer conjectures that it is finite if a and b are squares. Of course, $\Omega(a,b) = 1$ if a, b are each one less than a square. As an example, let us see why $\Omega(5,23) = 16$. If the triplets $(1,6,24)$ and $(4,9,27)$ are not all residues then 6 and 3 are not, and 2 must be a residue. If the triplets $(2,7,25)$ and $(13,18,36)$ are not all residues, then 7 and 13 must be nonresidues. Under these circumstances, $(r, r+5, r+23)$ are not all residues for $1 \le r \le 15$, but when $r = 16$, $(16,21,39)$ are residues.

Table 9 contains what are believed to be the (minimum) values of $\Omega(a,b)$. They provide good evidence for the conjectured finiteness in all cases except those already noted. Can an upper bound be obtained in terms of a and b?

What about patterns of four residues, $r, r+a, r+b, r+c$? Of course these won't necessarily occur if any of the four subpatterns of three residues aren't forced to do so. We need examine only $(a,b,c) = (2,5,6)$, $(1,6,7)$, $(1,4,9)$, $(5,6,9)$, $(1,6,10)$, $(1,7,10)$, ... where $\Omega(a,b)$, $\Omega(a,c)$, $\Omega(b,c)$ and $\Omega(b-a, c-a)$ are each known to be finite. Some corresponding values of $\Omega(a,b,c)$ are $\Omega(1,4,9) = 357$ (Peter Montgomery corrects an error in the first edition), $\Omega(1,4,15) = 675$, and of course $\Omega(3,8,15) = 1$.

But although $\Omega(1,6) = 24$, $\Omega(1,7) = 38$, $\Omega(5,6) = 49$ and $\Omega(6,7) = 57$, it appears that $\Omega(1,6,7) = \infty$. In fact the pattern r, $r+a$, $r+b$, $r+c$, $r+d$, with $(a,b,c,d) = (1,6,7,10)$ is such that, for each of the five subpatterns of four, $\Omega(1,6,7) = \Omega(1,6,10) = \Omega(1,7,10) = \Omega(5,6,9) = \Omega(6,7,10) = \infty$.

It is customary to define k**-th power residues** as numbers r for which $x^k \equiv r \bmod p$ has a solution, only with respect to those primes for which k divides $p-1$. In the same way that we remarked that every prime > 10 has a pair of consecutive quadratic residues not exceeding the pair $(9,10)$, Hildebrand has shown that for each k there is a fixed bound, $\Lambda(k,2)$ so that every sufficiently large prime has a pair of consecutive k-th power residues below this bound. There is no such bound for 3 consecutive quadratic or quartic, etc., residues. The argument consists in forcing primes of the form $3k+1$ to be residues and those of the form $3k+2$ to be nonresidues. Similarly by making 2 a residue and as many odd primes as necessary nonresidues, there is no such bound below which four consecutive k-th power residues must appear for any k. This leaves open only the question of three consecutive k-th powers for k odd. The case $k = 3$ was solved by the Lehmers, Mills & Selfridge.

Table 9. Values of $\Omega(a,b)$ for $a < b \leq 25$.

a	$b=4$	5	6	7	8	9	10	11	12	13	14
1	45	∞	24	38	∞	84	26	∞	∞	∞	∞
2	∞	25	20	∞	∞	∞	∞	70	30	∞	∞
3	174	39	∞	∞	1	∞	55	∞	∞	36	105
4		∞	∞	∞	∞	91	36	∞	∞	∞	∞
5			49	∞	∞	121	∞	25	4	∞	28
6				57	∞	33	30	∞	24	∞	42
7					∞	∞	75	∞	74	∞	∞
8						66	∞	∞	∞	∞	30
9							∞	54	∞	42	66
10								∞	60	85	∞
11									28	∞	119
12										∞	∞
13											∞

a	$b=15$	16	17	18	19	20	21	22	23	24	25
1	77	35	∞	∞	∞	∞	15	35	∞	21	69
2	54	∞	∞	∞	∞	25	98	∞	∞	∞	∞
3	1	∞	∞	18	36	95	∞	∞	∞	1	51
4	126	60	∞	38	168	∞	90	∞	∞	66	77
5	∞	∞	64	110	∞	100	4	∞	16	64	∞
6	60	36	38	∞	62	78	60	78	∞	45	∞
7	27	9	∞	∞	∞	∞	70	42	∞	∞	45
8	1	∞	∞	77	∞	48	∞	∞	42	1	∞
9	57	66	∞	36	27	16	72	∞	21	∞	119
10	55	∞	∞	32	102	∞	77	26	∞	28	39
11	49	∞	39	∞	∞	∞	64	∞	∞	25	∞
12	∞	65	98	∞	∞	36	4	∞	∞	∞	90
13	42	∞	∞	∞	36	∞	∞	∞	∞	36	∞
14	35	∞	∞	42	∞	52	56	∞	64	81	∞
15		66	27	69	∞	49	25	99	110	1	105
16			∞	∞	102	∞	169	95	∞	∞	56
17				∞	∞	76	64	∞	∞	∞	∞
18					81	∞	∞	∞	40	185	144
19						∞	33	∞	∞	36	96
20							74	∞	40	25	∞
21								93	∞	70	100
22									∞	∞	98
23										∞	∞
24											63

Alfred Brauer, Combinatorial methods in the distribution of kth power residues, in *Probability and Statistics*, **4**, University of North Carolina, Chapel Hill, 1969, 14–37.

Adolf Hildebrand, On consecutive k-th power residues, *Monatsh. Math.*, **102** (1986) 103–114; *MR* **88a**:11089.

Adolf Hildebrand, On consecutive k-th power residues II, *Michigan Math. J.*, **38** (1991) 241–253; *MR* **92d**:11097.

D. H. Lehmer & Emma Lehmer, On runs of residues, *Proc. Amer. Math. Soc.*, **13**(1962) 102–106; *MR* **25** #2035.

D. H. Lehmer, Emma Lehmer & W. H. Mills, Pairs of consecutive powerresidues, *Canad. J. Math.*, **15**(1963) 172–177; *MR***26** #3660.

D. H. Lehmer, Emma Lehmer, W. H. Mills & J. L. Selfridge, Machine proof of a theorem on cubic residues, *Math. Comput.*, **16**(1962) 407–415; *MR* **28** #5578.

René Peralta, On the distribution of quadratic residues and nonresidues modulo a prime number, *Math. Comput.*, **58**(1992) 433–440; *MR* **93c**:11115.

F7 A cubic analog of a Pell equation.

Hugh Williams observes that if $p \equiv 3 \bmod 4$, then the equation $x^2 - py^2 = 2$ has integer solutions just if the congruence $w^2 \equiv 2 \bmod p$ does, i.e. just if $\left(\frac{2}{p}\right) = 1$, and asks for a cubic analog in the cases where $p \not\equiv \pm 1 \bmod 9$: $x^3 + py^3 + p^2z^3 - 3pxyz = 3$ is solvable just if $w^3 \equiv 3 \bmod p$ is. Barrucand & Cohn have shown that this is true for $p \equiv 2$ or $5 \bmod 9$. What about $p \equiv 4$ or $7 \bmod 9$? This is a special case of a more general conjecture of Barrucand. If true, it would be useful in abbreviating the calculations needed to find the fundamental unit (regulator) of the cubic field $\mathbb{Q}\left(\sqrt[3]{p}\right)$.

P.-A. Barrucand & Harvey Cohn, A rational genus, class number divisibility and unit theory for pure cubic fields, *J. Number Theory*, **2**(1970) 7–21.

H. C. Williams, Improving the speed of calculating the regulator of certain pure cubic fields, *Math. Comput.*, **35**(1980) 1423–1434.

F8 Quadratic residues whose differences are quadratic residues.

Gary Ebert asks us to find the largest collection of quadratic residues $r_i \bmod p^n$, given $p^n \equiv 1 \bmod 4$, such that $r_i - r_j$ is a quadratic residue for all pairs (i, j).

F9 Primitive roots

A **primitive root**, g, of a prime p is a number such that the residue classes of g, g^2, ..., $g^{p-1} = 1$ are all distinct. For example, 5 is a primitive root of 23 because

$$\begin{array}{ccccccccccc} 5, & 5^2 \equiv 2, & 5^3 \equiv 10, & 4, & -3, & 8, & -6, & -7, & 11, & 9, & -1, \\ -5, & -2, & -10, & -4, & 3, & -8, & 6, & 7, & -11, & -9, & 1 \end{array}$$

all belong to different residue classes mod 23.

There is a famous conjecture of Artin, that for each integer $g \neq -1$, g not a square, there are infinitely many primes p with g as a primitive root. Hooley proved this assuming the extended Riemann hypothesis, and Gupta & Murty proved it unconditionally for infinitely many g. Heath-Brown has proved the remarkable theorem that, but for at most two exceptional primes p_1, p_2 the following is true: For each prime p there are infinitely many primes q with p a primitive root of q. For example, there are infinitely many primes q with either 2 or 3 or 5 as a primitive root.

Erdős asks: if p is large enough, is there always a prime $q < p$ so that q is a primitive root of p?

Given a prime $p > 3$, Brizolis asks if there is always a primitive root g of p and x $(0 < x < p)$ such that $x \equiv g^x \bmod p$. If so, can g also be chosen so that $0 < g < p$ and $g \perp (p-1)$?

Vegh asks whether, for all primes $p > 61$, every integer can be expressed as the difference of two primitive roots of p. W. Narkiewicz notes that there is an affirmative answer for $p > 10^{19}$, so that this can, in theory, be answered by computer.

If p and $q = 4p^2 + 1$ are both primes, Gloria Gagola asks if 3 is a primitive root of q for all $p > 3$; is $p = 193$ the only odd prime for which 2 is not a primitive root of q; is $p = 653$ the only prime for which 5 is neither a quadratic residue nor a primitive root of q; and is there a number, perhaps a function of p (such as $2p-1$ for large p), which is always a primitive root of q?

The Lehmers have checked that 6 is a primitive root for all primes p of the form $n^2 + 108$ for $p < 2 \cdot 10^8$.

Anton Dumitziu, Congruences du premier degré, *Rev. Roumaine Math. Pures Appl.*, **10**(1965) 1201–1234.

Rajiv Gupta & Maruti Ram Murty, A remark on Artin's conjecture, *Invent. Math.*, **78**(1984) 127–130; *MR* **86d**:11003.

D. R. Heath-Brown, Artin's conjecture for primitive roots, *Quart. J. Math. Oxford Ser.*(2) **37**(1986) 27–38; *MR* **88a**:11004.

C. Hooley, On Artin's conjecture, *J. reine angew. Math.*, **225**(1967) 209–220; *MR* **34** #7445.

Leo Murata, On the magnitude of the least primitive root, *Astérisque No.* **198-200**(1991) 253–257.

Maruti Ram Murty & Seshadri Srinivasan, Some remarks on Artin's conjecture, *Canad. Math. Bull.*, **30**(1987) 80–85; *MR* **88e**:11094.

Michael Szalay, On the distribution of the primitive roots of a prime, *J. Number Theory*, **7**(1975) 184–188; *MR* **51** #5524.

Emanuel Vegh, Pairs of consecutive primitive roots modulo a prime, *Proc. Amer. Soc.*, **19**(1968) 1169–1170; *MR* **37** #6240.

Emanuel Vegh, Primitive roots modulo a prime as consecutive terms of an arithmetic progression, *J. reine angew. Math.*,**235**(1969) 185–188; II, **244**(1970) 185–188; III, **256**(1972) 130–137; *MR* **39** #4086; **42** #1755; **46** #7137.

Emanuel Vegh, A note on the distribution of the primitive roots of a prime, *J. Number Theory*, **3**(1971) 13–18; *MR* **44** #2694.

F10 Residues of powers of two.

Graham asks about the residue of $2^n \bmod n$. There are no solutions of $2^n \equiv 1 \bmod n$ with $n > 1$. $2^n \equiv 2 \bmod n$ whenever n is a pseudoprime base 2 (see **A12**) or is a prime. The Lehmers have shown that the smallest solution of $2^n \equiv 3 \bmod n$ is $n = 4700063497 = 19 \cdot 47 \cdot 5263229$. Of course, n has to be composite, and it is not divisible by 2 or 3. In fact, Mąkowski (see reference at **B5**) notes that if $\left(\frac{2}{p}\right)$ and $\left(\frac{3}{p}\right)$ are of opposite sign, i.e. if $p = 24k \pm 7$ or ± 11, then n is not divisible by p.

Rotkiewicz (compare **A12**) notes that if m satisfies $2^m \equiv 3 \bmod m$, then $n = 2^m - 1$ is a solution of $2^{n-2} \equiv 1 \bmod n$.

Benkoski asked if there was a solution to $2^n \equiv 4 \bmod n$ which didn't end in 7 when written in decimal notation. Zhang Ming-Zhi gave the solutions where $n \equiv 1$ or $3 \bmod 10$ and asked if there were any with $n \equiv 9 \bmod 10$.

Victor Meally reports that $2^n \equiv -1 \bmod n$ for $n = 3^k$ and $2^n \equiv -2 \bmod n$ for $n = 2, 6, 66, 946, \ldots$. Schinzel notes that the existence of infinitely many n such that $2^n \equiv -2 \bmod n$ is proved in the remark to Exercise 4 on p. 235 of Sierpiński's *Elementary Theory of Numbers*, 2nd English Edition, 1987.

Zhang Ming-Zhi, A note on the congruence $2^{n-2} \equiv 1 \bmod n$ (Chinese. English summary), *Sichuan Daxue Xuebao*, **27** (1990) 130–131; *MR* **92b**: 11003 (where the wrong Benkoski reference appears to be given).

F11 Distribution of residues of factorials.

What is the distribution of $1!, 2!, 3!, \ldots, (p-1)!, p!$ modulo p? About p/e of the residue classes are not represented. Here are the missing ones for the first few values of p:

$p = 2$ or 3, none. $\quad p = 5$, $\{-2\}$. $\quad p = 7$, $\{-2, -3\}$.
$p = 11$, $\{-2, \pm 3, \pm 4\}$. $\quad p = 13$, $\{-3, 4, -5\}$.
$p = 17$, $\{4, 5, -6, -7, -8\}$. $\quad p = 19$, $\{3, -5, -6, \pm 7, \pm 8\}$.
$p = 23$, $\{-3, -4, -6, -7, -8, 10\}$.
$p = 29$, $\{-2, -4, 7, -8, -9, -10, -11, -12, 13, -14\}$.
$p = 31$, $\{\pm 3, 4, 8, \pm 10, 11, 12, 13, 14\}$.
$p = 37$, $\{3, 4, \pm 5, -9, 10, 11, -14, \pm 15, -18\}$.

Until we reach the last two entries we might be tempted to conjecture that there were always at least as many negative entries as positive ones. Are there infinitely many examples of each case? The value $p = 23$ is remarkable in that the only duplicates are ± 1.

In answer to a question of Erdős, Rokowska & Schinzel show that if the residues of $2!, 3!, \ldots, (p-1)!$ are all distinct, then the missing residue must be that of $-\frac{p-1}{2}!$, that $p \equiv 5 \bmod 8$, and that there are no such p with $5 < p \leq 1000$.

B. Rokowska & A. Schinzel, Sur une problème de *M.* Erdős, *Elem. Math.*, **15**(1960) 84–85.

R. Stauduhar, Problem 7, *Proc. Number Theory Conf.*, Boulder, 1963, p. 90.

F12 How often are a number and its inverse of opposite parity?

For each x $(0 < x < p)$ where p is an odd prime, define $\bar{x}$ by $x\bar{x} \equiv 1 \bmod p$ and $0 < \bar{x} < p$. Let N_p be the number of cases in which x and $\bar{x}$ are of opposite parity. E.g., for $p = 13$, $(x, \bar{x}) = (1,1)$, $\underline{(2,7)}$, $(3,9)$, $(4,10)$, $\underline{(5,8)}$, $\underline{(6,11)}$, $(12,12)$ so $N_{13} = 6$. D. H. Lehmer asks us to find N_p or at least to say something nontrivial about it. $N_p \equiv 2$ or $0 \bmod 4$ according as $p \equiv \pm 1 \bmod 4$.

p	3	5	7	11	13	17	19	23	29	31	37	41	43	47	53	59	61
N_p	0	2	0	4	6	10	4	12	18	4	14	18	20	16	30	32	30

F13 Covering systems of congruences.

A system of congruences $a_i \bmod n_i$ $(1 \leq i \leq k)$ is called a **covering system** if every integer y satisfies $y \equiv a_i \bmod n_i$ for at least one value of i. For example: $0 \bmod 2$; $0 \bmod 3$; $1 \bmod 4$; $5 \bmod 6$; $7 \bmod 12$. If $c = n_1 < n_2 < \cdots < n_k$ then Erdős offers \$500.00 for a proof or disproof of the existence of covering systems with c arbitrarily large. Davenport & Erdős, and Fried, found systems with $c = 3$; Swift with $c = 6$; Selfridge with $c = 8$; Churchhouse with $c = 10$; Selfridge with $c = 14$; Krukenberg with $c = 18$; and Choi with $c = 20$.

Erdős offers \$25.00 for a proof of the nonexistence of covering systems with all moduli n_i odd, distinct, and greater than one; while Selfridge offers \$900.00 for an explicit example of such a system. Berger, Felzenbaum & Fraenkel showed that the l.c.m. of the moduli of such a system must contain at least six prime factors. More generally, "odd" could be replaced by "not divisible by the first r primes." Simpson & Zeilberger showed that if the moduli are odd and squarefree then at least 18 primes are required.

Jim Jordan offers comparable rewards to those mentioned above for solutions to the analogous problems for Gaussian integers (**A15**).

Erdős noted that you can have a covering system with all moduli n_i distinct, squarefree, and greater than one by using the proper divisors of 210:

a_i	0	0	0	1	0	1	1	2	2	23	4	5	59	104
n_i	2	3	5	6	7	10	14	15	21	30	35	42	70	105

Krukenberg used 2 and squarefree numbers greater than 3. Selfridge asks if you can have such a system with $c \geq 3$ in place of $c = 2$. He observes that the n_i cannot all be squarefree with at most two prime factors, but the above example shows that you do not need more than three.

It is easy, but not trivial, to prove that, for a covering system with distinct moduli, $\sum_{i=1}^{k} 1/n_i > 1$. The sum can be arbitrarily close to 1 if $n_1 = 3$ or 4. Selfridge & Erdős conjecture that $\sum 1/n_i > 1 + c_{n_1}$ where $c_{n_1} \to \infty$ with n_1.

Schinzel has asked for a covering system in which no modulus divides another. This would not exist if a covering with odd moduli does not exist.

Simpson calls a covering system **irredundant** if it ceases to cover the integers when one of the congruences is removed. He has shown that if the l.c.m. of the moduli of such a system is $\prod p_i^{\alpha_i}$ then the system contains at least $\sum \alpha_i(p_i - 1)$ congruences.

Erdős conjectures that all sequences of the form $d \cdot 2^k + 1$ ($k = 1, 2, \ldots$), d fixed and odd, which contain no primes can be obtained from covering congruences (see **B21** for examples). Equivalently, the least prime factors of members of such sequences are unbounded.

Marc Aron Berger, Alexander Gersh Felzenbaum & A. S. Fraenkel, Necessary conditions for the existence of an incongruent covering system with odd moduli II, *Acta Arith.*, **48**(1987) 73–79.

S. L. G. Choi, Covering the set of integers by congruence classes of distinct moduli, *Math. Comput.*, **25**(1971) 885–895; *MR* **45** #6744.

R. F. Churchhouse, Covering sets and systems of congruences, in *Computers in Mathematical Research*, North-Holland, 1968, 20–36; *MR* **39** #1399.

Fred Cohen & J. L. Selfridge, Not every number is the sum or difference of two prime powers, *Math. Comput.*, **29**(1975) 79–81.

P. Erdős, Some problems in number theory, in *Computers in Number Theory*, Academic Press, 1971, 405–414; esp. pp. 408–409.

J. Haight, Covering systems of congruences, a negative result, *Mathematika*, **26**(1979) 53–61; *MR* **81e**:10003.

J. H. Jordan, Covering classes of residues, *Canad. J. Math.*, **19**(1967) 514–519; *MR* **35** #1538.

J. H. Jordan, A covering class of residues with odd moduli, *Acta Arith.*, **13**(1967-68) 335–338; *MR* **36** #3709.

C. E. Krukenberg, PhD thesis, Univ. of Illinois, 1971, 38–77.

A. Schinzel, Reducibility of polynomials and covering systems of congruences, *Acta Arith.*, **13**(1967) 91–101; *MR* **36** #2596.

R. J. Simpson, Regular coverings of the integers by arithmetic progressions, *Acta Arith.*, **45**(1985) 145–152; *MR* **86j**:11004.

R. J. Simpson & D. Zeilberger, Necessary conditions for distinct covering systems with squarefree moduli, *Acta Arith.*,**59**(1991) 59–70.

Zhang Ming-Zhi, A note on covering systems of residue classes, *Sichuan Daxue Xuebao* **26**(1989) 185–188; *MR* **92c**:11003.
Stefan Znám, A survey of covering systems of congruences, *Acta Math. Univ. Comen.*, **40-41**(1982) 59–79; *MR* **84e**:10004.

F14 Exact covering systems.

If a system of congruences is both covering and disjoint (each integer covered by just one conguence) it is called an **exact covering system**. Necessary, but not sufficient, conditions for a system to be exact are $\sum_{i=1}^{k} 1/n_i = 1$ and $(n_i, n_j) > 1$ for all i, j, where the notation is as in the first sentence of **F13**. There is a theorem, variously attributed to subsets of {Davenport, Mirsky, Newman, Rado}, that if a set of distinct numbers > 1 are the moduli of congruences, then either there is a number which is not in any of them or there is a number which is in more than one of them. The ingenious proof used a generating function and roots of unity. Combinatorial proofs were later given by Berger, Felzenbaum & Fraenkel and by Simpson.

Znám notes that $(n_1, n_2, \ldots, n_k) > 1$ is *not*, as stated in the first edition, a necessary condition, as is evinced by the example 0 (mod 6), 1 (mod 10), 2 (mod 15) and 3, 4, 5, 7, 8, 9, 10, 13, 14, 15, 16, 19, 20, 22, 23, 25, 26, 27, 28, 29 (mod 30). He confirmed a conjecture of Mycielski by further proving that if p is the least prime divisor of n_k, then $n_k = n_{k-1} = \ldots = n_{k-p+1}$. He conjectured that if there exist only *pairs* of equal moduli, then the moduli are all of the form $2^\alpha 3^\beta$, but later he and Burshtein & Schönheim and Joel Spencer each gave counter-examples, such as

	0, 1	2, 7	3, 8	13, 28	4, 9	14, 34	19, 39	59, 119
mod	5	10	15	30	20	40	60	120

Stein proved that if there is a single pair of equal moduli, the rest being distinct, then $n_i = 2^i$ $(1 \le i \le k-1)$, $n_k = 2^{k-1}$. Znám proved analogously that if there is a triple of equal moduli, the rest being distinct, then $n_i = 2^i$ $(1 \le i \le k-3)$, $n_{k-2} = n_{k-1} = n_k = 3 \cdot 2^{k-3}$. Beebee has extended Stein's result by showing that a system is exact just if

$$\sin \pi z = -2^{k-1} \prod_{i=1}^{k} \sin \frac{\pi}{n_i}(a_i - z).$$

Simpson extended work by Burshtein & Schönheim and showed that if the primes $p_1 < p_2 < \cdots < p_t$ are those dividing the moduli of an exact covering system in which no modulus occurs more than N times, then

$$p_t \le N \prod_{i=1}^{t-1} \frac{p_i}{p_i - 1}$$

The main outstanding problem is to characterize exact covering congruences.

Porubský asked if there is an "exactly m times covering system" which is *not* the union of m exact covering systems. More generally, call such a system S **reducible** if there is a partition $S = S_1 \cup S_2$ such that S_1 and S_2 are exactly l times and exactly $m-l$ times covering systems for some l, $0 < l < m$, and **irreducible** if there is no such partition. Zhang Ming-Zhi answers Porubský's question affirmatively by showing that for every $m > 1$ there is an irreducible exactly m times covering system. This had already been shown for $m = 2$ by S. L. G. Choi (Keszthely, 1973) and by Zeilberger, e.g.:

1(2); 0(3); 2(6); 0,4,6,8(10); 1,2,4,7,10,13(15); 5,11,12,22,23,29(30).

Infinite disjoint covering systems with all moduli distinct can exist. If the sum of the reciprocals of the moduli is 1, such systems exist with moduli $\{2, 2^2, 2^3, \ldots\}$ and with sets of moduli of shape $2^\alpha 3^\beta$. Fraenkel & Simpson conjecture that these are the only types. Lewis showed that the only possible exceptions had an infinite set of distinct primes dividing their moduli.

Questions can also be asked about covering systems of Beatty sequences (**E27**). Graham showed that if

$$\lfloor m\alpha_i + \beta_i \rfloor;\ m \in \mathbb{Z};\ 1 \le i \le k$$

is such a system with $k > 2$ and at least one α_i irrational, then some two α_i must be equal. This is not so if all the α_i are rational, since

$$\left\lfloor m\frac{2^k-1}{2^{k-i}} + 1 - 2^{i-1} \right\rfloor \qquad (1 \le i \le k)$$

are exactly covering systems. Fraenkel conjectures that the only such systems with distinct α_i are of this form.

There are connexions with pseudoperfect numbers (**B2**) and with Egyptian fractions (**D11**).

John Beebee, Examples of infinite, incongruent exact covers, *Amer. Math. Monthly*, **95**(1988) 121–123; errata **97**(1990) 412; *MR*, **89g**:11013, **91a**:11013.

John Beebee, Some trigonometric identities related to exact covers, *Proc. Amer. Math. Soc.*, **112**(1991) 329–338; *MR*, **91i**:11013.

John Beebee, Bernoulli numbers and exact covering systems, *Amer. Math. Monthly*, **99**(1992) 946–948; *MR* **93i**:11025.

Marc Aron Berger, Alexander Gersh Felzenbaum & A. S. Fraenkel, A non-analytic proof of the Newman-Znám result for disjoint covering systems, *Combinatorica*, **6**(1986) 235–243.

Marc Aron Berger, Alexander Gersh Felzenbaum, A. S. Fraenkel & R. Holzman, On infinite and finite covering systems, *Amer. Math. Monthly*, **98**(1991) 739–742; *MR* **92g**:11009.

N. Burshtein, On natural exactly covering systems of congruences having moduli occurring at most N times, *Discrete Math.*, **14**(1976) 205–214.

N. Burshtein & J. Schönheim, On exactly covering systems of congruences having moduli occurring at most twice, *Czechoslovak Math. J.*, **24(99)**(1974) 369–372; *MR* **50** #4521.

J. Dewar, On finite and infinite covering sets, in *Proc. Washington State Univ. Conf. Number Theory*, Pullman WA, 1971, 201–206.

P. Erdős, On a problem concerning systems of congruences (Hungarian; English summary), *Mat. Lapok*, **3**(1952) 122–128.

A. S. Fraenkel, The bracket function and complementary sets of integers, *Canad. J. Math.*, **21**(1967) 6–27.

A. S. Fraenkel, Complementing and exactly covering sequences, *J. Combin. Theory Ser. A*, **14**(1973) 8–20; *MR* **46** #8875.

A. S. Fraenkel, A characterization of exactly covering congruences, *Discrete Math.*, **4**(1973) 359–366; *MR* **47** #4906.

A. S. Fraenkel, Further characterizations and properties of exactly covering congruences, *Discrete Math.*, **12**(1975) 93–100; erratum 397; *MR* **51** #10276.

A. S. Fraenkel & R. Jamie Simpson, On infinite disjoint covering systems, *Proc. Amer. Math. Soc*, **119**(1993) 5–9; *MR* **93k**:11006.

R. L. Graham, Covering the positive integers by disjoint sets of the form $\{\lfloor n\alpha+\beta\rfloor : n=1,2,\ldots\}$, *J. Combin. Theory Ser. A*, **15**(1973) 354–358; *MR* **48** #3911.

R. L. Graham, Lin Shen & Lin Chio-Shih, Spectra of numbers, *Math. Mag.*, **51**(1978) 174–176; *MR* **58** #10808.

I. Korec, On a generalisation of Mycielski's and Znam's conjectures about coset decomposition of abelian groups, *Fundamenta Math.*, **85**(1974) 41–48.

I. Korec, On number of cosets in nonnatural disjoint covering systems, *Colloq. Math. Soc. János Bolyai* **51** (Number Theory, Vol. 1, Budapest, 1987), North-Holland, 1990, 265–278.

Ethan Lewis, Infinite covering systems of congruences which don't exist, *Proc. Amer. Math. Soc.*, (1993).

Ryozo Morikawa, Some examples of covering sets; On a method to construct covering sets; On eventually covering families generated by the bracket function, *Bull. Fac. Liberal Arts Nagasaki Univ.*, **21**(1981) 1–4; **22**(1981) 1–1; **23**(1982/83) 17–22; *MR* **84j**:10064; **84i**:10057; **84c**:10051.

Ryozo Morikawa, Disjointness of sequences $[\alpha_i n+\beta_i]$, $i=1,2$, *Proc. Japan Acad. Ser. A Math. Sci.*, **58**(1982) 269–271; *MR* **83m**:10096.

Morris Newman, Roots of unity and covering sets, *Math. Ann.*, **191**(1971) 279–282; *MR* **44** #3972 & err. p. 1633.

Břetislav Novák & Štefan Znám, Disjoint covering systems, *Amer. Math. Monthly*, **81** (1974) 42–45.

Štefan Porubský, On m times covering systems of congruences, *Acta Arith.*, **29**(1976) 159–169; *MR* **53** #2884.

Štefan Porubský, Results and problems on covering systems of residue classes, *Mitt. Math. Sem. Giessen*, **150**(1981)85 pp.; *MR* **83j**:10008.

R. J. Simpson, Disjoint covering systems of congruences, *Amer. Math. Monthly*, **94**(1987) 865–868; *MR* **89b**:11006.

R. J. Simpson, Exact coverings of the integers by arithmetic progressions, *Discrete Math.*, **59**(1986) 181–190.

R. J. Simpson, Disjoint covering systems of rational Beatty sequences, *Discrete Math.*, **92**(1991) 361–369.

Sherman K. Stein, Unions of arithmetic sequences, *Math. Ann.*, **134**(1958) 289–294; *MR* **20** #17.

Sun Zhi-Wei, On exactly m times covers, *Israel J. Math.*, **77**(1992) 345–348; *MR* **93k**:11007.

Charles Vanden Eynden, On a problem of Stein concerning infinite covers, *Amer. Math. Monthly*, **99**(1992) 355–358; *MR* **93b**:11004.

Doron Zeilberger, On a conjecture of R. J. Simpson about exact covering congruences, *Amer. Math. Monthly* **96**(1989) 243.

Zhang Ming-Zhi, Irreducible systems of residue classes that cover every integer exactly m times (Chinese, English summary), *Sichuan Daxue Xuebao*, **28**(1991) 403–408; *MR* **92j**:11001.

Štefan Znám, On Mycielski's problem on systems of arithmetical progressions, *Colloq. Math.*, **15**(1966) 201–204; *MR* **34** #134.

Štefan Znám, On exactly covering systems of arithmetic sequences, *Math. Ann.*, **180** (1969) 227–232; *MR* **39** #4087.

Štefan Znám, A simple characterization of disjoint covering systems, *Discrete Math.*, **12**(1975) 89–91; *MR* **51** #12772.

F15 A problem of R. L. Graham.

Szegedy won the prize that Graham offered for settling (affirmatively) the question: does $0 < a_1 < a_2 < \ldots < a_n$ imply that $\max_{i,j} a_i/(a_i, a_j) \geq n$? His proof, and that of Zaharescu, is for n sufficiently large. Cheng & Pomerance have given the specific bound 10^{4275} but there is still a fair amount of ground to be covered.

Cheng Yuanyou & Carl Pomerance, On Graham's conjecture, *Rocky Mountain J. Math.*, (1994) (to appear).

Paula A. Kemp, A conjecture of Graham concerning greatest common divisors, *Nieuw Arch. Wisk.*(4), **8**(1990) 61–62; *MR* **91e**:11003.

Rivka Klein, The proof of a conjecture of Graham for sequences containing primes, *Proc. Amer. Math. Soc.*, **95**(1985) 189–190; *MR* **86k**:11002.

J. W. Sander, On a conjecture of Graham, *Proc. Amer. Math. Soc.*, **102**(1988) 455–458; *MR* **89c**:11004.

R. J. Simpson, On a conjecture of R. L. Graham, *Acta Arith.*, **40**(1981/82) 209–211; *MR* **83j**:10062.

M. Szegedy, The solution of Graham's greatest common divisor problem, *Combinatorica*, **6**(1986) 67–71; *MR* **87i**:11010.

Alexandru Zaharescu, On a conjecture of Graham, *J. Number Theory*, **27** (1987) 33–40; *MR* **88k**:11009.

F16 Products of small prime powers dividing n.

Erdős defines $A(n,k)$ as $\prod p^a$ where the product is taken over primes p less than k with $p^a \| n$ and asks: is

$$\max_n \min_{1\le i\le k} A(n+i,k) = o(k) \qquad ?$$

He remarks that it is easy to show that it is $O(k)$. Is

$$\min_n \max_{1\le i\le k} A(n+i,k) > k^c$$

for every c and sufficiently large k? Is

$$\sum_{i=1}^{k} \frac{1}{A(n+i,k)} > c\ln k \qquad ?$$

F17 Series associated with the ζ-function.

Alf van der Poorten had asked for a proof that

$$\zeta(4)\left[= \sum_{n=1}^{\infty}\frac{1}{n^4} = \frac{\pi^4}{90}\right] = \frac{36}{17}\sum_{n=1}^{\infty}\frac{1}{n^4\binom{2n}{n}}$$

before he and others showed that

$$\frac{1}{2}\sum_{n=1}^{\infty}\frac{1}{n^4\binom{2n}{n}} = \int_0^{\frac{\pi}{3}} \theta\left(\ln 2\sin\frac{\theta}{2}\right)^2 d\theta = \frac{17\pi^4}{6480}.$$

It is also known that

$$\sum_{n=1}^{\infty}\frac{1}{\binom{2n}{n}} = \frac{2\pi\sqrt{3}+9}{27}, \qquad \sum_{n=1}^{\infty}\frac{1}{n\binom{2n}{n}} = \frac{\pi\sqrt{3}}{9},$$

$$\zeta(2)\left[= \sum_{n=1}^{\infty}\frac{1}{n^2} = \frac{\pi^2}{6}\right] = 3\sum_{n=1}^{\infty}\frac{1}{n^2\binom{2n}{n}},$$

$$2(\sin^{-1}x)^2 = \sum_{n=1}^{\infty}\frac{(2x)^{2n}}{n^2\binom{2n}{n}} \quad \text{and} \quad \zeta(3)\left[= \sum_{n=1}^{\infty}\frac{1}{n^3}\right] = \frac{5}{2}\sum_{n=1}^{\infty}\frac{(-1)^{n-1}}{n^3\binom{2n}{n}}$$

Some remarkable identities discovered by Gosper include

$$\sum_{k\ge1}\frac{30k-11}{4(2k-1)k^3\binom{2k}{k}^2} = \zeta(3), \qquad \sum_{n\ge1}\frac{2^{-n}}{1+x^{2^{-n}}} = \frac{1}{\ln x} + \frac{1}{1-x}$$

Louis Comtet, *Advanced Combinatorics*, D. Reidel, Dordrecht, 1974, p. 89.
John A. Ewell, A new series representation for $\zeta(3)$, *Amer. Math. Monthly*, **97**(1990) 219–220; *MR* **91d**:11103.
R. William Gosper, Strip mining in the abandoned orefields of nineteenth century mathematics, *Computers in Mathematics* (Stanford CA, 1986), *Lecture Notes in Pure and Appl. Math.*, Dekker, New York, **125**(1990) 261–284.
Leonard Levin, *Polylogarithms and Associated Functions*, North-Holland, 1981 [§7.62, and foreword by van der Poorten].
Alfred van der Poorten, A proof that Euler missed ... Apery's proof of the irrationality of $\zeta(3)$, *Math. Intelligencer*, **1**(1979) 195–203.
Alfred J. van der Poorten, Some wonderful formulas ... an introduction to polylogarithms, *Proc. Number Theory Conf.*, Queen's Univ., Kingston, 1979, 269–286; *MR* **80i**:10054.

F18 Size of the set of sums and products of a set.

If a_1, a_2, ..., a_n are n numbers (not necessarily integers), how big is the set of their sums and products in pairs?

$$¿ \qquad |\{a_i + a_j\} \cup \{a_i a_j\}| > n^{2-\epsilon} \qquad ?$$

Erdős & Szemerédi have proved that the cardinality is greater than n^{1+c_1} and less than $n^2 \exp(-c_2 \ln n / \ln \ln n)$.

P. Erdős, Some recent problems and results in graph theory, combinatorics and number theory, *Congress. Numer.*, **17** Proc. 7th S.E. Conf. Combin. Graph Theory, Comput., Boca Raton, 1976, 3–14 (esp. p. 11).
P. Erdős & E. Szemerédi, On sums and products of integers, *Studies in Pure Mathematics*, Birkhäuser, 1983, pp. 213–218; *MR* **86m**:11011.

F19 Partitions into distinct primes with maximum product.

In the first edition we asked: if n is large and written in the form $n = a + b + c$, $0 < a < b < c$ in every possible way, are all the products abc distinct, but Leech observed that **D16** answers this negatively. See also Kelly's paper. The analogous problem of maximizing the l.c.m. in place of product was studied algorithmically by Drago.

J. Riddell & H. Taylor asked if, among the partitions of n into *distinct primes*, the one having the maximum *product* of parts is necessarily one of those with the maximum *number* of parts, but Selfridge answered this negatively with the example

$$\begin{aligned} 319 &= 2+3+5+7+11+13+17+23+29+31+37+41+47+53 \\ &= 3+5+11+13+17+19+23+29+31+37+41+43+47 \end{aligned}$$

but the partition with the smaller number of parts gives the largest possible product. Is this the least counterexample? Can the cardinalities of the two sets differ by an arbitrarily large amount?

Antonino Drago, Rules to find the partition of n with maximum l.c.m., *Atti Sem. Mat. Fis. Univ. Modena*, **16**(1967) 286–298; *MR* **37** #180.

J. B. Kelly, Partitions with equal products, *Proc. Amer. Math. Soc.*, **15**(1964) 987–990.

F20 Continued fractions.

A number x may be expressed as a **continued fraction**

$$x = a_0 + \cfrac{b_1}{a_1 + \cfrac{b_2}{a_2 + \cfrac{b_3}{a_3 + \cdots}}}$$

which, out of kindness to the typesetter, is often written

$$x = a_0 + \frac{b_1}{a_1+}\frac{b_2}{a_2+}\frac{b_3}{a_3+}\cdots$$

When the numerators b_i are all 1 the continued fraction is called **simple**, and may be written

$$x = [a_0; a_1, a_2, a_3, \cdots]$$

It may be finite or infinite, but if x is rational it is finite. In this case there are two possible forms, one of which has its last **partial quotient**, a_k, equal to 1:

$$\frac{7}{16} = [0; 2, 3, 2] = [0; 2, 3, 1, 1]$$

Zaremba conjectured that given any integer $m > 1$, there is an integer a, $0 < a < m$, $a \perp m$ such that the simple continued fraction $[0; a_1, \cdots, a_k]$ for a/m has $a_i \le B$ for $1 \le i \le k$ where B is a small absolute constant (say $B = 5$). He was only able to prove $a_i \le C \ln m$.

T. W. Cusick, Zaremba's conjecture and sums of the divisor function, *Math. Comput.*, **61**(1993) 171–176.

S. K. Zaremba, La méthode des "bons treillis" pour le calcul des intégrales multiples, in *Applications of Number Theory to Numerical Analysis* (*Proc. Symp. Univ. Montréal*, 1971) Academic Press, 1972, 93–119 esp. 69 & 76; *MR* **49** #8271.

F21 All partial quotients one or two.

Not every number n can be expressed as the sum of two positive integers $n = a + b$ so that the continued fraction for a/b has all its partial quotients equal to 1 or 2. For 11, 17 and 19 we can have

$$\frac{4}{7} = [0; 1, 1, 2, 1], \qquad \frac{5}{12} = [0; 2, 2, 2], \quad \text{and} \quad \frac{7}{12} = [0; 1, 1, 2, 2]$$

but 23 can't be so expressed. However Leo Moser conjectured that there is a constant c such that every n can be so expressed with the *sum* of the partial quotients, $\sum a_i < c \ln n$.

Bohuslav Divis asked for a proof that in any real quadratic field there is always an irrational number whose simple continued fraction expansion has all its partial quotients 1 or 2. He also asks the same question with 1 and 2 replaced by any pair of distinct positive integers.

F22 Algebraic numbers with unbounded partial quotients.

Is there an algebraic number of degree greater than two whose simple continued fraction has unbounded partial quotients? Does *every* such number have unbounded partial quotients? Ulam asked particularly about the number $\xi = 1/(\xi + y)$ where $y = 1/(1 + y)$.

Littlewood observed that if θ has a continued fraction with bounded partial quotients a_n, then $\liminf n|\sin n\theta| \leq A(\theta)$, where $A(\theta)$ is not zero (though it is for almost all θ). He also asks if

$$\liminf n|\sin n\theta \sin n\phi| = 0$$

for all real θ and ϕ ? It is for almost all θ and ϕ. Cassels & Swinnerton-Dyer treat a dual problem and show incidentally that $\theta = 2^{1/3}$, $\phi = 4^{1/3}$ does *not* provide a counterexample. Davenport suggested that a computer might help with proving that

$$|(x\theta - y)(x\phi - z)| < \epsilon$$

has solutions for *every* θ, ϕ when, for example, $\epsilon = \frac{1}{10}$ or $\frac{1}{50}$.

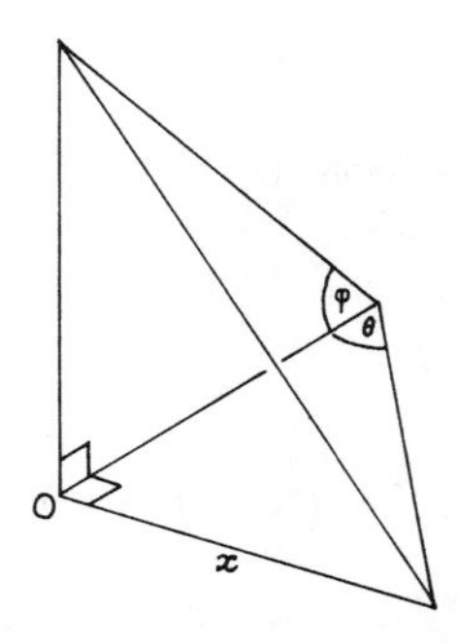

Figure 18. Rectangular Tetrahedron.

J. W. S. Cassels & H. P. F. Swinnerton-Dyer, On the product of three homogeneous linear forms and indefinite ternary quadratic forms, *Philos. Trans. Roy. Soc. London Ser. A*, **248**(1955) 73–96; *MR* **17**, 14.

Harold Davenport, Note on irregularities of distribution, *Mathematika*, **3** (1956) 131–135; *MR* **19**, 19.

John E. Littlewood, *Some Problems in Real and Complex Analysis*, Heath, Lexington MA, 1968, 19–20, Problems 5, 6.

F23 Small differences between powers of 2 and 3.

Problem 1 of Littlewood's book asks how small $3^n - 2^m$ can be in comparison with 2^m. He gives as an example

$$\frac{3^{12}}{2^{19}} = 1 + \frac{7153}{524288} \approx 1 + \frac{1}{73}$$

(the ratio of D$^\sharp$ to E$^\flat$).

The first few convergents to the continued fraction (see **F20**)

$$1 + \frac{1}{1+}\,\frac{1}{1+}\,\frac{1}{2+}\,\frac{1}{2+}\,\frac{1}{3+}\,\frac{1}{1+}\cdots$$

for $\log 3$ to the base 2 are

$$\frac{1}{1},\ \frac{2}{1},\ \frac{3}{2},\ \frac{8}{5},\ \frac{19}{12},\ \frac{65}{41},\ \frac{84}{53},\ \ldots$$

so Victor Meally observed that the octave may conveniently be partitioned into 12, 41 or 53 intervals, and that the system of temperament with 53 degrees is due to Nicolaus Mercator (1620–1687; not Gerhardus, 1512–1594, of map projection fame).

Ellison used the Gel'fond-Baker method to show that

$$|2^x - 3^y| > 2^x e^{-x/10} \quad \text{for} \quad x > 27,$$

and Tijdeman used it to show that there is a $c \geq 1$ such that $|2^x - 3^y| > 2^x/x^c$.

Croft asks the corresponding question for $n! - 2^m$. The first few best approximations to $n!$ by powers of 2 are

5!	20!	22!	24!	61!	63!	90!
2^7	2^{61}	2^{70}	2^{79}	2^{278}	2^{290}	2^{459}
−1.34	+0.13	−0.10	+0.046	+0.023	−0.0017	−0.0007

where the third row is the percentage error in the exponent.

In the "Stellingen" that accompanied Benne de Weger's PhD thesis he observed that if the primes $p_1, \ldots, p_t$ are given, then there is an effectively computable constant C, depending only on the p_i, such that for all n, k_1, ..., k_t with $n! \neq p_1^{k_1} \cdots p_t^{k_t}$ it is true that

$$\left| n! - p_1^{k_1} \cdots p_t^{k_t} \right| > \exp(Cn/\ln n).$$

There is some experimental support for the conjecture that the right side could be replaced by $\exp(C'n \ln n)$. For a fixed m, the methods of his thesis will determine all solutions of

$$n! - p_1^{k_1} \cdots p_t^{k_t} = m.$$

Erdős believes the conjecture. He also observes that $n! = 2^a \pm 2^b$ only when $n = 1$, 2, 3, 4 and 5.

F. Beukers, Fractional parts of powers of rationals, *Math. Proc. Cambridge Philos. Soc.*, **90**(1981) 13–20; *MR* **83g**:10028.

A. K. Dubitskas, A lower bound on the value of $\|(3/2)^k\|$ (Russian), *Uspekhi Mat. Nauk*, **45**(1990) 153–154; translated in *Russian Math. Surveys*, **45**(1990) 163–164; *MR* **91k**:11058.

W. J. Ellison, Recipes for solving diophantine problems by Baker's method, *Sém. Théorie Nombres*, 1970–71, Exp. No. 11, C.N.R.S. Talence, 1971.

R. Tijdeman, On integers with many small factors, *Compositio Math.*, **26** (1973) 319–330.

F24 Squares with just two different decimal digits.

Sin Hitotumatu asks for a proof or disproof that, apart from 10^{2n}, $4 \cdot 10^{2n}$ and $9 \cdot 10^{2n}$, there are only finitely many squares with just two different decimal digits, such as $38^2 = 1444$, $88^2 = 7744$, $109^2 = 11881$, $173^2 = 29929$, $212^2 = 44944$, $235^2 = 55225$ and $3114^2 = 9696996$.

F25 The persistence of a number.

In the sequence 679, 378, 168, 48, 32, 6, each term is the product of the decimal digits of the previous one. Neil Sloane defines the **persistence** of a number as the number of steps (five in the example) before the number collapses to a single digit. The smallest numbers with persistence 1, 2, ..., 11 are 10, 25, 39, 77, 679, 6788, 68889, 2677889, 26888999, 3778888999, 277777788888899. There is no number less than 10^{50} with persistence greater than 11. Sloane conjectures that there is a number d such that no number has persistence greater than d.

In base 2 the maximum persistence is 1. In base 3 the second term is zero or a power of 2. It is conjectured that all powers of 2 greater than 2^{15}

contain a zero when written in base 3. This is true up to 2^{500}. The truth of this conjecture would imply that the maximum persistence in base 3 is 3.

Sloane's general conjecture is that there is a number $d(b)$ such that the persistence in base b cannot exceed $d(b)$.

Erdős modifies the problem by letting $f(n)$ be the product of the *non-zero* decimal digits of n, and asks how fast one reaches a one-digit number, and for which numbers is the descent slowest. He says that it is easy to prove that $f(n) < n^{1-c}$, so that at most $c \ln\ln n$ steps are needed.

N. J. A. Sloane, The persistence of a number, *J. Recreational Math.*, **6**(1973) 97–98.

F26 Expressing numbers using just ones.

Let $f(n)$ be the least number of ones that can be used to represent n using ones and any number of + and × signs (and parentheses). For example,

$$80 = (1+1+1+1+1) \times (1+1+1+1) \times (1+1+1+1)$$

so $f(80) \le 13$. It can be shown that $f(3^k) = 3k$ and $3\log_3 n \le f(n) \le 5\log_3 n$ where the logs are to base 3. Does $f(n) \sim 3\log_3 n$?

Daniel Rawsthorne has shown that $f(n) = 2a + 3b$ when n is of the form $2^a 3^b$ and not greater than 3^{10}. Is this true for larger such n ?

Is it always true that for a prime p, $f(p) = 1 + f(p-1)$? And that $f(2p) = \min\{2 + f(p), 1 + f(2p-1)\}$?

J. H. Conway & M. J. T. Guy, π in four 4's, *Eureka*, **25**(1962) 18–19.

Richard K. Guy, Some suspiciously simple sequences, *Amer. Math. Monthly*, **93**(1986) 186–190; and see **94**(1987) 965 & **96**(1989) 905.

K. Mahler & J. Popken, On a maximum problem in arithmetic (Dutch), *Nieuw Arch. Wiskunde*, (3) **1**(1953) 1–15; *MR* **14**, 852e.

Daniel A. Rawsthorne, How many 1's are needed? *Fibinacci Quart.*, **27**(1989) 14–17.

F27 Mahler's generalization of Farey series.

The **Farey series** of order n consists of all positive rational numbers in their lowest terms, with numerators and denominators not exceeding n, arranged in order of magnitude. For example, the Farey series of order 5 is

$$\frac{1}{5}\ \frac{1}{4}\ \frac{1}{3}\ \frac{2}{5}\ \frac{1}{2}\ \frac{3}{5}\ \frac{2}{3}\ \frac{3}{4}\ \frac{4}{5}\ \frac{1}{1}\ \frac{5}{4}\ \frac{4}{3}\ \frac{3}{2}\ \frac{5}{3}\ \frac{2}{1}\ \frac{5}{2}\ \frac{3}{1}\ \frac{4}{1}\ \frac{5}{1}$$

The determinant formed from the numerators and denominators of two adjacent fractions is -1. Mahler regards the members of the sequence as positive real roots of linear equations whose coefficients have g.c.d. 1

and do not exceed n, and obtained the following apparent generalization to quadratic equations. List the coefficients (a, b, c) of the quadratic equations

$$ax^2 + bx + c = 0, a \geq 0, (a, b, c) = 1, b^2 \geq 4ac, \max\{a, |b|, |c|\} \leq n$$

which have positive real roots, in order of size of the roots. Then the third-order determinant (see **F28**) formed from any three consecutive rows of a, b, c appeared always to take the value 0 or ± 1. In the first edition, Table 10 illustrated this for $n = 2$, with initial and final entries $(0, 1, 0)$ and $(0, 0, 1)$, corresponding to roots 0 and ∞, just as the Farey series could include the terms $\frac{0}{1}$ and $\frac{1}{0}$. We also adopted a suggestion of Selfridge of duplicating rational roots to avoid trivial exceptions. The present Table 10 is an excerpt from the generalized series for $n = 3$. The entry in the last column is the value of the determinant formed from that row and its immediate neighbors.

Table 10. Segment of Generalized Farey Series of Order 3.

a	b	c	root	determinant
0	1	-1	1	
3	-1	-3	$(1 + \sqrt{37})/6$	0
3	-2	-2	$(1 + \sqrt{7})/3$	1
2	0	-3	$\sqrt{6}/2$	-1
3	-3	-1	$(3 + \sqrt{21})/6$	-1
2	-1	-2	$(1 + \sqrt{17})/4$	0
1	1	-3	$(\sqrt{13} - 1)/2$	-1
2	-2	-1	$(1 + \sqrt{3})/2$	1
3	-2	-3	$(1 + \sqrt{10})/3$	0
1	0	-2	$\sqrt{2}$	-1
3	-3	-2	$(3 + \sqrt{33})/6$	1
0	2	-3	$3/2$	

The conjecture was verified for $n \leq 5$, but Lambertus Hesterman of Canberra discovered counterexamples when $n = 7$; for example

a	b	c	root	determinant
2	-7	-7	$(7 + \sqrt{105})/4$	
1	-3	-6	$(3 + \sqrt{33})/2$	-2
1	-6	7	$3 + \sqrt{2}$	

Can this be rescued, or is it yet another example of the Strong Law of Small Numbers? Lewis Low proved that the absolute value of the determinant cannot exceed n. Can this bound be substantially reduced?

What can be said about the fourth-order determinants associated with cubic equations?

H. Brown & K. Mahler, A generalization of Farey sequences: some exploration via the computer, *J. Number Theory*, **3**(1971) 364–370; *MR* **44** #3959.

Lewis Low, Some lattice point problems, PhD thesis, Univ. of Adelaide, 1979; *Bull. Austral. Math. Soc.*, **21**(1980) 303–305.

Kurt Mahler, Some suggestions for further research, Res. Report No. 20, 1983, Math. Sci. Res. Centre, Austral. Nat. Univ.

F28 A determinant of value one.

The **third order determinant**

$$\begin{vmatrix} a_1 & a_2 & a_3 \\ a_4 & a_5 & a_6 \\ a_7 & a_8 & a_9 \end{vmatrix}$$

may be defined as $a_1(a_5a_9 - a_6a_8) - a_2(a_4a_9 - a_6a_7) + a_3(a_4a_8 - a_5a_7)$.

Find whole numbers a_1, a_2, ... a_9, none of them 0 or ± 1, so that

$$\begin{vmatrix} a_1 & a_2 & a_3 \\ a_4 & a_5 & a_6 \\ a_7 & a_8 & a_9 \end{vmatrix} = 1 = \begin{vmatrix} a_1^2 & a_2^2 & a_3^2 \\ a_4^2 & a_5^2 & a_6^2 \\ a_7^2 & a_8^2 & a_9^2 \end{vmatrix}$$

In the first edition we attributed this to Basil Gordon. It was asked by Molnar, who only required that $a_i \neq \pm 1$, and did not restrict the order of the determinants to 3. A topological significance was given, with references to Hilton. Solutions by Morris Newman, Peter Montgomery, Harry Applegate, Francis Coghlan and Kenneth Lau appeared, some of orders greater than 3, including several parametric families, for example

$$\begin{vmatrix} -8n^2 - 8n & 2n+1 & 4n \\ -4n^2 - 4n & n+1 & 2n+1 \\ -4n^2 - 4n - 1 & n & 2n-1 \end{vmatrix}$$

Richard McIntosh gave examples with a high proportion of Fibonacci numbers:

$$\begin{vmatrix} 1167 & 2 & 5 \\ 1698 & 3 & 8 \\ 2866 & 5 & 13 \end{vmatrix} \qquad \begin{vmatrix} 610 & 5 & 13 \\ 1054 & 8 & 21 \\ 1665 & 13 & 34 \end{vmatrix}$$

Rudolf Wytek restricted himself to integers > 1 and in the closing days of 1987 used a computer to find

$$\begin{vmatrix} 2 & 3 & 2 \\ 4 & 2 & 3 \\ 9 & 6 & 7 \end{vmatrix} \quad \begin{vmatrix} 2 & 3 & 5 \\ 3 & 2 & 3 \\ 9 & 5 & 7 \end{vmatrix} \quad \begin{vmatrix} 2 & 3 & 6 \\ 3 & 2 & 3 \\ 17 & 11 & 16 \end{vmatrix} \quad \begin{vmatrix} 5 & 7 & 6 \\ 6 & 4 & 7 \\ 17 & 16 & 20 \end{vmatrix} \quad \begin{vmatrix} 8 & 7 & 8 \\ 12 & 11 & 7 \\ 17 & 15 & 16 \end{vmatrix} \quad \begin{vmatrix} 10 & 7 & 12 \\ 4 & 2 & 7 \\ 17 & 12 & 20 \end{vmatrix}$$

the second of which had been found earlier by Kenneth Lau. The others were new and are not special cases of any of the parametric solutions. Solutions are evidently more numerous than might at first be thought.

Dănescu, Vâjâitu & Zaharescu solve the problem for determinants of any order in which all the elements are $\geq k$ for any given k.

Will the problem extend to cubes?

Alexandru Dănescu, Viorel Vâjâitu & Alexandru Zaharescu, Unimodular matrices whose components are squares of unimodular one.

P. J. Hilton, On the Grothendieck group of compact polyhedra, *Fundamenta Math.*, **61**(1967) 199–214.

P. J. Hilton, General Cohomology Theory & K-Theory, *L.M.S. Lecture Notes*, **1**, Cambridge University Press, 1971, p. 58.

E. A. Molnar, Relation between wedge cancellation and localization for complexes with two cells, *J. Pure Appl. Alg.*, **3**(1973) 141–158.

E. A. Molnar, A matrix problem, *Amer. Math. Monthly*, **81**(1974) 383–384; and see **82**(1975) 999–1000; **84**(1977) 809 and **94**(1987) 962.

Sadao Saito, Third-order determinant: E. A. Molnar's problem, *Acta Math. Sci.*, **8**(1988) 29–34; *MR* **89j**:15031.

F29 Two congruences, one of which is always solvable.

Given a prime p, find pairs of functions $f(x)$, $g(x)$ such that one of the congruences $f(x) \equiv n$, $g(x) \equiv n \bmod p$ is solvable for all integers n. A trivial example is $f(x) = x^2$, $g(x) = ax^2$ where a is a quadratic nonresidue (**F5**) of the odd prime p. Mordell gives the further example $f(x) = 2x + dx^4$, $g(x) = x - 1/4dx^2$, where d is any integer prime to p and $1/z$ is defined as $\bar{z}$, where $z\bar{z} \equiv 1 \bmod p$.

F30 A polynomial whose sums of pairs of values are all distinct.

It was noted in **D1** that no nontrivial solution of $a^5 + b^5 = c^5 + d^5$ is known. In fact x^5 is a likely answer to the following unsolved problem of Erdős. Find a polynomial $P(x)$ such that all the sums $P(a) + P(b)$ $(0 \leq a < b)$ are distinct.

F31 An unusual digital problem.

Express the integers in base 4, using the digits 0, 1, 2 and $\bar{1}$ $(= -1)$. Let L be the set of integers which can be written in this way using the digits 0, 1 and $\bar{1}$, but not 2. Can every odd integer be written as the quotient of two elements of L? Loxton & van der Poorten show that, given an odd k, there is indeed a multiplier m such that m and km are both in L, but their analysis is ineffective in the sense that they do not know how to estimate the smallest such m. It may be that there is an absolute constant C such that there is always a multiplier less than $|k|^C$. Examples requiring large multipliers are $k = 133 = 2011_4$, $m = 333 = 111\bar{1}1_4$ and $k = 501 = 20\bar{1}11_4$, $m = 2739 = 1\bar{1}\bar{1}\bar{1}\bar{1}1\bar{1}_4$.

John Selfridge & Carole Lacampagne ask if every $k \equiv \pm 1 \bmod 3$ can be written as the quotient of integers which can be represented in base 3 using the digits 1 and $\bar{1}$, but not 0. Experiments suggest that the answer is yes. If we allow the digits 0 and 1, but not 2, then which integers can be written as such a quotient?

F. M. Dekking, M. Mendès France & A. J. van der Poorten, Folds! *Math. Intelligencer*, **4**(1982) 130–138, 173–181, 190–195; *MR* **84f**:10016abc.

D. H. Lehmer, K. Mahler & A. J. van der Poorten, Integers with digits 0 and 1, *Math. Comput.*, **46**(1986) 683–689; *MR* **87e**:11017.

J. H. Loxton & A. J. van der Poorten, An awful problem about integers in base four, *Acta Arith.*, **49**(1987) 193–203; *MR* **89m**:11004.

Index of Authors Cited

The names appearing here are those of authors whose works are referred to in this volume. References coccur at the end of each problem (e.g. D11, pp.158–166); in the Introduction, I (pp. 1–2) and at the beginning of Sections A (pp. 3–4), D (p. 139), E (p. 199) and F (p. 240). Mentions unsupported by references are listed in the General Index.

Aaltonen, M., D9
Abbott, Harvey L., B2, C9, C14, D3, E1, E10, E11, E12, E28
Abel, Ulrich, A17
Abouabdillah, Driss, B24
Acland-Hood, F4
Adams, William W., A, A17
Adena, Michael A., F4
Adleman, Leonard M., A
Agarwal, P. K., E20
Aiello, Walter, C6
Ajtai, Miklós, C9
Alanen, Jack, B4, B6
Alaoglu, Leon, B2
van Albada, P. J., D11
Alex, Leo J., D10
Alexander, L. B., B2
Alford, W. Red, A13
Alladi Krishnaswami, B3, B22
Alles, Peter, E38
Allouche, J.-P., E16
Almering, J. H. J., D19, D21
Alon, Noga, E12, F3
Alter, Ronald, B37, D27
Althoen, Stephen C., D3
Altwegg, M., D20
Anderson, David Brent, F4
Andrews, George E., C20, E30
Ang, D. D., D20
Anglin, W. S., D3
Ankeny, Nesmith C., B16
Ansari, A. R., A17
Anshel, Michael, D11
Antoniadis, Jannis A., D26
Applegate, David, E16
Archibald, Raymond Clare, A3
Arno, Steven, A12
Arshon, S., E21
Artin, Emil, B16
Artuhov, M. M., B2, B4
Atanassov, Krassimir T., B41
Atkin, A. Oliver L., A3, A8
Aull, C. E., B2
Avanesov, È. T., D3

Babcock, W. C., C11
Bailey, D. F., B31
Baillie, Robert, A13, B21, B36
Baker, T., D15
Balasubramaniam, R., E1
Ball, W. W. Rouse, F4
Ballew, David, B36
Balog, Antal, A9
Bang, Thøger, A17
Baragar, Arthur, D12
Baranov, Valery Ivanov, A17
Barbeau, Edward J., D11
Barbette, E., D3
Barrucand, P.-A., F7
Bastien, L., D27
Bateman, Paul T., A3, A17, B2, B25, C20
Battany, David M., D9
Battiato, S., B4
Baumert, Leonard D., C10, E11
Baumgartner, James E., E29
Baxa, Christoph, A17
Bays, Carter, A4
Beach, B. D., B16

Bear, Robert, B2
Beatty, S., E27
Beck, Walter E., B2, B5
Bedocchi, E., A17
Beebee, John, F14
Beeckmans, Laurent, D3, D11
Beeger, N. G. W. H., A13
Beeler, Michael D., E10
Behrend, F. A., E10
Bell, D. I., D20
Bencze, Mihály, B12
Benkoski, S. J., B2
Bennett, Michael, D9
Berend, Daniel, D25
Berge, Claude, C18
Berger, Marc Aron, F13, F14
Berlekamp, Elwyn Ralph, C7, E10, E26, E27
Berman, Paul, C18
Bernstein, Daniel J., E16
Bernstein, Leon, D8, D11
Berry, T. G., D19
Besicovitch, Abram Samoilovitch, D20, E5
Best, M. R., C10, D7
Beukers, Frits, F23
Beutelspacher, Albrecht, E11
Beyer, Ö, C7
Beyer, W. A., E31
Bini, U., D1
Birch, Bryan J., D1, D27
Bleicher, M. N., D11
Blundon, W. J., A6, D5, D18
Bode, Dieter, B9
Boesch, F. T., C10
Bohman, Jan, C1, C20
Bombieri, Enrico, A8
Borho, Walter, B4, B7
Borning, Alan, A2
Borwein, Peter B., B14
Bose, R. C., C9, C11
Boshernitzan, Michael, E27
Bosma, Wieb, A3, B21
Bosznay, Á. P., C16
Boyarsky, A., D5
Boyd, David, E16
Brakemeier, W., C15
Brandstein, Michael S., B1
Bratley, P., B4
Brauer, Alfred T., B28, C6, C7, F5, F6
Braun, J., A17
Braunholtz, C. H., E21
Bremner, Andrew, D1, D2, D3, D8, D18, D19
Brenner, Joel L., D10
Brent, Richard P., A3, A8, B1
Brenton, Lawrence, D28
Brestovansky, Walter, E11
Breusch, Robert, B14, D11
Briggs, William E., C3
Brillhart, John David, A3
Brindza, B., D25
Brocard, H., D25
Bromhead, H., D18
Brouwer, Andreas E., C10
Browkin, Jerzy, B19, E25
Brown, Alan L., B2
Brown, B. H., B4
Brown, Ezra, B2, D18
Brown, H., F27
Brown, J. L., D11
Brown, Martin Lawrence, D24
Brown, Thomas Craig, E10, E21
Brudno, Simcha, D1
Bruen, A., C18
Brzeziński, Juliusz, B19
Buchholz, Ralph Heiner, D21, D22
Buck, R. Creighton, A17
Buell, Duncan A., A3, B21
Bugulov, E. A., B2
Buhler, Joseph P., A2, D2
Burbacka, E., B34
Burnside, William, D18
Burshtein, N., F14
Buschman, Robert G., C3

Cadwell, J. H., A8
Calkin, Neil J., E32
Callan, David, B2
Cameron, Peter J., E32
Campbell, Paul J., D11
Cao Zhen-Fu, D3, D9, D23, D28
Carlitz, Leonard, B44
Carmichael, Robert Daniel, A12, A13, B2, B39
Case, Janell, B36
Cassels, John William Scott, C20, D5, D12, D27, F, F22
Catalan, E., B6
Catlin, P. A., B41
Cattaneo, Paolo, B2
Cauchy, Augustin-Louis, C15
Cayley, Arthur, E37
Challis, M. F., C12
Chan Wah-Keung, D28
Chandra, Ashok K., E10
Chang, S. J., D11
Chein, E. Z., D18
Chellali, M., D2
Chen, Sheng, C11
Chen Jing-Run, A4, A8, C1, F1

Chen Wen-De, C11
Cheng Yuanyou, F15
Chinburg, Ted, B25
Choi, S. L. G., B26, C14, E11, F13
Choudhry, Ajai, D1
Chowla, Sardomon D., A6, B16, C9, D4, E28
Churchhouse, R. F., F13
Chvátal, Váslav, E10
Cilleruelo, Javier, C9, E1, F1
Cipolla, M., A12
Cockayne, Ernest T., C18
Cohen, Fred, A19, F13
Cohen, Graeme L., B1, B2, B3, B37
Cohen, Robert, D11
Cohn, Harvey, D12, F7
Cohn, J. H. E., D6, D26
Cole, George Raymond, D21
Colman, W. J. A., D18
Colquitt, W. N., A3
Comtet, Louis, F17
Conway, John Horton, A17, C7, C8, C10, D18, E17, E26, E27, E31, F26
Cook, T. J., B3
Coppersmith, Don, D2, E30
Córdoba, Antonio, C9
Cormack, G. V., B21
van der Corput, J. G., C1
Corzatt, C. E., F4
Costello, Patrick, B4
Cottrell, A., C6
Coxeter, Harold Scott MacDonald, D18, F4
Craggs, D., F4
Cramér, Harald, A8
Crandall, R. E., A2, D2, E16
Cremona, John E., D3
Crews, Philip L., B2
Crittenden, Richard B., E23
Crocker, R., A19
Croft, Hallard T., F4
Cross, J. T., B2
Cucurezeanu, Ion, D3
Čudakov, N. G., C1
Curtis, Frank, C7
Curtiss, D. R., D11
Curtz, T. B., D27
Cusick, Thomas W., D12, F20

Dănescu, Alexandru, F28
Dardis, J. A., D1
Davenport, Harold, A8, C15, D4, E20, F5, F22
Davidson, J. L., E16
Davis, James A., E10, E33
Daykin, David E., D20
Dean, Richard A., E21
DeBoer, Jennifer L., B3
Dekking, F. M., E21, F31
Delorme, Jean-Joël, D1
Dem'janenko, V. A., D1
Deshouillers, Jean-Marc, A19, C1
Deuber, Walter, E14
Devitt, John Stanley, B6
Dewar, James, F14
Diananda, Palahenedi Hewage, C14
Dias da Silva, J. A., C15
Dickson, Leonard Eugene, A17, B6, D12, D18, D27, E32
Ding Xia-Xi, C1
Dirksen, P. H., E20
Dixmier, Jacques, C7
Dixon, R., C18
Djawadi, Mehdi, C12
Dodge, Clayton W., D19
Doig, Stephen K., B49
Dolan, Stanley W., D5
Dove, Kevin L., D22
Downey, Peter J., E31
Drago, Antonino, F19
Drazin, David, B16
Drucker, Daniel S., D28
Dubitskas, A. K., F23
Dubner, Harvey, A2, A3, A13
Dubouis, D1
Dudeney, Henry Ernest, D3, E34, F4
Dudley, Underwood, A17
Duke, W., B16
Duparc, H. J. A., A13

Ecker, Michael W., B36
Eckert, Ernest J., D21
Ecklund, Earl F., B23, B33
Edgar, Hugh Maxwell Wallace, D8
Edwards, Harold M., D2
Eggleston, Harold Gordon, D21
Eggleton, Roger Benjamin, B23, D19, E27
Ehrman, John R., A3
Eliahou, Shalom, E16
Elkies, Noam D., B19, C8, D1
Elliott, D. D., A17
Elliott, Peter D. T. A., A5
Ellis, David, A17
Ellison, W. J., D5, F23
Engel, Marcus, D18
Entringer, Roger C., E10, E21, E33
Erdős, Pál, A5, A6, A9, A11, A12, A13, A18, A19, B2, B4, B6, B8, B9, B10, B14, B16, B18, B21, B22, B23, B24, B26, B27, B30, B31,

B32, B33, B35, B36, B39, B40, B41, B42, B46, C3, C6, C7, C8, C9, C10, C12, C15, C16, C17, D2, D3, D3, D4, D7, D11, D17, D25, E, E1, E2, E3, E4, E5, E9, E10, E11, E21, E24, E28, E31, F2, F13, F14, F18
Ernvall, Reijo, A17, D2
Escott, E. B., B4
Estermann, Theodore, C1
Evans, Ronald J., B28, C10, D19
Evdokimov, A. A., E21
Everett, C. J., E16
Everts, F., E10
Ewell, John A., C5, C20, F17

Fabrykowski, J., B18
Falconer, Kenneth J., F4
Fässler, Albert, D21
Faulkner, Marilyn, B33
Fejes Tóth, László, F
Felzenbaum, Alexander Gersh, F13, F14
Fife, Earl Dennet, E21
Filipponi, P., E16
Filz, Antonio, C1
Finch, Steven R., C4, E32
Fine, Nathan J., D19, E25
Finkelstein, Raphael, D3
Flammenkamp, Achim, B7, F4
Flatto, Leopold, E18
Flood, P. W., D18
Forbes, A., B5
Forman, William, E19
Forman, Robin, A17
Foster, Lorraine L., B39, D10, D16
Fouvry, É., A8
Fraenkel, Aviezri S., C5, D8, E27, F13, F14
Franceschine, Nicola, D11
Franqui, Benito, B2
Fredricksen, Harold, E11, E12
Freiman, G. A., C15
Freud, Róbert, C9, E30
Frey, H. A. M., B3
Friedlander, John B., A8, A9, A12
Friedman, Charles N., D11
Frobenius, G., D12
Froberg, Carl-Erik, C1, C20
Fuchs, W. H. J., C9
Funar, Louis, C14
Fung, Gilbert W., A1, A17
Furedi, Zoltan, E20
Furstenberg, H., E10

Gale, David, E15, E17
Gallagher, Patrick X., A19
Gallian, Joseph A., C13
Gamble, B., C18
Gandhi, J. M., A17
García, Mariano, B2, B4
Gardiner, Verna L., C3, D5
Gardner, Martin, A1, A2, C8, D3, D18, D21, F4
Gardy, D., E20
Garner, Lynn E., E16
Garrison, Betty, A17
Gelfond, A. O., D9
Genocchi, A., D27
Gérardin, A., D1, D27
Gerver, Joseph L., A5, E10
Gibbs, Richard A., C10
Giese, R. P., C6
Gill, C., D15
Gillard, P., B36
Gillies, Donald B., A3
Gilmer, Robert, B16
Gioia, A. A., B4, C6
Giuga, Giuseppe, A17
Gloden, A., D1
Godwin, Herbert James, A2, D27
Goetgheluck, P., A17
Golay, Marcel J. E., C10
Goldfeld, Dorian, D11
Goldstein, Richard, F4
Goldston, Daniel L., A8, C1
Golomb, Solomon W., A2, A17, B42, C18, D11
Golubev, V. A., A5
Goodstein, Reuben Louis, A17
Gordon, Basil, C5
Gordon, Daniel M., A12
Gorzkowski, Waldemar, D4
Gosper, R. William, E16, F17
Gostin, Gary B., A3
Gottschalk, W. H., E21
Gouyou-Beauchamps, D., E20
Gould, Henry W., A17
Graham, Ronald Lewis, A3, B23, B33, C7, C9, C10, C20, D11, E2, E10, E25, E27, E30, E31, E33, F14
Graham, Sidney W., A4, B3, C11
Granlund, Torbjorn, D18
Granville, Andrew, A4, A9, A13, A19, B33, B41, C1, C20, D2, F1
Greenberg, Harold, C7
Greenwell, Raymond N., E10
Grimm, Charles A., B32
Grinstead, Charles M., B22, D3
Griswold, Ralph E., E31
Grosswald, Emil, A5, C20
Grupp, F., A8

Gryte, D. G., D26
Guillaume, D., A13
Güntsche, R., D22
Guo Zhi-Tang, D10
Gupta, Rajiv, F9
Gupta, Hansraj, B33, C8
Gurak, S., A12
Guy, Andrew William Peter, B6
Guy, Michael John Thirian, F26
Guy, Richard Kenneth, A, A3, A17, B6, B8, B13, C1, C7, C8, C20, D1, D19, E26, E27, E37, F2, F4, F4, F26
Györy, K., D7

Haentzschel, E., D22
Hagis, Peter, A5, B1, B2, B3, B4, B5, B8, B37, B39
Hahn, H. S., D11
Hahn, Liang-Shin, D11
Haight, J., F13
Hajela, D., C11
Hajnal, Péter, E20
Hajós, G., E10
Halberstam, Heini, A5, C9, E
Hales, Alfred W., E10
Hall, Marshall, C10, E25
Hall, Richard R., B36, B41, F4
Halter-Koch, Franz, C20
Hamidoune, Yahya Ould, C15
Hammer, Joseph, F
Hämmerer, N., C12
Hansche, B., C18
Hansen, W., C6
Hanson, Denis, D3, E10, E11
Harborth, Heiko, B28, D3, D20, F4
Hardy, B. E., B2
Hardy, Godfrey Harold, A1, A17, D4
Harman, Glyn, A8
Harris, Vincent C., A17
Hart, S., E20
Härtter, E., A17
Haugland, Jan Kristian, A8
Hausman, Miriam, B41
Hawkins, David, C3, E21
Heath-Brown, D. Roger, A1, A4, A5, B16, B18, D5, F9
Hebb, Kevin, C6
Hedetniemi, Stephen T., C18
Hedlund, Gustav A., E21
Heilbronn, Hans, C15
Helm, Martin, C11
Hendy, M. D., B2
Henriksen, Melvin, B25
Hensley, Douglas, A9
Heppner, E., E16
Herschfeld, Aaron, D9
Herstein, I. N., E16
Herzberg, Norman P., D12
Heuer, Karl W., F4
Higgins, Olga, A17
Higgins, Robert N., B36
Hildebrand, Adolf, A9, B18, C20, F6
Hill, Jay Roderick, A13
Hilton, Peter J., F28
Hindman, Neil, E29
Hoffmann, H., B4
Hofmeister, Gerd, C12, D11
Hofstadter, Douglas R., E31
Hoggatt, Verner E., D26
Holton, Derek A., F4
Holzman, R., F14
Hooley, Christopher, A19, B39, B40, F9
Horn, Roger A., A17
Hornfeck, B., B2
Hudson, Richard H., A4, F5
Hudson, W. H., D19
Huenemann, Joel, A13
Huff, G. B., C18, D20
Hunsucker, John L., B9, B13
Hurwitz, Adolf, D12
Huxley, Martin, A8, F1

Ibstedt, Henry, E15
Il'in, A. M., C6
Inkeri, K., D2, D9
Irving, R. W., E11
Isenkrahe, C., A17
Ivić, Aleksandar, B16, F1
Iwaniec, Henryk, A1, A8, A9, B16, B39, F1

Jabotinsky, Eri, C3
Jackson, D. E., E21
Jackson, T. H., F4
Jaeschke, Gerhard, A12, A13, B21
Jerrard, R. P., B2
Jewett, R. I., E10
Jia Chao-Hua, A8, B16
Jia Xing-De, C9, C11, C12
Johnson, Allan William, D11
Johnson, G. D., A3
Johnson, Wells, B2, D2
Joint, W. Howard, D18
Jones, James P., A6, A17
Jones, Patricia, B36
Jönsson, Ingemar, B20
Joó, I., A12, A13, D11
Jordan, James H., A16, D21, F13
Judd, J. S., A

Justin, J., E21
Jutila, Matti, A4

Kac, Mark, B14, E31
Kahan, Steven, A17
Kalbfleisch, J. G., C10
Kalyamanova, K. È., D22
Kang Ji-Ding, D7
Kanold, Hans-Joachim, B2, B4, B9
Kaplansky, Irving, D1, E16
Karst, Edgar, A5, A17
Kashihara, Kenji, D23
Katayama, Shin-ichi, D23
Kato, H., C6
Katz, M., C17
Kay, David C., E16
Keller, Ott-Heinrich, F
Keller, Wilfrid, A, A3, A13, B20, B21
Kellogg, O. D., D11
Kelly, John B., D1, F19
Kelly, Patrick A., F4
Kemnitz, Arnfried, D19, D20
Kemp, Paula A., F15
Kendall, David G., B2
Kenney, Margaret J., C1
Khinchin, A. Y., E10
Killgrove, Raymond B., A10
Kim Su-Hee, A12
Kingsley, R. A., D26
Kirfel, Christoph, C12, C12
Kishore, Masao, B1, B2, B37
Kiss, Péter, A12, E31
Klamkin, Murray S., B2
Klarner, David A., C18, E36
Klee, Victor L., B39
Klein, Rivka, F15
Kleitman, Daniel J., E12
Klotz, Walter, C12
Kløve, Torliev, C11, F4
Knapowski, Stanisław, A4
Knödel, W., A13
Knopfmacher, John, A17
Knuth, Donald Ervin, A3, C6
Ko, Chao, D5, D11, D13
Kobayashi, Masaki, A17
Kolesnik, G. A., A1
Kolsdorf, H., C12
Komjáth, Péter, E20
Komlós, János, C9, E11, F4
Korselt, A., A13
Koyama, Kenyi, D5
Kraitchik, Maurice, A3, C18, D18, D25
Krasikov, I., E16
Kravitz, Sidney, A2, A3, B2
Krückeberg, F., C9
Krukenberg, C. E., F13
Kubiček, Jan, D1
Kuipers, Lauwerens, A17
Kurepa, Đuro, B44

Laatsch, Richard, B2
LaBar, Martin, D15
Laborde, M., A8
Lacampagne, Carole, B31, D3
Lagarias, Jeffery C., A8, A17, E8, E16, E30
Lagrange, Jean, D4, D15, D18, D20, D27
Lal, Mohan, A6, B5, B8, B36, D5, D18
Lalout, Claude, A7
Lambek, Joachim, E27
Lander, Leon J., D1
Landman, Bruce M., E10
Lang, Serge, B19
Langevin, Michel, B32, D9
Langmann, Klaus, A17
Laub, Moshe, D3
Lazarus, R. B., C3, D5
Le Mao-Hua, B38, D9, D10
Lebensold, Kenneth, B24
Lee, Elvin J., B4
Leech, John, A4, A9, C10, D1, D18, D20, E21
Lehmer, Derrick Henry, A3, A7, A12, A13, A17, B6, B29, B37, F1, F6, F31
Lehmer, Derrick Norman, D20
Lehmer, Emma, B24, B45, D2, F6
Leitmann, D., A1
Lekkerkerker, Cornelius Gerrit, F
Lenstra, Arjen K., A, A3
Lenstra, Hendrik Willem, A, A3, B6
Levin, Leonard, F17
Levine, Eugene, E28
Lewin, Mordechai, C7
Lewis, Ethan, F14
Li An-Ping, C11
Li De-Lang, D11
Li Xiao-Ming, C10
Lieuwens, E., B37
Lin Chio-Shih, E27, F14
Lin Shen, C20, E27, F14
Lind, D. A., D3
Lindström, Bernt, C8, C9, C11
Linnik, U. V., A4
van Lint, Jacobus H., D11
Lioen, Walter M., A3, D5
Littlewood, John Edensor, A1, D4, F22
Liu, A. C., E10
Liu Hong-Quan, A1, A8, B16
Liu Jian-Min, A4
Liverance, Eric, A17

Ljunggren, W., D3, D6
Löh, Günter, A7, A13
Lord, Graham, B5, B9
Lossers, O. P., D3
Lou Shi-Tuo, A8
Low, Lewis, C15, F27
Loxton, John H., F31
Lucas, Edouard, D3
van de Lune, J., C1, D7
Lunnon, W. Fred, B4, C8, C12, E21
Luo Ming, D26
Lyness, Robert Cranston, D18

Ma De-Gang, D3
Macdonald, Shiela Oates, E20
MacMahon, Percy A., E30
MacWilliams, F. Jessie, C10
Madachy, Joseph S., B4, E34
"Mahatma", D18
Mahler, Kurt, D4, E18, F26, F27, F31
Maier, Helmut, A8, A9, B9, B36, E3
Mąkowski, Andrzej, A12, A15, B2, B5, B9, B16, B19, B25, B42, D5, D10
Mallows, Colin L., E31
Malm, Donald E. G., A19
Manasse, Mark S., A, A3
Mann, Henry B., A17, C10, C15
Mansfield, Richard, C15
Marcus, Daniel, E25
Markoff, A., D12
Masai, Pierre, B39
Mason, T. E., B2
Massias, J.-P., A17
Mathieu, D1
Matiyasevich, Yuri V., A17
Matthews, K. R., E16
Mattics, L. E., C20, D3
Mauldon, James G., D16, D19
McDaniel, Wayne L., A12, B2, B16, B49
McIntosh, Richard John, B31
McKay, John H., B4
McLaughlin, Philip B., A3
Meeus, Jean, A7
Mendelsohn, Nathan S., B21
Mendès France, Michel, F31
Metropolis, N., C3
Metsänkylä, T., D2
Mian, Abdul Majid, E28
Miech, R. J., B37
Mientka, Walter E., A19, B13, E21
Mignotte, Maurice, A1, D9, D23
Mijajlović, Ž., B44
Miller, G. L., A
Miller, Jeffrey Charles Percy, C10, D5
Miller, Kathryn, B36
Miller, V. S., A17
Mills, William H., A17, B2, B41, D13, E20, F6
Minoli, Daniel, B2
Mirsky, Leon, B18, E11
Misiurewicz, M., D24
Mittelbach, F., D11
Mo De-Ze, D10
Moews, David, B7
Moews, Paul C., B7
Mohanty, Shreedhara Prasada, D3
Möller, Herbert, E16
Mollin, Richard A., A17, B16
Molnar, E. A., F28
Monsky, Paul, D27
Montgomery, Hugh L., A9, B40, C1
Montgomery, Peter Lawrence, A, A3
Moore, Eliakim Hastings, D18
Moran, Andrew, A5
Mordell, Louis Joel, B16, D6, D18, D20, D27
Moree, Pieter, C14, D7
Morikawa, Ryozo, F14
Moroz, B. Z., B16
Morse, Marston, E21
Morton, Patrick, D2
Moser, Leo, A17, B33, C2, C12, C14, C17, D7, E1, E10, E11, E27
Mossige, Svein, C12, E32
Motzkin, Theodor S., C17
Mozzochi, C. J., A8, F1
Mrose, Arnulf, C12
Muir, Thomas, E37
Müller, Helmut, E16
Muller, P., C4
Mullin, Ronald C., C10, E20
Murata, Leo, F9
Murdeshwar, M. G., C17
Murty, Maruti Ram, F9

Nagell, Trygve, D9, D11
Nair, M., B18
Najar, Rudolph M., B2, B5
Nakamura, Shigeru, B2
Namboodiripad, K. S., A17
Narkiewicz, Władysław, A19, C1
Nash, John C. M., C11
Nathanson, Melvyn B., C9, C12, C15
Naur, Thorkil, A3
Nebb, Jack, B13
Neill, T. B. M., A17
Nelson, Harry L., A6
Newberry, R. S., B3
Newman, David, E31
Newman, Donald J., C6, D11

Newman, Morris, F14
Nicol, Charles A., B41, D2
Nicolas, Jean-Louis, D15
Niebuhr, Wolfgang, A13
Nitaj, Abderrahmane, B19
Niu Xue-Feng, C7
Niven, Ivan, A17, B41, D11
Noda, Kazunari, D27
Noll, Landon Curt, D20
Norrie, R., D1
Novák, Břetislav, F14
Nowakowski, Richard Joseph, E37

Obláth, R., D2, D11, D25
Odda, Tom, E33
Odlyzko, Andrew M., A10, A17, B21, C10, E8, E10, F4
Oertel, Philipp, F4
Oltikar, Sham, B49
Olsen, John E., C15
Ondrejka, Rudolf, A3
O'Neil, Patrick E., E10
Oppenheim, Alexander, D8
Ore, Oystein, A17, B2, B4
Ortega Costa, Joaquin, A17
Osgood, Charles F., D25
Ostmann, H., E
O'Sullivan, Joseph, E28
Overhagen, T., D25, F1
Owings, J. C., D11

Pajunen, Seppo, B2
Palamà, G., D11
Pall, Gordon, C20
Pan Cheng-Dong, C1
Pan Cheng-Tung, A4
Papadimitriou, Makis, A17
Parady, Bodo K., A8
Parkin, T. R., A6, D1
Patterson, Cameron Douglas, C20
Paxson, G. Aaron, B6
Peeples, W. D., D20
Pelikán, Jozsef, E38
Penk, M. A., A2
Penney, David E., A2
Peralta, René, F6
Perelli, A., C1
Perrine, Serge, D12
Peterkin, C. R., E20
Peterson, Blake E., D21
Peterson, Ivars, B49
Phillips, Steven, E30
Phong Bui-Ming, A12, A13
Piekarczyk, J., B34
Pillai, S. Sivasankaranarayana, B28, D9
Pil'tjai, G. Z., A8
Pinch, Richard G. E., A13
Pinner, C., B18
Pintér, Ákos, D3
Pintz, János, C1, F4
Pitman, Jane, C15
Plaksin, V. A., C20
Pleasants, P. A. B., E21
Pocklington, H. C., D18
de Polignac, A., A19
Pollack, Richard M., A3, D25
Pollard, John M., A, C15
Pomerance, Carl, A, A1, A4, A5, A8, A12, A13, B2, B4, B9, B18, B32, B36, B37, B39, B41, B42, B46, C1, D19, E, E10, F4, F15
Popken, J., F26
Porta, Horacio A., E16
Porubský, Štefan, C20, E30, F14
Posner, Edward C., D9
Post, Karel, E36
Potler, Aaron, A8
Poulet, Paul, B2, B4, B6
Pounder, J. R., C12
Powell, Barry J., B45
Prachar, K., A4
Prellberg, Thomas, F4
Pritchard, Paul, A5
Prodinger, Helmut, E21
Propp, James Gary, E10
Proth, F., A10
Prothro, E. T., D21
Purdy, George B., C20
Pyateckii-Šapiro, I. I., A1

Queneau, Raymond, C4

Rabung, John R., A16, E10
Rado, Richard, E10, E14, E36
Raitzin, Carlos, A17
Ralston, Kenneth E., A10, C17
Ramachandra, K., B32
Ramaré, Olivier, B33
Ramsey, L. Thomas, A5, E10
Rangamma, M., B37
Rankin, Robert Alasdair, E10
Rassias, Themistocles Michael, E25
Rathbun, Randall L., D18, D21
Rawsthorne, Daniel A., E16, F26
Recamán, Bernardo, C4
Reddy, D. Ram, B3
Regimbal, Stephen, A17
Reidlinger, Herwig, B2
Remak, Robert, D12
Rennie, B. C., E20

Rényi, Alfred, A6
Rhemtulla, A. H., E12
Ribenboim, Paulo, A, D2
Ribet, Kenneth, D2
Richards, Ian, A9
Richert, Hans-Egon, A5, C20
Rickert, N. W., A3, A8
Rickert, U.-W., C15
Riddell, James, C12, E10
Rieger, G. J., B4
te Riele, Herman J. J., A3, B1, B2, B4, B6, B8, C1, D5, D7
Riesel, Hans, A, B20, C20
Rivat, J., A1
Rivest, R., A
Robbins, Herbert E., E21
Robbins, Neville, B2, D26
Roberts, J. B., C7
Roberts, S., D27
Robin, G., A17
Robins, Gabriel, D16
Robinson, Raphael M., B21, E15
Rödne, Arne, C12
Rødseth, Øystein J., C7, C12, C15
Rogers, Claude Ambrose, F
Rohrbach, H., C12
Rokowska, B., F11
Root, S. C., A5
Rosati, L. A., D11
Roselle, David P., E20
Rosenberger, Gerhard, D12
Rosenstiel, C. R., D1
Rosenstiel, E., D1
Ross, P. M., C1
Rosser, J. Barkley, A17
Roth, Klaus F., C9, E, E10, F4
Rothschild, Bruce L., E10
Rotkiewicz, Andrzej, A3, A12, A13, D10
Rubinstein, Michael, A17
Ruderman, Harry D., B31
Rumney, Max, B15
Rumsey, Howard, D9
Russell, W., D5
Ruzsa, Imre Z., B33, C9, C15, E1
Ryavec, Charles, C15

Saito, Sadao, D28, F28
Salem, R., E10
Salié, H., E10
Sander, J. W., B33, D11, E16, F15
Sándor, József, B3, B37, B42
Sárközy, András, A12, B18, B19, B33, C16, E, E2, E9
Sastry, K. R. S., D21
Sato, Daihachiro, A17
Satyanarayana, M., B2
Sawyer, Walter Warwick, D18
Sayers, M. D., B1
Scarowsky, W., D5
Schatz, J. A., E21
Scheidler, Renate, B33
Schinzel, Andrzej, A, B9, B18, B25, B27, B34, B36, B42, B47, D5, D10, D11, D23, E20, F11, F13
Schlafly, Aaron, B39
Schmidt, Peter Georg, B16
Schmidt, Wolfgang M., E10, F4
Schnitzer, F., C17
Schoenfeld, Lowell, A17
Scholz, Arnold, C6
Schönheim, Johanan, C10, E11, F14
Schrandt, R. G., E31
Schroeppel, Rich, E16
Schubert, H., D21
Schuh, Fred., B37
Schultz, O., D22
Schur, Issai, B33, E11
Schwarz, Hermann Amandus, D11
Scott, Reese, D9
Sebastian, J. D., C18
Sedláček, Jiří, D11
Segal, David, B31
Segal, Sanford L., B37, D8
Selfridge, John Lewis, A3, A12, A19, B6, B21, B23, B27, B31, B32, C5, C17, C18, C19, D1, D2, D17, F6, F13
Selmer, Ernst S., C7, C12, E32
Selvik, Björg Kristin, C12
Sentance, W. A., B16
Serf, P., D27
Shallit, Jeffery, E16
Shamir, A., A
Shanks, Daniel, A, A1, A3, A4, A8, A17, B13, D2
Shantaram, R., E10
Shapiro, Harold N., B2, B41, D25, E19
Sharir, Micha, E20
Shearer, James B., E8
Shelah, Saharon, E10
Shen, Mok-Kong, A12, C18
Shen Tsen-Pao, C18
Sheng, T. K., D20
Shepherd, B., C18
Shiu, Peter, B16, B18
Shor, P., E20
Shorey, T. N., B32, D10
Siebert, Hartmut, A17
Sierpiński, Wacław, A, A1, A5, A17, B2, B18, B21, B36, D5, D8, D11,

D18, D22, D23, E25
Silverman, Joseph H., D1, D12
Silverman, Robert D., A3
Simmons, Gustavus J., D25, E10, E33
Simpson, R. Jamie, E10, E23, F13, F14, F15
Singer, J., C9
Singer, M., A17
Singmaster, David Breyer, D3, D11
Sinisalo, Matti K., C1
Sirota, E. R., A1
Sitaramachandra Rao, R., B16
Siva Rama Prasad, V., B3, B37
Skinner, Christopher M., D10
Slater, M., C18
Slater, Peter J., C10
Sloane, Neil J. A., C9, C10, F25
Smith, Herschel F., A9
Smith, Joel F., A8
Smith, Paul, C8
Smyth, C. J., D1
Solovay, R., A
Somos, Michael, E15
Sompolski, R. W., D2
Sonntag, Rolf, C6
Sorenson, Jonathan, A
Sós, Vera Turán, C9, E9
Soundarajan, K., E1
Spencer, D. C., E10
Spencer, Joel H., E2, E10
Spencer, P. H., C18
Spiro, Claudia, B41
Spohn, W. G., D18
Sprague, Roland Percival, C20
Srinivasan, B. R., A17
Srinivasan, Seshadri, F9
Stanley, Richard P., E10
Stanton, Ralph G., B21, C10, E20
Stark, Harold M., D26
Starke, Emory P., D24
Stauduhar, Richard, F11
Stechkin, Boris S., A17
Steen, Adolf, E37
Stein, M. L., C1
Stein, P. R., C1, D5
Stein, Sherman K., D11, F14
Steinberg, Robert, C14
Steiner, Ray P., D6, D26, E16
Stemmler, Rosemarie M., B25
Stemple, J. G., B4
Stephens, A. J., B16, B25
Stephens, Nelson M., D27
Stevens, R. S., E10
Stewart, Bonnie M., D11
Stewart, C. L., B19
Stöhr, Alfred, C12, E
Stolarsky, Kenneth B., C6, E16, F4
Stoll, Peter, D11
Strassen, Volker, A
Straus, Ernst G., B2, B17, B33, C5, C6, C16, D11, E24
Street, Anne Penfold, C14, E11, E12
Stroeker, Roelof Jacobus, D11
Styer, Robert, D10
Subba-Rao, K., D1
Subbarao, Mathukumalli V., B2, B3, B17, B18, B37, B39, C6
Sudbery, A., F4
Sugunamma, M., B15, C6
Sulyok, Miklós, E11
Summers, T., B8
Sumner, John L., D22
Sun, Chi, D11
Sun Qi, B47, D5, D13
Suryanarayana, D., B2, B3, B9, B16, B17
Suyama, Hiromi, A3
Swierczkowski, S., C17
Swinnerton-Dyer, Henry Peter Francis, D1, D9, F22
Sylvester, James Joseph, B33, C7, D11
Szabó, Zoltán István, E10
Szalay, Michael, F9
Szegedy, Márió, F15
Szekeres, Esther, E11
Szekeres, George, B16, B31, E11
Szemerédi, Endre, C9, C15, D4, E, E2, E10, E11, E20, F4, F18
Szymiczek, Kazimierz, A12

Tait, Peter Guthrie, E37
Tallman, M. H., D26
Tanner, Jonathan W., D2
Tanny, Stephen M., E31
Taylor, Herbert, C10, C18
Tee, Garry J., A17
Temperley, Nicholas, B2
Templer, Mark, A2
Tenenbaum, Gérald, D11, E3
Terras, Riho, E16
Tetali, Prasad, C9
Teuffel, E., A17
Thatcher, Alfred R., D15
Thiele, Torsten, F4
Thompson, John G., A17
Thouvenot, J. P., E10
Thue, Axel, E21
Thurber, Edward G., C6
Tijdeman, Robert, B19, B32, D7, D9, D10, D11, E18, F23
Tiller, G., B8

Tovey, Craig A., D3
Trigg, Charles W., A9
Trost, E., D24
Trusov, Ju. D., D1
Tsangaris, P. G., A17
Tunnell, Jerrold B., D27
Turán, Pál, A4, A5, A11, C9, E, E10
Turgeon, Jean M., B24
Turk, Jan, B30
Tzanakis, Nikos, D3, D6

Uchiyama, Saburô, A8, A19, D3
Ulam, Stanislas M., C3, C4, E31
Urbanek, Friedrich J., E21
Urbanowicz, Jerzy, D7
Utz, W. R., B33, C6

Vaidya, A. M., B4
Vâjâitu, Viorel, F28
Valette, Alain, B39
Vanden Eynden, Charles L., A17, B16, E23, F14
van der Poorten, Alfred J., A13, D2, F17, F31
Vandiver, Harold S., D2
Vantieghem, E., A17
Vardi, Ilan, C18, E16, E25
Varnavides, P., C14
Vaughan, Robert C., A9, A19, B40, C1, D11
Vegh, Emanuel, F9
Venturini, G., E16, E17
Vinogradov, I. M., C1
Vogt, R. L., B13
Voorhoeve, M., D7
Vucenic, W., C18
Vulah, L. Ja., D12

Wada, Hideo, A17, D27
van der Waerden, B. L., E10
Wagon, Stanley, B39, E16
Wagstaff, Samuel S., A3, A12, A13, D2
Wakulicz, A., B36
Wald, Morris, C1
Walker, David T., B16
Wall, Charles R., B2, B2, B3, B8, B36, B41, D26
Wall, David W., B37
Wallis, Jennifer Seberry, E11
Wallis, Walter Denis, E11
Walsh, P. Gary, B16
Wang Du-Zheng, D9
Wang, Edward T. H., C1, C14, E12
Wang Tian-Ze, A8, C1
Wang Yan-Bin, D23
Wang Yuan, C1
Wang Wei3, A4
Ward, Morgan, D1
Warren, L. J., B3
Watson, George Neville, D3
Watts, A. M., E16
Webb, William A., D11
Weber, J. M., B3
de Weger, Benne M. M., D3, D10, D26
Weinberger, Peter, A3
Weintraub, Sol, A5
Weiss, A., E16
Weitzenkamp, Roger C., A19
Welsh, L., A3
Wen Zhang-Zeng, D28
Werebrusow, A. S., D1
Whitehead, Earl Glen, E11
Wichmann, B., C10
Wiens, Douglas, A17
Wilansky, Albert, B49
Wild, K., F4
Wilf, Herbert S., D11
Willans, C. P., A17
Williams, E. R., E14
Williams, Hugh Cowie, A, A1, A3, A13, A17, B16, B21, B33, D2, D26, F7
Winter, David, F4
Winter, Dik, A3
Wirsing, Edward, B2
Witsenhausen, Hans S., E2
Witt, E., E10
Woods, Dale, A13, B4
Wooldridge, K. R., B39
Woollett, M. F. C., D5
Wormell, C. P., A17
Wright, Edward Maitland, A17, C20
Wu, Jie, A8
Wu Yun-Fei, D10
Wunderlich, Marvin C., B6, B8, C3, C4, D8
Wyburn, C. T., C6

Xie Sheng-Gang, A9
Xu Yi-jing, F1
Xu Z.-Y., D3

Yamada, Masaji, E16
Yamamoto, Koichi, D11
Yao, Qi, A8
Yates, Samuel, A3, B49
Yau, Stephen S.-T., F1
Yip L.-W., B39
Yokota, Hisashi, D11
Yorinaga, Masataka, A12, A13, A19, B36

Yoshitake, Motoji, B2
Young, Jeff, A3, A8, B21
Yu Kun-Rui, B19
Yuan Ping-Zhi, B16

Zaccagnini, Alessandro, A8
Zachariou, Andreas, B2
Zachariou, Eleni, B2
Zagier, Don B., D12
Zaharescu, Alexandru, F15, F28
Zajta, Aurel J., D1, D9
Zarankiewicz, Kazimierz, A5
Zarantonello, Sergio E., A8
Zaremba, Stanisław Krystyn, F20
Zarnke, C. R., B16
Zay, Béla, E31
Zeilberger, Doron, F13, F14
Zeitlin, David, E31
Zhang Ming-Zhi, A13, D28, F10, F13, F14
Zhang Zhen-Xiang, B47, E4, E28
Zhou Guo-Fu, D7
Zhu Yi-Liang, C20
Złotkowski, W., D10
Znám, Štefan, E11, E13, E16, F13, F14
Zun, Shan, B37
Zwillinger, Dan, C1

General Index

Names appear here if their mention in the relevant section is unsupported by references. Single letter entries refer to the Introduction (pp. 1–2) and to the beginning of Sections A (pp. 3–4), E (p. 199) and F (p. 240).

A-sequence, E28
ABC conjecture, A3, B19, D2
Abe, Nobuhisa, D25
abundance, B2
abundant, B2
addition chain, C6
additive basis, C12
additive sequence, C4, E32
admissible partition, E11
Alanen, Jack, B10
algebraic number, F22
aliquot cycle, B7
aliquot parts, B, B4
aliquot sequence, B6
almost perfect, B2
amicable numbers, B4
amicable triples, B5
Anderson, Claude, D13
Andrica, Dorin, A8
Applegate, Harry, F28
arithmetic progression, A, A5, A6, A19, E10, E23, E33
Ashbacher, Charles, D21
associates, A16
asymptotic, A1
asymptotic density, A17

B_2-sequence, C9, E28
Baker's method, D10
Baker, Alan, C4
ballot numbers, B33
Bang, Thøger, B46
Baragar, Arthur, D23
barrier, B8
basis, E32
Beatty sequences, E27, F14
Benkoski, S. J., F10
Bergmann, Horst, D18
Bergum, Gerald E., D19
Bernoulli numbers, A17, D2
betrothed numbers, B5
Betsis, Dimitrios, A9
binomial coefficient, B31, B33, D3, D17
Bond, Reginald, B43
Bowen, Rufus, D7
Brauer chain, C6
Brauer number, C6
Brauer, Alfred, E10
Brizolis, Demetrios, F9
Browkin, Jerzy, B36
Brun's method, B42
Brun–Titchmarsh theorem, B39
Buckley, M. R., D27
Burr, Stefan, C4

Cameron, Peter J., C9
Carmichael number, A12, A13, B37, D1
Carmichael's conjecture, B39
Catalan conjecture, B19, D9
Catalan numbers, B33
chain, E17
champion, A8
Choi, S. L. G., F14
Choudhry, Ajai, D13
Chowla, Sardomon D., A4, B18, D8
Chudnowsky, Gregory V., D9
class-number formula, F5
cluster, D20
coin problem, C7
Collatz, Lothar, E16
collinear integers, E10
composite number, A

congruent, A4
congruent number, D27
continued fraction, D11, F20, F21, F22, F23
Conway, John Horton, A8, B19, B20, B33, D19, E20
coprime, A
Costas array, C18
Coughlan, Francis, F28
covering congruences, B21, E3, F13, F14
covering problem, C12, C13
covering system, F13
Cramer's conjecture, A2
Croft, Hallard T., A10, F23
cube, C13
Cullen numbers, B20
Cunningham chain, A7, A18
cycle, D18, E17
Czipszer, J., C17

Davenport-Schinzel sequence, E20
Dedekind's function, B8, B37, B38, B41
deficiency, B31
deficient, B2
density, B2, E, E26, E36
derived cuboid, D18
determinant, F27, F28
Dickerman, Mitchell R., B6
Dickson, Leonard Eugene, D22
digital problems, F24, F25, F31
Dirichlet's theorem, A
Dirichlet, Peter G. Lejeune, F5
Divis, Bohuslav, F21
Dressler, R. E., C20
Dubner, Harvey, A8
Dudeney, Henry Ernest, D19
Düntsch, Ivo, B18

e-multiperfect, B17
e-perfect, B17
Easter, Michael, B23
Ebert, Gary, F8
Echevarria, Javier, C1
Edgar, Hugh Maxwell Wallace, D9, D10
Eggleton, Roger Benjamin, B18
Egyptian fractions, D11, F14
Eisenstein-Jacobi integers, A16
Eisenstein-Jacobi primes, A16
Elliott-Halberstam conjecture, B41
elliptic curve, D1, D2
elliptic functions, D3
Epstein, Richard, E26
Erdős, Pál, A2, A15, B13, B15, B19, B34, B38, C1, C2, C11, D9, D13, D15, E7, E8, E12, E29, E30, E33, F3, F9, F16, F23, F25, F30
error-correcting codes, C10
Euler numbers, B45
Euler pseudoprime, A12
Euler's constant, B41
Euler's function, D11
Euler, Leonhard, A1, B48, D1, D9, D17, D18, D20, D21, D22
exact covering system, F14
exponential divisor, B17
extremal basis, C12

factorial, A2, B22, B23, B43, D25, F11
fan, C13
Farey series, D11, F27
Feit, Walter, B25
Fermat number, A3, A12, B21
Fermat problem, B19
Fermat's (little) theorem, B20
Fermat, Pierre de, D17, D27
Fibonacci number, A12, D26
Fibonacci sequence, A3
figurate number, D3
Finucane, Daniel M., B41
Folkman, Jon, E11
fortunate primes, A2
Fortune's conjecture, A2
Frénicle de Bessy, Bernard, D1
friendship graph, C13
frieze pattern, D18
Frobenius, G., C7

Gagola, Gloria, F9
Gallyas, K., D27
Gandhi, J. M., D2
Gardner, Martin, A6, B4, D19
Gaussian integer, A16, F13
Gaussian prime, A16
Gauß's problem, F1
Gauß, Carl Friedrich, F5
Gel'fond-Baker method, F23
Gessel, Ira, B33
Gilbreath's conjecture, A10
Göbel, Fritz, E15
Godwin, Herbert James, B6
Goldbach conjecture, B10, B19, C1
golden number, E25
Golomb ruler, C10
Golomb, Solomon W., E25
good primes, A14
Goormaghtigh, R., B25
Gordon, Basil, A16, B37, F28
Gorzkowski, Waldemar, B20
graceful graph, C13

graceful labelling, C10
Graham, Ronald Lewis, B2, B30, B37, D17, D23, D25, E8, E29, E33, F10, F15
Graham, Sidney W., A13
Granville, Andrew, B19
greatest common divisor, A, E2
Grecu, Dan, A8
greedy algorithm, E10, E28, E32
Grimm's conjecture, B32
Gupta, Hansraj, D9, E22
Guy, Michael John Thirian, D19, E17

Hajós, G., B37
Hall, Richard R., C19
Hansen chain, C6
happy number, E34
Harborth, Heiko, D21
Hardy, Godfrey Harold, F1
harmonic mean, B2
harmonic number, B2
harmonious graph, C13
harmonious labelling, C10
Heath-Brown, D. Roger, B1
Heilbronn, Hans, F4
Helenius, Fred, B2
Heron triangle, D22
Hesterman, Lambertus, F27
hexagonal number, C20
hexahedra, D21
Hickerson, Dean, B23, C10
Hitotumatu, Sin, F24
Hoey, Dan, D21
Hoffman, Fred, B42
Hooley, Christopher, D4
Hunter, J. A. H., D19, D27
Hurwitz equation, D12
hyperperfect number, B2

independent, C19
infinitary divisor, B3
infinitary multiperfect number, B3
infinitary perfect number, B3
irrationality sequence, E24
irreducible semiperfect, B2
irregular prime, D2, D7
Isaacs, Rufus, B5

Jacobi symbol, A12, F5
Jacobsthal, E., B40
Jones, James P., B31

k-th power residues, F6
Kalsow, William, D20
Kanapka, Joe, B19
Kaplansky, Irving, B4
Keller, Wilfrid, B43
Kelly, Blair, D1
Kemnitz, Arnfried, D21
Khare, S. P., B31, B33
Kimberling shuffle, E35
Klarner, David, C14
Klarner-Rado sequence, E36
Kleitman, Daniel J., B24
Kolba, Z., E20
Kummer surface, D19
Kummer, Ernst Eduard, B33, D2

Lacampagne, Carole, F31
Lagrange, Jean, D14
Lam, Clement W. H., C15
Landau, Edmund, D4, F1
Lander, Leon J., A8
lattice point, D20, E10, F, F1, F2, F3, F4
Lau, Kenneth, F28
least common multiple, B26, E2, F19
Leech, John, A16, B33, C20, D2, D9, D11, D14, D17, D19, D21, D22, D25
left factorial, B44
Legendre symbol, F5
Lehmer, Derrick Henry, A8, B2, B36, D1, D11, F9, F10, F12
Lehmer, Emma, B36, F9, F10
Lenstra, Hendrik Willem, E15
Leonardo of Pisa, D27
Levine, Eugene, B33, E11
Linnik's constant, A4
van Lint, Jacobus H., E25
Littlewood, John Edensor, A4, E18, F23
Longyear, Judith, D28
loopy game, E26
Lucas number, D26
Lucas's problem, D3
Lucas-Lehmer sequence, A3
Lucas-Lehmer test, A3
lucky numbers, C3

MacMahon, Percy A., E31
magic square, A6
Mahler, Kurt, B46, C4, D1
Mąkowski, Andrzej, B8, B11, B13, B48, C20, F10
Markoff equation, D12, D23
Markoff number, D12
Marsias, J. P., E8
max sequence, E27
McIntosh, Richard, F28
McKay, John H., B25
Meally, Victor, A8, B37, C18, D11, F10, F23

Mercator, Nicolaus, F23
Mersenne prime, A3, B1, B5, B9, B11, B15, B38
mex, E27
mex sequence, E27
minimum overlap problem, C17
modular Mousetrap, E37
monotone A.P., E33
Montgomery, Peter Lawrence, B33, D1, D11, F6, F28
Mordell, Louis Joel, D, D1, D3, D4, D11, D17, D28, F29
Moser, Leo, B11, C1, C5, D14, D15, F21
Motzkin, Theodor S., A16, B19, E22
Mousetrap, E37
multigrade equations, D1
multiperfect, B2
multiply perfect, B2

$\mathcal{N}$-position, E26
Nagashima, Takahiro, D28
Napier's rules, D18
Narkiewicz, Władysław, A15, B16, F9
Newman, Morris, F28
nim addition, E27
nim-like games, E27
no-three-in-line problem, F4
Noll, Landon Curt, A15
nonaveraging set, C16
noncototients, B36
nondividing set, C16
nontotients, B36

octahedron, C13
octal game, E27
Odlyzko, Andrew, A8
ordinary point, D18
Ore number, B2
Oursler, Clellie, D28
Owens, Frank, C4

$\mathcal{P}$-position, E26
packing problem, C9
paradox, E17
Parker, Ernest T., B25
Parkin, T. R., A8
partial quotient, F20, F21, F22
Pascal triangle, B33
Pell equation, B16, D3, D17
Pell sequence, A3
Penney, David E., B8, B41
pentagonal number, C20
perfect cuboid, D18
perfect difference set, C10
perfect number, A3, B1, B5
permutation sequence, E17
persistence, F25
Petersen graph, C13
Platonic solid, C13
Pomerance, Carl, A3, B8, B19, B47, D17, E30
postage stamp problem, C12
power residue, F6
powerful number, B16, D2
prime circle, C1
prime desert, A8
prime factorial, A8
prime number graph, A5, A14
prime number race, A4
prime numbers of measurement, E30
prime pyramid, C1
primitive abundant, B2
primitive part, A3
primitive pseudoperfect, B2
primitive root, F9
primitive sequence, E4
pseudoperfect number, B2, D11, F14
pseudoprime, A12, F10
Purdy, George B., F3
Pythagorean ratio, D17
Pythagorean triple, D18, D21

quadratic reciprocity law, F5
quadratic residue, F5, F6, F8, F9
quadri-amicable, B5
quasi-amicable, B5
quasi-perfect, B2
queens problem, C18

Ramanujan-Nagell equation, D23
Ramsey number, E11
Rathbun, Randall L., D1
rational box, D18, D21
rational simplex, D22
recurrence relation, A3
regular equation, E14
regular prime, D2, D7
Rényi, Alfred, B9
repunit, A3
residue class, A
Rhind papyrus, D11
Riddell, James, F19
Riemann ζ-function, B48
Riemann hypothesis, A4, A8, A19, F9
Robinson, Raphael M., A3
Rosenburg, Bryan, D20
Rote, Günter, E20
Roth, Klaus F., E9
Rotkiewicz, Andrzej, F10
Ruderman, Harry D., B47

Rudnick, Carl, D9
ruler function, E22

Säfholm, Sten, A9
Sastry, K. R. S., D3, D17, D19
Scher, Bob, D1
Scherk, Peter, C17
Schinzel's conjecture, A2
Schinzel, Andrzej, B37, D3, D9, D16, D24, E11
Schroeppel, Rich, B2, B42
Schur number, E11
Schur's conjecture, F5
Scott, Reese, D10
self-contained integer, E16
Selfridge, John Lewis, A2, A9, A18, B2, B8, B22, B28, B30, B33, B42, B46, B47, C14, C15, D11, F19, F27, F31
Shallit, Jeffery, B23
Shapiro, Harold N., D11
Shedd, Charles L., D21
Sidon sequence, C9
Sierpiński, Wacław, B13, D24, F10
sieve of Eratosthenes, C3
Silverman, David, B48, E8
Simmons, Chuck, A15
Singer, J., C10
singular point, D1
singular solution, D12
Slavić, Dušan V., B44
Sloane, Neil J. A., B33
Smith numbers, B49
sociable numbers, B7
Sós, Vera Turán, E14
span, E10
sphere, F1
spherical triangle, D18
Spiro, Claudia, B18
Sprague-Grundy theory, E27
square pyramid, D3
squarefree, A3, B33, B44
Stanley, Richard P., B16
Stechkin, Boris S., A17
Stewart, Bonnie M., A2
Stirling's formula, B22
Stöhr sequence, E32
Stong, Richard, D11
Straus, Ernst G., B19, B22, E22
Stroeker, Roelof Jacobus, D23
Strong Law of Small Numbers, A3, B20, D23, F5, F27
strong pseudoprime, A12
strongly independent, C19
strongly sum-free, E13
Struppeck, Thomas, B6
Styer, Robert, B19
Sulyok, Miklós, C9
sum-free, E12
sum-free sequence, E32
sum-free set, C14
superperfect numbers, B9
Szemerédi's theorem, A5, E10
Szpiro, Lucien, D25

table of primes, A
Tarry–Escott problem, D1
Taylor, Herbert, F19
tetrahedroid, D19
tetrahedron, D3, D22, F1
Thatcher, Alfred R., D18
Thompson, John G., B25
Thue sequence, E21
Thue–Siegel theorem, B30
Tijdeman, Robert, B46, E7
totient function, B11, B36, B41, B42, D11
triangular number, C20, D3, D21
triperfect number, B2
Turán, Pál, C20, E13
twin primes, A8, B19, B38

U-numbers, C4
Ulam, Stanislas M., F22
unique factorization, A16
unit, A
unit fractions, D11
unitary aliquot sequence, B8
unitary amicable numbers, B4
unitary divisor, B3
unitary multiperfect, B3
unitary perfect number, B3
unitary sociable numbers, B8
units, A16
untouchable numbers, B10
Upton, Leslie J., D19

Vandemergel, Stephane, A4, B18, D1, D11, D16
Vélez, William Yslas, B47
Viète, François, D1
Vojta, P., D1

van der Waerden's theorem, E10
Wagon, Stanley, A2
Wagstaff, Samuel S., B43, B48
Wall, Charles R., B38
Waring's problem, D4
weakly independent, C19
de Weger, Benne M. M., D23, F23
Weintraub, Sol, A8

weird, B2
wheel, C13
Whiteman, Albert Leon, E19
Wiethaus, Holger, B4
Wiles, Andrew J., D2
Wilson's theorem, A17
Windecker, C12
windmill, C13
Wolstenholme's theorem, B31
Woods, Alan R., B29
Wytek, Rudolf, F28
Wythoff pair, E27

Yang Yuan-Sheng, C10
Yuanhua, Ren, B7

Z-number, E18
Zeitlin, David, C4
Zhang, Ming-Zhi, A12

Problem Books in Mathematics *(continued)*

Theorems and Problems in Functional Analysis
by *A.A. Kirillov and A.D. Gvishiani*

Problem-Solving Through Problems
by *Loren C. Larson*

A Problem Seminar
by *Donald J. Newman*

Exercises in Number Theory
by *D.P. Parent*